ELEMENTARY
BUSINESS
STATISTICS:
the modern approach

ELEMENTARY
BUSINESS
STATISTICS:
the modern approach

SECOND EDITION

JOHN E. FREUND
Professor of Mathematics,
Arizona State University

FRANK J. WILLIAMS
Professor of Statistics,
San Francisco State College

PRENTICE-HALL, INC., Englewood Cliffs, N.J.

CONTENTS

PREFACE

In the Preface to the first edition of this book we observed that, in the past few decades, the development and application of new mathematical, statistical, and computer techniques had brought about radical changes in virtually all areas of business. As for statistics itself, we said that recent years had seen its shift from a backward-looking process to one concerned with current affairs, and also with future operations and their consequences. "In today's statistics," we wrote, "experiments are designed, samples selected, data collected and analyzed with reference to decisions that must be made, controls that must be exercised, judgments entailing action that must be taken, and so on." Now, almost eight years after the first Preface was written, we find that this forward- and outward-looking attitude— what we call the "modern approach"—is pretty much the accepted way of life both in the practice and the teaching of statistics.

Our aim in writing the first edition of this book was to describe, as best we could, the modern approach to decision making in the face of uncertainty. That is again our aim, and in writing this new edition we now have years of valuable classroom experience to draw on. Over these years we, and hundreds of others, have discussed both formally and informally with thousands of students our earlier ideas on how modern business statistics could best be organized and presented at an elementary level of mathematical difficulty. Literally dozens of these teachers and hundreds of students have generously taken time to share their experiences and thoughts with us. The great extent to which we have relied on these friends will not be evident to every reader, but their counsel and advice are reflected everywhere in this book.

In this present version, we have rearranged the basic material somewhat, added some new material, and given substantially greater attention to personal probability and the Bayesian approach to business decision making. To anyone familiar with the first edition, all this should be evident from an examination of the table of contents which precedes this Preface. In our judgment, this book contains enough material for a full year's work (six semester hours, or nine quarter hours), and it thus permits a good deal of latitude in the selection of topics for shorter courses. We will again resist the temptation to tell anyone specifically what chapters, sections, or subjects he should study or teach, but we do want to make the following brief comments: Many topics (no matter how practically important or intellectually stimulating they might be) can be omitted without loss of continuity. It is possible to present a one-quarter or one-semester course stressing the more traditional topics (for example, frequency distributions, index numbers, time series, and regression and correlation) without becoming formally involved in the combinatorial and probability and decision work. Of course, there is some loss involved in not doing this work, but the loss is more in the understanding of probability itself (which many feel should be part of everyone's general education) than in the understanding of classical statistical inference and methodology. On the other hand, there is ample material here for those who wish to stress the probability and decision material, and they can omit the more traditional topics without difficulty. Actually, the number of possibilities for a sound and useful introduction to statistics seems to be limited only by one's imagination. In any case, we have included over 600 numbered exercises to accompany the various topics we have discussed. Most of them are drawn from actual problems, but many of them have been modified and scaled down somewhat to simplify the computational burden.

We are greatly indebted to our many faculty colleagues and students who have contributed so much to this book; to them, and to Raymond Ewer who worked all the exercises, we extend our special thanks.

We are also indebted to the reviewers—Professors Richard H. Haase of Drexel University, Francis B. May of the University of Texas at Austin, George Miaoulis, Jr. of Fordham University, Henry W. Nace of the Lawrence Institute of Technology, Benjamin M. Perles of the University of Alaska, and Othmar W. Winkler of Georgetown University—who offered many helpful suggestions.

Finally, the authors would like to express their appreciation and indebtedness to the Literary Executor of the late Sir Ronald A. Fisher, F. R. S., to Dr. Frank Yates, F. R. S. and to Oliver & Boyd, Edinburgh, for their permission to reprint parts of Tables II, III, and VI from their books *Statistical Methods for Research Workers* and *Statistical Tables for Biological, Agricultural and Medical Research;* to Professor E. S. Pearson and the Biometrika trustees for permission to reproduce the material in Tables IVa, IVb, Va, and Vb from their *Biometrika Tables for Statisticians;* to Donald B. Owen and Addison-Wesley, Inc. for permission to reproduce parts of the Table of Random Numbers from their *Handbook of Statistical Tables.*

John E. Freund
Frank J. Williams

ELEMENTARY
BUSINESS
STATISTICS:
the modern approach

1

INTRODUCTION

MODERN STATISTICS

One of the most remarkable phenomena of the past few decades has been the growth of statistical methods and statistical ideas. For many years, statistics was concerned mostly with the collection of data and their presentation in tables and charts; today, it has evolved to the extent that its impact is felt in almost every area of human endeavor. This is because modern statistics is looked upon as encompassing a process as old as history itself, that of *decision making in the face of uncertainty*. Needless to say, there are uncertainties wherever we turn—when we predict the weather or the outcome of an election, when we toss a coin or roll a pair of dice, prospect for gold, experiment with a new paint, market a new product, indeed, even when we order a meal, step on an airplane, or pick up the phone.

Thus, the most important feature of the recent growth of statistics has been the shift in emphasis from methods which merely *describe* to methods which serve to make *generalizations*, or in other words, a shift in emphasis from *descriptive statistics* to *statistical inference*. By descriptive statistics we mean any treatment of data which is designed to summarize or describe some of their important features without attempting to infer anything that goes beyond the data. For instance, if someone compiles the necessary data and reports that in a given year a company's 1,200 employees received raises totaling $390,000, his work belongs to the domain of descriptive statistics. This is true also if he determines that the average raise received by the employees is $\dfrac{390,000}{1,200} = \$325$, but *not* if he uses the data to estimate average raises for subsequent years or for the employees of other companies.

Descriptive statistics is an important branch of statistics and it continues to be widely used in business and other areas of activity. In most cases, however, statistical information arises from samples (from observations made only on some of a large set of items), or from observations made on past happenings. Time, cost, or the *impossibility* of doing otherwise usually requires such a procedure, even though our real interest lies in the whole large set of items from which the sample was obtained, and not in the past but the future. Since generalizations of any kind lie outside the scope of descriptive statistics, we are thus led to the use of statistical inferences in making both short- and long-range plans and in solving many problems of day-to-day operations. To mention but a few examples, the methods of statistical inference are required to estimate the 1980 population of San Diego County, to predict the operating lifespan of an electronic computer, to forecast the 1985 military requirements for certain raw materials, or to decide upon an effective dose of a new antibiotic.

It must be understood, of course, that when we make a statistical inference, that is, a generalization which goes beyond the limits of our data, we must proceed with considerable caution. We must decide carefully how far we can go in generalizing from a given set of data, whether such generalizations are at all reasonable or justifiable, whether it might be wise to wait until we have more data, and so forth. Indeed, the most important problem of statistical inference is to appraise the risks to which we are exposed by making generalizations from sample data, the probabilities of making wrong decisions or incorrect predictions, and the chances of obtaining estimates which do not lie within permissible limits of error. These various possibilities may seem somewhat frightening, but they cannot be eliminated; *so long as we have to live with uncertainties, we simply must learn how to live with them intelligently.*

STATISTICS IN BUSINESS MANAGEMENT

There is hardly any area in which the impact of statistics has been felt more strongly than in business. Indeed, it would be hard to overestimate the contributions statistical methods have made to the effective planning and control of business activities of all sorts. In the past 25 to 30 years the application of statistical methods has brought about drastic changes in all the major areas of business management: general management, research and development, finance, production, sales, advertising, and the like. Of course, not all problems in these areas are of a statistical nature, but the list of those which can be treated either partly or entirely by statistical methods is very long. To illustrate, let us mention but a few which might face a large manufacturer.

In the *general management* area, for example, where long-range planning is of great concern, population trends must be forecast and their effects on consumer markets must be analyzed; in *research and engineering*, costs must be estimated for various projects, and manpower, skill, equipment, and time requirements must be anticipated; in the area of *finance*, the profit potential of capital investments must be determined, overall financial requirements must be projected, and capital markets must be studied so that sound long-range financing and investment plans can be developed. Although we cannot illustrate at this time how statistics is actually used in these areas of application, let us point out that they will be touched upon in the examples in the text and the exercises following the various sections.

In *production*, problems of a statistical nature arise in connection with plant layout and structure, size and location, inventory, production scheduling and control, maintenance, traffic and materials handling, quality assurance, and so forth. Enormous strides have been made in recent years in the application of statistics to the last area, that is, to sampling inspection and quality control. In the area of *sales*, many problems are encountered which require statistical solutions. For instance, sales must be forecast for both present and new products for existing as well as new markets, channels of distribution must be determined, requirements for sales forces must be estimated, and so on. Building a successful *advertising* campaign is also a troublesome task; budgets must be determined, allocations must be made to various media, and the effectiveness of the campaign must be measured (or predicted) by means of survey samples of public response and other statistical techniques.

So far we have been speaking of problems of a statistical nature which might typically be encountered by a large manufacturer. However, similar problems are faced, say, by a large railroad trying to make the best use of its thousands of freight cars, by a large rancher trying to decide how to feed his cattle so that nutritional needs will be met at the lowest possible cost, or by an investment company trying to decide which stocks and bonds to include in its portfolio.

It is not at all necessary to refer to large organizations to find business applications of statistics. For the smaller businesses, problems usually differ more in degree than in kind from those of their large competitors. Neither the largest supermarket nor the smallest neighborhood grocery store, for example, has unlimited capital or shelf space, and neither can afford to tie these two assets up in the wrong goods. The problem of utilizing capital and shelf space most effectively is as real for the small store as for the large, and it is extremely shortsighted to think that modern management tools (including modern statistical techniques) are of value only to big business. In fact, they could hardly be needed more anywhere else than in small business, where each year thousands of operating units fail and many of the thousands of new units entering the field are destined to fail because of inadequate capital, overextended credit, overloading with the wrong stock, and, generally speaking, no knowledge of the market or the competition.

Although, in this text, our attention will be directed largely toward business statistics, and our specific goal is to acquaint the reader with the statistical concepts and methods with which any student of business should be familiar, it should be understood that *the formal notions of statistics as a way of making rational decisions ought to be part of any thoughtful person's equipment.* After all, business managers are not the only persons who must make decisions involving uncertainties and risks; *everyone* has to make decisions of this sort professionally or as part of his everyday life. It is true that many of the choices we have to make entail only matters of taste and personal preference, in which case there is, of course, no question of a decision's being right or wrong. On the other hand, many choices we have to make between alternatives can be definitely wrong in the sense that there is an actual loss or penalty of some sort—possibly only a minor annoyance, perhaps something as serious as loss of life, or anything between these two extremes. The methods of modern statistics deal with problems of this kind and they do so not only in business, industry, and in the world of everyday life, but also in such fields as medicine, physics, chemistry, agriculture, economics, psychology, government, education, and so on. Although the examples and exercises used in this book will be mostly from the area of business, we shall not hesitate to refer to these other areas from time to time. In this way, the reader will be reminded of the fact that although specialized techniques exist for handling particular kinds of problems, the underlying principles and ideas are identical regardless of the field of application.

OPERATIONS RESEARCH

In recent years we have seen the birth of a new technology called *operations research* which is partly mathematics, partly statistics, partly engineering, partly industrial know-how, and partly philosophy (a new outlook and a new approach). At the present time there is no generally accepted definition of operations research, but it can be described as dealing with *the application of modern scientific techniques to problems involving the operations of a "system" looked upon as a whole: the conduct of a war, the management of a firm, the manufacture of a product, and so forth.*

In view of our objectives, we shall devote the last chapter of this book to some of the methods of operations research which are directly related to probability, statistics, and decision making in general. Unfortunately, even the simplest of these methods are fairly difficult mathematically, and we shall, therefore, have to be satisfied with giving only some of the basic language, concepts, and tools. After all, a person in the management ranks does not have to be a scientist (and vice versa), but he should be familiar enough with the new concepts and methodology to know where, when, and how they can be applied.

A WORD OF CAUTION

The amount of statistical information that is disseminated to the public for one reason or another is almost beyond comprehension, and what part of it is "good" statistics and what part is "bad" statistics is anybody's guess. Certainly, all of it cannot be accepted uncritically. Another important consideration is that sometimes entirely erroneous conclusions are based on sound data. For instance, a certain city once claimed to be the "nation's healthiest city," since its death rate was the lowest in the country. Even if we go along with their definition that healthy means "not dead," there is another factor that was not taken into account: since the city had no hospital its citizens had to be hospitalized elsewhere, and their deaths were recorded in the cities in which death actually occurred. The following are some other examples of *non sequiturs* based on otherwise sound statistical data: *"Statistics show that there were fewer airplane accidents in 1920 than in 1970; hence, flying was safer in 1920 than in 1970." "Since there are more automobile accidents in the daytime than there are at night, it is safer to drive at night." "Recent statistics show that the average income per person in a certain area is $1,600; thus, the average income for a family of five is $8,000."*

Sometimes, identical data are made the basis for directly opposite conclusions, as in collective-bargaining disputes when the same data are used by one side to show that employees are getting rich and by the other side to show that they are on the verge of starvation. In view of examples like these, it is understandable that some persons are inclined to feel that figures can be made to show pretty much what one wants them to show. Sadly enough, this is uncomfortably close to the truth, unless we carefully distinguish between "good" statistics and "bad" statistics, between statistics properly applied and statistics shamefully abused, and between statistics correctly analyzed and statistics unintentionally or intentionally perverted. We shall repeatedly remind the reader of this problem in special sections titled "A Word of Caution," which are given at the end of each chapter.

It is also important to realize that the sound statistical treatment of a problem consists of a good deal more than merely making a few observations on some conveniently available data, performing a few calculations, and reaching a conclusion. Questions as to how the data were collected and how the whole experiment or survey was planned are of prime importance. As elsewhere, we get "nothing for nothing" in statistics, and unless proper care is taken in all phases of an investigation—from the conception and statement of the problem to the planning and design, through the stages of data collection, analysis, and interpretation—no useful or valid conclusion whatever may be reached. *Generally speaking, no amount of fancy mathematical or statistical manipulation can salvage poorly designed surveys or experiments.* Indeed, professional statisticians insist that even the simplest of sampling studies be rigidly conducted according to certain well-defined rules. There is no more justification for calling a study which does not conform to these rules "statistical" than there is for calling a barnacle a ship.

2

SUMMARIZING DATA: FREQUENCY DISTRIBUTIONS

INTRODUCTION

In recent years, business decisions have come to depend more and more on the analysis of very large sets of data. This includes the small businessman who may need information about income patterns in the area which he serves, the market research analyst who may have to deal with the views expressed by thousands of shoppers, the government statistician who must handle, treat, and analyze census data which can only be described as voluminous, and the head of a large corporation who must consider information which would overwhelm him if it were not presented in a compact and usable form. This trend in the use of mass data is due partly to the increasing availability of high-speed computers; indeed, many current applications of statistical methods would have been practically impossible before the advent of modern data processing techniques. The trend is also due partly to an increasing awareness of the need for scientific methods in business management. Of course, we do not always deal with very large sets of data; there are instances where they are costly and hard to obtain, but the problem of putting mass data into a usable form is so important that it deserves special attention.

When dealing with large sets of data, we can often gain much information and obtain a good overall picture by *grouping* the data into a number of classes, as is illustrated in the following two examples:

Weekly earnings (dollars)	Number of secretaries
Under 80	21
80– 89	114
90– 99	182
100–109	290
110–119	204
120–129	159
130–139	88
140–149	74
150–159	45
160 and over	39
	1,216

The data which have been grouped here are the weekly earnings of 1,216 secretaries in the Phoenix, Arizona, metropolitan area as of March 1969; they constitute part of an area wage survey conducted by the *Bureau of Labor Statistics*. This kind of table is called a *frequency distribution* (or simply a *distribution*): It shows the frequencies with which the various incomes are distributed among the chosen classes. Tables of this sort, in which the data are grouped according to numerical size, are called *numerical* or *quantitative* distributions. In contrast, tables like the one given below (taken from the same source), in which the data are sorted according to a number of categories, are called *categorical* or *qualitative* distributions.

Industry of last job held by unemployed persons in the United States	Number of persons (thousands)
Agriculture	83
Construction	257
Manufacturing	763
Transportation and public utilities	128
Wholesale and retail trade	438
Finance and service industries	503
Public administration	86
No previous work experience	293
	2,551

Although frequency distributions present data in a relatively compact form, give a good overall picture, and contain information which is adequate for many purposes, there are evidently some things which can be obtained from the original data that cannot be obtained from a distribution. For instance, referring to the first of the above tables, we can find neither the exact size of the lowest and highest of the weekly earnings nor, for that matter, the exact average of the weekly earnings of the 1,216 secretaries. Nevertheless, frequency distributions present *raw* (unprocessed) data on which they are based in a more usable form, and the price which we must pay, the loss of certain information, is usually a fair exchange.

Data are sometimes grouped solely to facilitate the calculation of further statistical descriptions. We shall go into this briefly in Chapter 3, but it is worth noting that this function of frequency distributions is diminishing in importance in view of the ever increasing availability of high-speed computing equipment.

FREQUENCY DISTRIBUTIONS

The construction of a numerical distribution consists essentially of three steps: (1) choosing the classes into which the data are to be grouped, (2) sorting (or tallying) the data into the appropriate classes, and (3) counting the number of items in each class. Since the last two of these steps are purely mechanical, we shall concentrate on the first, the problem of choosing suitable classifications. (If the data are recorded on punch-cards, the sorting and counting can be done automatically in a single step.)

The two things we shall have to consider in the first step are determining the *number of classes* into which the data are to be grouped, and the *range of values* each class is to cover, that is, "from where to where" each class is to go. Although both of these choices are essentially arbitrary (they depend largely on the ultimate purpose the distribution is to serve), the following rules are usually observed:

(a) We seldom use fewer than 6 or more than 15 classes; the exact number used in a given situation will, of course, have to depend on the nature, magnitude, and range of the data.

(b) We *always* choose classes which are such that all of the data can be accommodated.

(c) We *always* make sure that each item belongs to only one class; in other words, we avoid overlapping classes, namely, successive classes having one or more values in common.

(d) Whenever possible, we make the class intervals of *equal length;* that is, we make them cover equal ranges of values. It is also generally desirable to make these ranges multiples of 5, 10, 100, and so on (or other numbers that are easy to work with), to facilitate the reading and the use of the resulting table.

Note that the first three, but not the fourth, of these rules were observed in the construction of the numerical distribution on page 8, assuming that the weekly earnings were all rounded to the nearest dollar. (Had these figures been given to the nearest cent, earnings of, say, $99.45 could not have been accommodated, for this value would have fallen *between* the third and fourth class.) The fourth rule was violated in connection with the first and last classes, which are said to be *open*—the first class has no specified lower limit and the last class has no specified upper limit. In general, the term "open" is applied to any class of the "less than," "or less," "more than," or "or more" variety. If a set of data contains a few values that are much greater (or much smaller) than the rest, open classes can help to simplify the overall picture by reducing the number of required classes; otherwise, open classes should be avoided as they can make it difficult, if not impossible, to give certain further descriptions of the data.

As we have pointed out, the appropriateness of a classification may depend on whether the data are rounded to the nearest dollar or the nearest cent. Similarly, it may depend on whether data are given to the nearest inch or the nearest hundredth of an inch, whether they are given to the nearest per cent or the nearest tenth of a per cent, and so on. Thus, if we want to group the sizes of all sales made by a college bookstore on a given day, we might use the classification

Size of sale
(dollars)

0.00– 4.99
5.00– 9.99
10.00–14.99
15.00–19.99
20.00–24.99
etc.

assuming that the data are given to the nearest cent. Similarly, for the price-earnings ratios of stocks given to the nearest tenth, we might use the classification

Price-earnings ratio

10.0–11.9
12.0–13.9
14.0–15.9
16.0–17.9
etc.

and for the number of calls received per hour at a department store's switchboard, we might use the classification

Number of calls

0– 99
100–199
200–299
300–399
etc.

We shall now construct a frequency distribution, or frequency *table*, from the following set of data showing the number of minutes 100 customers occupied their seats in a college cafeteria.

29	67	34	39	23	66	24	37	45	58
51	37	45	26	41	55	27	96	22	43
73	48	63	37	19	31	38	68	22	35
31	58	35	82	28	35	44	40	41	34
15	31	34	56	45	27	54	46	62	29
51	31	56	43	39	35	23	28	45	48
47	41	34	47	30	54	49	34	53	61
82	45	26	35	67	73	30	16	52	35
46	40	41	56	37	51	33	92	70	63
72	35	62	28	38	61	33	49	59	36

Since the smallest of these values is 15 and the largest is 96, it seems reasonable to choose the nine classes going from 10 to 19, from 20 to 29, from 30 to 39, . . . , and from 90–99. Performing the actual tally and counting the number of items in each class, we obtain the following frequency distribution:

Seat occupancy time (minutes)	Tally	Frequency
10–19	///	3
20–29	𝖭𝖫 𝖭𝖫 ////	14
30–39	𝖭𝖫 𝖭𝖫 𝖭𝖫 𝖭𝖫 𝖭𝖫 ////	29
40–49	𝖭𝖫 𝖭𝖫 𝖭𝖫 𝖭𝖫 //	22
50–59	𝖭𝖫 𝖭𝖫 ////	14
60–69	𝖭𝖫 𝖭𝖫	10
70–79	////	4
80–89	//	2
90–99	//	2
		100

The numbers shown in the right-hand column of this table, giving the number of items falling into each class, are called the *class frequencies.* Also, the smallest and largest values that can go into any given class are referred to as its *class limits;* thus, the limits of the nine classes are 10 and 19, 20 and 29, 30 and 39, and so on. More specifically, 10, 20, 30, . . . , and 90 are referred to as the *lower class limits,* while 19, 29, 39, . . . , and 99 are referred to as the *upper class limits* of the respective classes.

Numerical distributions such as this also have "class marks," "class intervals," and "class boundaries." *Class marks* are simply the midpoints of the classes, and they are easily obtained by averaging the class limits, that is, by adding the two limits of a class and dividing by two. Thus, the class marks of the seat-occupancy-time distribution are 14.5, 24.5, 34.5, . . . , and 94.5. A *class interval* is simply the length of a class, or the range of values it can contain, and when we are dealing with *equal* class intervals, their length is given by the difference between any two successive class marks. Thus, the above distribution has class intervals of 10, or as it is customary to say, it has a class interval of 10. Note that the class intervals of this distribution are *not* given by the respective differences between the upper and lower class limits, which would equal 9 instead of 10.

Using the terminology just introduced, we can now restate our comment on page 10 by saying that *the choice of the class limits depends on the extent to which the numbers we want to group are rounded off.* If we are dealing with prices rounded to the nearest dollar, the class $10–$19 actually contains all prices between $9.50 and $19.50; similarly, if we are dealing with measurements rounded to the nearest tenth of an inch, the class 1.5–1.9 actually contains all values between 1.45 and 1.95. Referring again to the seat-occupancy-time example, we thus observe that the first class actually contains all values falling between 9.5 and 19.5 minutes, the second class contains all values falling between 19.5 and 29.5 minutes, the third class contains all values falling between 29.5 and 39.5 minutes, and so forth. It is customary to refer to these values as *class boundaries;* sometimes, they are referred to as the "real" class limits. Note that the difference between the two boundaries of a class equals its class interval; in fact, for distributions having classes of unequal length, the difference between the two boundaries of a class serves to *define* its interval.

It is important to remember that class boundaries must always be "impossible" values, that is, values which cannot occur among the data we want to group. To make sure of this, we have only to observe the extent to which the data are rounded; for instance, for the size-of-sale classification on page 10 the class boundaries are the impossible values −0.005, 4.995, 9.995, 14.995, and so on.

Sometimes it is preferable to present data in what is called a *cumulative frequency distribution*, or simply a *cumulative distribution*, which shows directly how many of the items are less than (or, if preferred, greater than) various values. Successively adding the frequencies which we obtained for the seat-occupancy-time data, we get the following *cumulative "or less" distribution:*

Seat occupancy time (minutes)	Cumulative frequency
19 or less	3
29 or less	17
39 or less	46
49 or less	68
59 or less	82
69 or less	92
79 or less	96
89 or less	98
99 or less	100

Note that in this table we could have written "less than 20" instead of "19 or less," "less than 30" instead of "29 or less," "less than 40" instead of "39 or less," . . . , and we would then have referred to the distribution as a *cumulative "less than" distribution.* If we successively add the frequencies starting at the other end, we get a corresponding "or more" or "more than" distribution, depending on whether we refer to the classes as "10 or more," "20 or more," "30 or more," . . . , or as "more than 9," "more than 19," "more than 29," . . . .

Sometimes, it is preferable to show what percentage of the items falls into each class instead of the class frequencies. To convert a frequency distribution (or a cumulative distribution) into a corresponding percentage distribution, we have only to divide each class frequency by the total number of items and multiply by 100. For instance, referring to the secretarial earnings distribution on page 8, we find that the first class contains $\frac{21}{1216} \cdot 100 = 1.7$ per cent of the data, that the second class contains $\frac{114}{1216} \cdot 100 = 9.3$ per cent of the data, and so on.

So far we have discussed only numerical distributions, but the general problem of constructing categorical (or qualitative) distributions is very much the same. Again we must decide how many classes (categories) to use and what kind of items each category is to contain, making sure that all of the items are accommodated and that there are no ambiguities. Since the categories must often be selected before any data are actually obtained, sound practice is to include a category labeled "others."

When dealing with categorical distributions, we do not have to worry about such mathematical details as class limits, class boundaries, and class marks; on the other hand, we now have a more serious problem with ambiguities and we must be careful and explicit in defining what each category is to contain. For this reason, it is often advisable to use standard categories developed by the Bureau of the Census and other government agencies. (For references to such lists see the book by P. M. Hauser and W. R. Leonard in the Bibliography at the end of the book.)

EXERCISES

1. Decide for each of the following quantities whether it can be determined on the basis of the weekly earnings distribution on page 8; if possible, give a numerical answer:

   (a) The number of secretaries with weekly earnings of at least $130.
   (b) The number of secretaries with weekly earnings of more than $130.
   (c) The number of secretaries with weekly earnings of more than $200.
   (d) The number of secretaries with weekly earnings of less than $100.
   (e) The number of secretaries with weekly earnings of at most $100.
   (f) The number of secretaries with weekly earnings of at most $139.

2. The following is the distribution of the actual shelf weight (in ounces) of a sample of 60 "one-pound" sacks of trisodium phosphate, which were filled from bulk stock by a part-time clerk in a hardware store:

Weight	Number of sacks
15.60–15.79	2
15.80–15.99	6
16.00–16.19	11
16.20–16.39	16
16.40–16.59	12
16.60–16.79	7
16.80–16.99	3
17.00–17.19	2
17.20–17.39	1

Find (a) the class limits of each class, (b) the class marks, (c) the class boundaries, and (d) the class interval.

3. A customs official groups the declared values of a number of packages mailed from foreign countries into a frequency distribution with the classes $0.00–$9.99, $10.00–$19.99, $20.00–$29.99, $30.00–$39.99, $40.00–$49.99, $50.00–$59.99, and $60.00 and over. Decide for each of the following quantities whether it can be determined on the basis of the resulting distribution:

(a) The number of packages valued at less than $40.00.

(b) The number of packages valued at $40.00 or less.

(c) The number of packages valued at more than $30.00.

(d) The number of packages valued at $30.00 or more.

4. A set of data consisting of the sizes of orders in dollars and cents is grouped into a table whose classes have the limits $0.00–$24.99, $25.00–$49.99, $50.00–$99.99, and $100.00–$199.99. Find (a) the limits of each class, (b) the class boundaries, (c) the class marks, and (d) the class intervals.

5. From the birth dates given by a number of persons responding to a magazine advertisement for a newly-formed record club, the respondents' ages are grouped into a table having the classes 10–14, 15–19, 20–24, 25–29, and 30–34. What are the class boundaries of this distribution and what are its class marks?

6. A set of measurements of the lengths of a large number of pieces of scrap metal given to the nearest tenth of an inch are grouped into a table whose classes have the boundaries 4.95, 6.95, 8.95, 10.95, 12.95, 14.95, and 16.95. What are the lower and upper limits of each class?

7. The weights of 435 applicants for positions in the fire department of a large city are given to the nearest tenth of a pound, with the lowest being 153.2 pounds and the highest being 223.7 pounds. Give the class limits of a table with eight equal classes into which these weights might be grouped.

8. The year-end bonuses paid by a large manufacturer to 140 members of its managerial personnel are all multiples of $100, with the smallest being $8,200 and the largest being $24,700. Show both the class limits and the class boundaries of a table with six equal classes into which these figures might be grouped.

9. The following are the number of years which 100 of the many filing cabinets owned by a chain of drugstores have been in service:

4.2	5.8	6.1	9.1	4.7	5.8	6.9	4.0	6.4	7.7
6.9	8.6	8.2	4.0	5.5	5.6	9.4	7.2	2.5	6.3
8.2	8.1	5.2	8.8	9.2	6.6	7.8	6.9	5.8	7.5
5.3	5.6	5.7	3.5	6.0	5.4	2.6	9.3	5.2	2.3
6.1	5.3	2.0	5.4	9.8	6.0	5.2	8.6	5.1	9.2
5.6	4.3	3.8	6.5	7.5	4.5	4.2	3.7	6.1	5.4
6.2	6.3	10.4	6.7	7.8	3.9	7.1	5.6	3.3	6.7
5.0	6.5	5.0	5.8	5.7	4.8	8.5	6.3	7.5	3.1
7.5	3.7	10.6	5.8	6.8	7.4	3.0	9.7	8.4	5.9
2.6	5.9	6.8	5.1	5.0	5.8	5.5	5.2	4.1	6.8

(a) Group these figures, which are rounded to the nearest tenth of a year, into a distribution having the classes 2.0–2.9, 3.0–3.9, 4.0–4.9, . . . , and 10.0–10.9.

(b) Convert the distribution obtained in part (a) into a cumulative "less than" distribution.

10. With reference to Exercise 9, suppose that the chain of drugstores is offered by one firm a trade-in of $60 for each filing cabinet which (rounded to the nearest tenth of a year) is less than 4 years old, $30 for each filing cabinet which (rounded to the nearest tenth of a year) is at least 4 years old but less than 8 years old, and $5 for each older filing cabinet. Use the cumulative distribution obtained in part (b) of Exercise 9 to determine whether this firm offers a higher total trade-in on the 100 filing cabinets than another firm which offers a flat $2,500.

11. A group of 50 employees from the accounting department of a large company is given an intensive course in computer programming. Of the various exercises assigned during the course, the following are the number of exercises satisfactorily completed by the members of the group:

13	9	5	11	14	6	5	8	11	13
10	16	15	3	19	18	9	9	5	12
13	12	15	9	18	12	16	7	12	13
11	18	15	9	21	9	11	6	12	12
10	16	2	14	10	17	8	15	11	12

Group these figures into a table having the classes 2–4, 5–7, 8–10, . . . , and 20–22, and also construct the corresponding percentage and cumulative "or more" percentage distributions.

12. During the instruction period referred to in Exercise 11, the study group was given a test on number systems and Boolean algebra, and the following are the grades which the members of the group obtained:

73	65	82	70	45	50	70	54	32	75
75	67	65	60	75	87	83	40	72	64
58	75	89	70	73	55	61	78	89	93
43	51	59	38	65	71	75	85	65	85
49	97	55	60	76	75	69	35	45	63

Group these grades into a table having the classes 30–39, 40–49, 50–59, . . . , and 90–99, and also construct the corresponding cumulative "or less" distribution.

13. A stock brokerage firm published a list of 50 stocks which it considered to be "fully priced" at the time. On this list it also gave the price-earnings ratios of these stocks (based on a recent price and latest 12-month earnings) as

19.3	16.4	29.5	15.2	20.9	12.2	25.9	37.9	15.8	59.1
16.1	10.8	29.2	16.3	57.3	18.3	26.4	15.0	19.7	20.0
22.8	17.7	52.5	15.6	27.2	22.4	23.0	19.3	25.4	45.7
15.2	51.9	17.9	48.7	24.3	19.3	40.3	40.8	34.4	35.8
30.3	67.7	33.1	15.0	25.4	22.4	42.9	28.8	29.0	26.1

Group these price-earnings ratios into a table having the eight equal classes 8.0–15.9, 16.0–23.9, 24.0–31.9, . . . , 64.0–71.9.

14. With reference to the list of Exercise 13, the last full-year earnings (in dollars per share) were given as

3.27	2.71	1.15	2.94	2.22	2.37	1.46	3.89	4.60	2.70
4.02	2.99	1.26	1.05	0.90	1.95	1.11	2.14	3.67	1.45
3.12	1.77	1.85	1.42	1.60	1.18	1.98	1.52	1.29	3.06
2.41	2.92	1.87	1.91	1.55	1.61	3.59	0.91	3.84	2.71
2.10	1.03	1.77	2.14	2.73	3.09	3.19	0.46	2.44	1.35

Group these figures into a table having the classes $0.00–$0.99, $1.00–$1.99, $2.00–$2.99, $3.00–$3.99, and $4.00–$4.99, and also construct the corresponding percentage and cumulative "or less" percentage distributions.

15. In its training program for inspectors, a company uses sheets of paper with random arrays of the digits 0, 1, 2, 3, 4, 5, 6, 7, 8, and 9. There are 1,000 digits on each sheet, among which the fives and sixes, of which there are altogether 60, are considered to be defective parts. To simulate the inspection of manufactured products, a trainee is given 120 "lots" of 1,000 "parts" to "inspect" in an allotted time. The following are the number of "defectives" one trainee recorded for each lot:

59	53	61	63	58	57	60	59	56	62	59	61
50	62	57	56	60	60	62	58	48	58	52	53
58	60	60	60	54	61	60	54	58	60	59	54
55	59	46	55	57	60	58	56	61	61	57	57
63	60	58	60	59	51	59	61	62	67	55	56
60	56	60	62	60	56	54	59	57	58	63	62
59	57	62	51	64	58	62	60	60	59	60	59
58	48	60	59	60	60	58	57	59	61	49	58
53	60	53	60	56	61	60	52	60	57	58	57
61	58	60	54	59	64	57	60	55	58	60	54

Group these figures into a frequency distribution showing how many times each value occurs, and also construct the corresponding "or more" cumulative distribution.

16. The following are the net wages earned during one week by 90 temporary employees placed by an employment agency specializing in such personnel:

68.10	26.15	63.70	34.10	71.75	48.66	79.51	35.18	28.10
49.24	30.18	32.15	29.90	60.12	47.11	53.33	40.26	31.17
29.66	35.01	58.56	31.24	52.02	41.63	39.54	69.40	33.09
32.05	26.70	44.40	83.74	37.20	25.65	46.42	45.89	47.29
30.09	33.81	40.10	73.78	30.33	50.12	59.39	33.55	39.19
38.07	48.62	38.69	55.17	41.10	65.74	46.82	50.11	45.25
25.09	39.71	26.42	43.71	36.55	61.18	33.72	54.60	55.00
25.48	42.70	37.37	44.49	30.52	52.89	42.63	36.10	43.11
34.15	31.51	32.50	77.98	39.35	62.91	31.71	56.05	27.12
30.70	43.62	28.55	45.62	28.69	35.40	27.16	47.50	36.10

Group these wages into a distribution having the seven equal classes $20.00–$29.99, $30.00–$39.99, . . . , and $80.00–$89.99.

17. To show how the choice of different classes can alter the overall appearance of a distribution, regroup the wages of Exercise 16 into a table having the six equal classes $25.00–$34.99, $35.00–$44.99, . . . , and $75.00–$84.99.

18. Use a daily newspaper listing prices on the New York Stock Exchange and construct a table showing how many of the R, S, and T stocks traded on a certain day showed a net increase, a net decrease, or no change in price.

19. Take that part of the classified ads of a large daily newspaper where individuals (not dealers) advertise cars for sale, and construct a distribution showing how many of these cars are station wagons, how many are sedans, how many are convertibles, and so on.

20. Choose a local television station and construct a table showing how many of the programs it broadcasts during one week are situation comedies, adventure stories, children's programs, educational programs, and so forth. (Disregard differences in the lengths of the programs.)

GRAPHICAL PRESENTATIONS

When frequency distributions are constructed primarily to condense large sets of data and display them in an "easy to digest" form, it is usually advisable to present them graphically, that is, in a form that appeals to the human power of visualization. The most common among all graphical presentations of statistical data is the *histogram*, an example of which is shown in Figure 2.1. A histogram is constructed by representing the measurements of observations that are grouped (in Figure 2.1, the seat occupancy times in minutes) on a horizontal scale, the class frequencies on a vertical scale, and drawing rectangles whose bases equal the class interval and whose heights are determined by the corresponding class frequencies. The markings on the horizontal scale can be the class limits, as in Figure

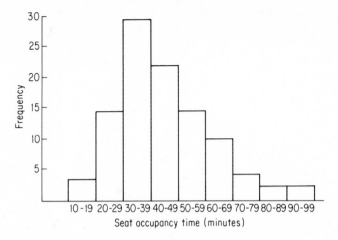

FIGURE 2.1   Histogram of seat-occupancy-time distribution

2.1, the class boundaries, the class marks, or arbitrary key values. For easy readability, it is generally preferable to indicate the class limits, although the rectangles actually go from one class boundary to the next. Histograms cannot be used in connection with frequency distributions having open classes and they must be used with extreme care if the classes are not all equal.

An alternative, though less widely used, form of graphical presentation is the *frequency polygon* (see Figure 2.2). Here the class frequencies are

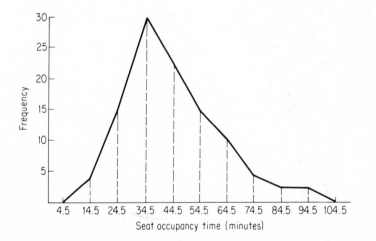

FIGURE 2.2   Frequency polygon of seat-occupancy-time distribution

plotted at the class marks and the successive points are connected by means of straight lines. If we apply this same technique to a cumulative distribution, we obtain a so-called *ogive*. Note, however, that now the cumulative frequencies are *not* plotted at the class marks—it stands to reason that the frequency corresponding, say, to "29 or less" should be plotted at 29 or preferably at the class boundary of 29.5, since "29 or less" actually includes everything up to 29.5. Figure 2.3 shows an ogive corresponding to the "or less" distribution of the seat-occupancy-time data.

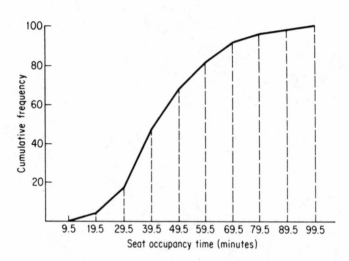

FIGURE 2.3   Ogive of seat-occupancy-time distribution

Although the visual appeal of histograms, frequency polygons, and ogives exceeds that of frequency tables, there are ways in which distributions can be presented even more dramatically and probably also more effectively. Two examples of such pictorial presentations (often seen in newspapers, magazines, and reports of various kinds) are given in Figures 2.4 and 2.5; they are, respectively, a *pictogram* and a *pie chart*.

EXERCISES

1. Draw a histogram of the frequency distribution constructed in Exercise 11 on page 16 and an ogive of the cumulative "or more" percentage distribution.

2. Draw a histogram of the frequency distribution constructed in Exercise 12 on page 16 and an ogive of the cumulative "or less" distribution.

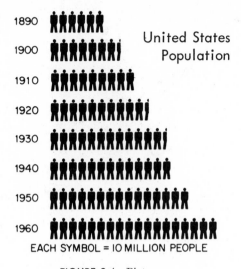

EACH SYMBOL = 10 MILLION PEOPLE

FIGURE 2.4 Pictogram

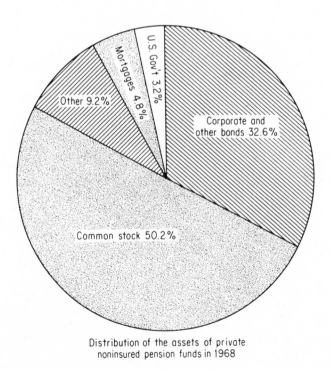

Distribution of the assets of private
noninsured pension funds in 1968

FIGURE 2.5 Pie chart

3. Draw a histogram and a frequency polygon of the distribution constructed in Exercise 13 on page 17.

4. Draw a histogram of the frequency distribution constructed in Exercise 14 on page 17, a frequency polygon of the percentage distribution, and an ogive of the cumulative "or less" percentage distribution.

5. Draw a histogram of the frequency distribution constructed in Exercise 15 on page 17 and an ogive of the "or more" cumulative distribution.

6. Draw histograms of the distributions constructed in Exercises 16 and 17 on page 18 and compare their shapes.

7. The following is a distribution of the number of days which elapsed between the day on which 90 commercial property owners in a large city petitioned for tax hearings and the day on which formal acknowledgments of the petitions were received from the appeals board:

Number of days	Number of petitioners
20–24	2
25–29	6
30–34	13
35–39	18
40–44	22
45–49	10
50–54	9
55–59	5
60–64	3
65–69	2
	90

Construct a histogram of this distribution and also draw a histogram of the modified distribution obtained by combining the cases from 30 through 39 into one class (making the adjustment indicated on page 24).

8. With reference to Exercise 2 on page 14 draw (a) a histogram of the shelf weight of the 60 sacks of trisodium phosphate, and (b) an ogive of the corresponding "or less" cumulative percentage distribution.

9. A pictorial presentation which is very similar to a histogram is a *bar chart* like that of Figure 2.6; the lengths of the bars are proportional to the class frequencies, but there is no pretense of having a continuous (horizontal) scale.

    (a) Draw a bar chart of the distribution constructed in Exercise 13 on page 17.

    (b) Draw a bar chart of the distribution constructed in Exercise 16 on page 18.

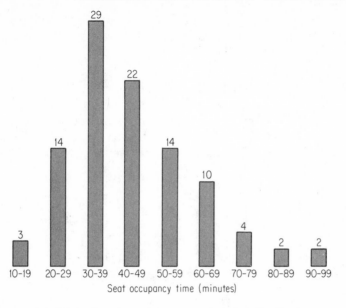

FIGURE 2.6 Bar chart of seat-occupancy-time distribution

10. Categorical distributions are often presented as *pie charts* like the one of Figure 2.5, where a circle is divided into sectors which are proportional in size to the frequencies of the categories they represent. To obtain the proper central angle for each sector, note that 1 per cent is represented by a central angle of 3.6 degrees.

(a) Construct a pie chart conveying the information that (according to the most recently available figures) of the total number of bottles of wine consumed in the United States, 69 per cent came from California, 18 per cent came from other parts of the country, 5 per cent was imported from France, and the rest was imported from other countries.

(b) Draw a pie chart to display the information that a hospital in a large city accounts for its expenses as follows: 73 per cent for salaries, professional medical fees, and employee benefits; 13 per cent for medical and surgical supplies and equipment; 8 per cent for maintenance, food, and power; and 6 per cent for administrative costs.

(c) Of each dollar currently spent by the state of California, 12.5 cents goes for higher education, 28.0 cents for other education, 22.5 cents for health and welfare, 14.3 cents for property tax relief and shared revenue, 11.2 cents for transportation, 2.9 cents for correction, 2.3 cents for resources, 1.8 cents for business and commerce, 1.4 cents for administration and fiscal management, and 3.1 cents for other activities. Draw a pie chart to display this information.

A WORD OF CAUTION

Intentionally or unintentionally, frequency tables, histograms, and other pictorial presentations can be very misleading. Suppose, for instance, that in the seat-occupancy-time illustration we had combined the two classes 50–59 and 60–69 into the one class 50–69. This new class has a frequency of 24, but in Figure 2.7, where we still use the *heights* of the

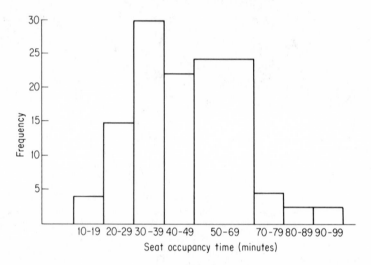

FIGURE 2.7   Incorrectly modified histogram of seat-occupancy-time distribution

rectangles to represent the class frequencies, we get the erroneous impression that this class contains more than a third of the items. This is due to the fact that when we compare the sizes of rectangles, triangles, and other plane figures, we instinctively compare their *areas* and not their sides. All this does not matter when the class intervals are equal, but in Figure 2.7 the class 50–69 is *twice as wide* as the others, and we will have to compensate for this by *dividing the height of the rectangle by two*. This was done in Figure 2.8, and we now get the correct impression that the class 50–69 contains about as many items as the class 40–49. (This is as it should be, for the respective class frequencies are 24 and 22.) *Note that in Figure 2.8 the vertical scale (which has lost its significance) is omitted.* The whole idea of representing class frequencies or proportions by means of *areas* is very important, because this is precisely what we shall have to do in Chapter 7 in connection with continuous distributions.

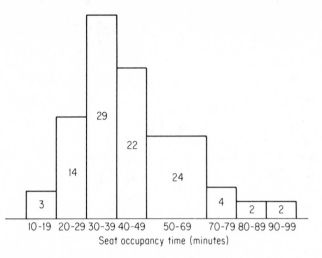

FIGURE 2.8 Correctly modified histogram of seat-occupancy-time distribution

The same difficulty arises in the construction of so-called *pictograms*, where the sizes of various kinds of objects are supposed to illustrate and emphasize differences among the data. Suppose, for instance, we want to

16 billion
dollars in 1953

48 billion
dollars in 1968

FIGURE 2.9 Misleading pictogram

dramatize the fact that the value of industrial bonds held by U.S. life insurance companies has increased from about $16 billion in 1953 to roughly $48 billion in 1968. Since the amount has *tripled,* we might be tempted to draw a picture like the one of Figure 2.9, where the height and the width of the bond certificate representing 1968 are both three times the respective height and width of the bond certificate representing 1953. Unfortunately, this gives the erroneous impression that the amount is multiplied by *nine instead of three.* To avoid this we must make the *area* of the second bond certificate three times that of the first, and we accomplished that in Figure 2.10 by making each side of the second bond certificate $\sqrt{3} = 1.73$ times the corresponding side of the first.

16 billion
dollars in 1953

48 billion
dollars in 1968

FIGURE 2.10 Pictogram

There are many other ways in which tabular and graphical presentations can give misleading and erroneous impressions; some of these are treated in an amusing fashion in *How to Lie with Statistics,* the book by D. Huff referred to in the Bibliography at the end of the book.

3

SUMMARIZING DATA: STATISTICAL DESCRIPTIONS

SAMPLES AND POPULATIONS

Descriptions of statistical data can be quite brief or quite elaborate, depending partly on the nature of the data themselves and partly on the purpose for which they are to be used. In some instances we might even describe the same set of data in several different ways. To draw an analogy, a San Francisco department store might describe itself to the fire department by giving the number of exits, the total floor space, the number of sprinklers, and the number of employees; on the other hand, it might describe itself to someone calling by phone as a "full-line department store with a fine men's department on the second floor." Both of these descriptions may serve the purposes for which they are designed, but they would hardly satisfy the New York Stock Exchange in passing on the department store's application for the listing of its stock. This would require detailed information on the nature of the store's business, its history, various kinds of financial statements, and so on. In any event, when we describe collections of data, it is always desirable to say neither too little nor too much. Thus, on some occasions it may be satisfactory to present data simply as they are, in *raw* form, and let them speak for themselves; at other times it may be best to group a set of data and present its distribution in tabular or graphical form. Usually, however, it is necessary to summarize data further by means of statistical measures, which are designed to give appropriate descriptions. In the next two sections we shall discuss two special kinds of descriptive measures, called *measures of location* and *measures of variation*, reserving mention of some other measures until later.

When we said that the choice of a statistical description depends partly on the nature of the data themselves, we were referring among other things to the following fundamental distinction: *If a set of data consists of all conceivably possible (or hypothetically possible) observations of a certain phenomenon, we refer to it as a population; if a set of data contains only part of these observations, we refer to it as a sample.* The qualification "hypothetically possible" was added here to take care of such clearly hypothetical situations as where we look at the outcomes (heads or tails) of 10 flips of a coin as a *sample* from the *population* which consists of the outcomes of all possible flips of the coin, or where we look at five measurements of the length of a steel shaft as a *sample* from the *population* of all possible measurements of its length. In fact, we often look at the results obtained in an experiment as a sample of what we might obtain if the experiment were repeated over and over again.

In this book, we shall always denote the number of items in a *finite* population (the population *size*) by the letter N, and denote the number of items in a sample (the sample size) by the letter n.* Although we are free to call any group of items a population, in actual practice this depends on the context in which they are to be viewed. Suppose, for instance, we are offered a lot of 1,000 one-inch pine boards, which we may or may not buy depending on their moisture content. If we measure the moisture content of 5 of the boards to estimate the average moisture content of all the boards, these 5 measurements constitute a sample from the population which consists of the moisture content of the $N = 1,000$ boards. In another context, however, if we were considering entering into a long-term contract for the delivery of 10 million such boards, we would look at the measurements of the moisture content of the 1,000 boards only as a sample. Similarly, the complete set of figures for a recent year, giving the elapsed time between the application for and the issuance of residential building permits in Atlanta, can be looked upon as a sample or as a population. If we were interested only in the city of Atlanta and that particular year, we would consider these data to constitute a population; on the other hand, if we wanted to generalize about the speed with which residential building permits

---

* The distinction between finite and infinite populations is discussed in some detail in Chapter 8.

are issued in the entire United States, or in some other year, the given data would constitute only a sample.

As we have been using it here, the word "sample" has very much the same meaning as it has in everyday language. A newspaper considers the attitudes of 100 readers toward a proposed city-county sports complex to be a sample of the attitudes of all its readers toward this complex, and a housewife considers a box of a new detergent to be a sample of all the boxes of this detergent. Later, we shall use the word "sample" only in connection with sets of data which can reasonably serve as the basis for valid generalizations about the population from which they came, and in this more technical sense many sets of data which are popularly called samples are not samples at all.

It must also be evident from this discussion that the word "population" is not used in its everyday sense; neither is the word "universe" which is sometimes used in its place. In statistics, both terms refer to the actual or hypothetical totality of measurements or observations with which we are concerned in a given situation and not to collections of human beings or animals, or even to the cosmos as a whole.

MEASURES OF LOCATION

It is often necessary to represent a set of data by means of a single number which, in its way, is descriptive of the entire set. Exactly what sort of number we choose depends on the particular characteristic of the set we wish to describe. In one study, for example, we might be interested in the extreme (smallest and largest) values among the data; in another, in the value which is exceeded by only 10 per cent of the values; and in still another, in the total of all the values. In this section we shall be concerned with so-called *measures of central location*, namely, measures which somehow describe the "center" or "middle" of a set of data.

Of the different measures of central location, by far the best known and the most widely used is the *arithmetic mean*, or simply, the *mean*. The mean is widely referred to as the "average," a term which in popular usage has a rather loose connotation and different meanings (as, for example, in batting average, average day, average task, or feeling average). Thus, we shall sometimes use the terms "average" or "arithmetic mean," but mostly we shall use the term "mean," to refer to the measure of central location which we now define:

The mean of a set of values is the sum of the values divided by their number.

Since we shall have occasion to calculate the means of many different sets of data, it will be convenient to develop a simple formula that is applicable to any set of data. This requires that we represent the figures (measurements or observations) to be averaged by some general symbol such as x, y, or z. Choosing, say, the letter x, we can refer to the n values in a sample as x_1 (which reads *"x sub-one"*), x_2, x_3, . . . , and x_n and write

$$\text{sample mean} = \frac{x_1 + x_2 + x_3 + \cdots + x_n}{n}$$

This formula is perfectly general and it will take care of any set of sample data, but it can be made more compact with the use of the Σ (*capital sigma*) notation. In this notation we let $\Sigma\, x$ stand for "the sum of the x's," namely, for $x_1 + x_2 + \cdots + x_n$. Thus, the mean, $\bar{x}$, of a set of sample values x_1, x_2, . . . , x_n is given by the formula

$$\bar{x} = \frac{\Sigma\, x}{n} \qquad\qquad \bigstar$$

Here $\bar{x}$ reads *"x bar,"* and if we referred to the measurements as y's or z's, we would correspondingly write their mean as $\bar{y}$ or $\bar{z}$. In the above formula, the term $\Sigma\, x$ does not state explicitly which values of x are to be added; let it be understood, however, that $\Sigma\, x$ always refers to the sum of *all* the x's under consideration in a given situation. In Technical Note 1 on page 61 the use of the sigma notation will be discussed in more detail.

The mean of a population of N items is defined in the same way: It is the sum of the N items, $x_1 + x_2 + x_3 + \cdots + x_N$, or $\Sigma\, x$, divided by N. Giving this mean the special symbol μ (*mu*, the Greek letter for m), we can thus write

$$\mu = \frac{\Sigma\, x}{N} \qquad\qquad \bigstar$$

with the reminder that in this formula $\Sigma\, x$ means the sum of all N values of x which constitute the population.

To illustrate the calculation of a mean, suppose that we want to know the mean weight of a production lot (considered to be a population) of

---

★ Formulas marked with a star are actually used for practical computations. This will make it easier for the reader to distinguish between formulas used for calculations and those given primarily for definitions or as part of derivations.

$N = 5,000$ drums of a wax-base floor cleaner. As it may be too time consuming and too costly to weigh all the drums and then calculate μ, a solution, widely used in practice, is to take a sample, calculate the sample mean $\bar{x}$, and use this quantity to estimate the population mean μ. If $n = 5$ and the actual measurements are, respectively, 248, 250, 252, 250, and 251 pounds, we have

$$\bar{x} = \frac{\Sigma\, x}{n} = \frac{248 + 250 + 252 + 250 + 251}{5} = 250.2 \text{ pounds}$$

and if these weights constitute a sample in the technical sense (that is, a set of data from which valid generalizations can be made), we can *estimate* the mean weight μ of all the 5,000 drums to be 250.2 pounds. Moreover, if we assume that the sample is a particular kind of sample called a "random sample," we can apply techniques we shall study later and say that we are "practically certain" that this estimate is in error by at most 2.5 pounds.

In the same way we might estimate the mean lifetime of a population of 5 million 40-watt light bulbs (for which $\mu = \dfrac{\Sigma\, x}{N}$ could not feasibly be determined in practice since life tests are *destructive*) by taking a sample of, say, $n = 4$. If the lifetimes of these four bulbs are, respectively, 504, 485, 501, and 494 hours, we get

$$\bar{x} = \frac{504 + 485 + 501 + 494}{4} = \frac{1,984}{4} = 496 \text{ hours}$$

and if the four lifetimes constitute a random sample, we can *estimate* μ to be 496 hours. Furthermore, using techniques to be learned in Chapter 9, we can assert that we are "reasonably sure" that the actual mean lifetime is not less than 469.3 hours nor more than 522.7 hours.

In order to distinguish between descriptions of populations and of samples, statisticians use different symbols such as μ and $\bar{x}$. Furthermore, descriptions of populations are referred to as *parameters* and they are usually denoted by Greek letters; in contrast, descriptions of samples are called *statistics*. Thus, in both of our examples we used the statistic $\bar{x}$ to estimate the parameter μ.

The widespread use of the mean to describe the "middle" of a set of data is not just accidental. Aside from the fact that it is a simple, familiar measure, the mean has the following desirable properties: (1) it can be calculated for any set of numerical data, so it always exists; (2) a set of numerical data has only one mean, so it is always unique; (3) it lends itself to further statistical manipulation (for instance, the means of several sets of data can be combined by using an appropriate formula into the grand mean of all the data); and (4) it is relatively *reliable* in the sense that the means of many samples drawn from the *same* population do not vary (fluctuate) as widely as some other estimators of the population mean μ, and this is of fundamental importance in statistical inference.

There is another characteristic of the mean which, on the surface, seems desirable, but may not be so: The mean takes into account each individual item. Sometimes samples contain very small or very large observations, which are so far removed from the main body of the data that the appropriateness of including them in a sample is questionable. When such values, called "outliers," are averaged in with the other values, they can affect the mean to such an extent that its value as a meaningful description of the "middle" of the data becomes debatable. (Outliers resulting from such things as gross errors in calculations, gross errors in recording data, malfunctions of equipment, and contamination can sometimes be identified as to their source and simply eliminated from the data before they are averaged.)

Instead of omitting outliers and calculating a sort of "modified" mean in an attempt to avoid the difficulty described, we could describe the "middle" of our data by means of another measure of location called the *median*. By definition, *the median of a set of data is the value of the middle item (or the mean of the two middle items) when the data are arrayed or ordered, that is, arranged in an increasing or decreasing order of magnitude.* Unlike the mean, the median is not easily affected by extreme values.

If there is an odd number of items in a set of data, there is always a middle item whose value is the median. The median of the five numbers 5, 3, 2, 8, and 1 is 3 (and not 2, since the numbers must first be arrayed), and the median of the nine numbers 5, 3, 7, 11, 500, 8, 2, 10, and 7 is 7. Generally speaking, the median of a set of n items, where n is *odd*, is the *value* of the $\frac{n+1}{2}$th largest item. Therefore, the median of 25 numbers is the value of the $\frac{25+1}{2} = 13$th largest item, the median of 99 numbers is the value of the $\frac{99+1}{2} = 50$th largest, and so on.

For a set containing an even number of items, there is no single middle item, and the median is defined as the mean of the middle two items, but the formula $\frac{n+1}{2}$ still serves to locate the median. If $n = 6$, for instance, $\frac{6+1}{2} = 3.5$, and we interpret this as "halfway between the values of the third and fourth items." Thus, the median of the numbers 3, 5, 8, 10, 12, and 15 is 9, and the median of the numbers 14.7, 14.9, 15.1, and 16.6 is 15.0.

For the last set of numbers of the preceding paragraph the median is 15.0 and the mean is 15.3, yet each of these values is an "average" of the four items—each describes the middle of the data in its own way. The median is typical (central or average) in the sense that *it splits the data into two parts so that the values of half the items are less than or equal to the median, while the values of the other half are greater than or equal to the median.* The mean, on the other hand, is typical in the sense that *if all the values were the same size (while their total remains unchanged) they would all be equal to the mean.* (Since $\bar{x} = \dfrac{\Sigma\,x}{n}$, it follows that $n \cdot \bar{x} = \Sigma\,x$, and hence, that n values, each of size $\bar{x}$, have the same total $\Sigma\,x$ as the actual n values of x.)

The median, like the mean, always exists and is unique for any set of data. Furthermore, the median can be used to define the middle of a number of objects, properties, or qualities which do not permit a quantitative description. It is possible, for instance, to rank a number of tasks according to their difficulty and then describe the middle (or median) one as being of "average" difficulty. On the less desirable side, we find that ordering large sets of data manually can be a very tedious job, and what is more serious from the standpoint of statistical inference, a sample median is generally *not so reliable* an estimate of a population mean as the (arithmetic) mean of the same data. The medians of many samples drawn from one and the same population usually vary more widely than the corresponding sample means.

So far as symbolism is concerned, we shall write the median of a sample as $\tilde{x}$ when the sample values are denoted by the letter x. The population median is usually written as $\tilde{\mu}$, but there will be no need for this notation in this book. In most problems of estimation we assume that the population mean and median coincide; in fact, our main interest in the median is in problems where sample medians are used to estimate population means.

Another measure which is sometimes used to describe the center of a set of data is the *mode*, which is defined simply as the value that occurs with the highest frequency. Thus, if more students in a class make 70 on a quiz than any other grade, then 70 is the *modal grade* (that is, the mode is 70); if at some time more banks in a region pay 6 per cent on large time deposits than any other rate, then 6 per cent is the modal rate; and if more turret lathes in a shop are 7 years old than any other age, then 7 is their *modal age*. The principal advantage of the mode is that it can be used to "average" qualitative as well as quantitative data. For instance, if most of the handbags sold by a large retailer are white, then white is the *modal color* of these bags. Two definite disadvantages of the mode are that for some sets of data it may not exist and for others it may not be unique. For instance, there is no mode of the ages 19, 23, 29, 20, 31, 25, 22, and 24 (which are all different), and there are two modes (9 and 14) of the dress sizes 7, 10, 14, 9, 9, 14, 9,

18, 16, 12, 11, 14, 14, 14, 9, 20, 9, and 11. The fact that a set of data has more than one mode is sometimes indicative of a lack of homogeneity in the data (for instance, the list of 18 dress sizes given above includes the sizes worn by nine mothers and their nine teenage daughters).

Besides the mean, the median, and the mode, there are various other measures which can be used to describe in some way the center, middle, or average, of a set of data. Some of these are used fairly widely in special situations, while others are used hardly at all. Three other fairly important measures of central location will be given in Exercises 12, 15, and 16 below.

EXERCISES

1. Suppose that we have, for all 45 manufacturers of automotive parts located in a certain area, the total dollar value of their past-due accounts as of a certain date. Give one example each of a situation where these data might be looked upon (a) as a sample and (b) as a population.

2. Suppose we are given complete information about the total sales tax collected by the 138 restaurants in a city during January, 1971. Give one illustration each of a problem in which we would consider these data to constitute (a) a sample and (b) a population.

3. One of the best measures of overall railroad operating profitability is the percentage of revenues brought down to net railway operating income before Federal income taxes. Calculate the mean of the following percentages given for all the railroads operating in a certain district in 5 successive years: 10.4, 11.6, 12.3, 8.1, and 9.6.

4. The "cut-out" syrup density in canned fruits is the percentage (or "degree") by weight of sugar in the syrup solution when a can is opened. A check on samples of eight cans each of three grades of dried prunes yielded the following cut-out densities (measured in degrees on the Brix scale):

> *Fancy grade:* 33, 35, 32, 32, 35, 30, 33, 34
> *Choice grade:* 24, 27, 25, 30, 30, 28, 28, 30
> *Standard grade:* 23, 22, 18, 18, 20, 24, 20, 20

Calculate the mean, the median, and (if it exists) the mode of each of the three samples.

5. On the same day, a sample of 15 grocery stores in a large city posted the following prices (in cents) for a pint of Brand A sour cream, 8 ounces of Brand B cream cheese, and 9-inch Brand C fruit pies:

Sour cream: 59, 58, 60, 61, 59, 59, 59, 62, 59, 58, 62, 59, 59, 61, 61
Cream cheese: 28, 29, 29, 28, 29, 30, 29, 31, 29, 29, 32, 28, 31, 29, 30
Fruit pies: 59, 59, 60, 62, 62, 62, 63, 58, 62, 62, 60, 60, 61, 59, 62

Calculate the mean, the median, and the mode of the prices of each item.

6. Calculate the mean, the median, and the mode of the following percentage interest rates quoted at the same time on conventional mortgage loans by 10 lending institutions: $8\frac{1}{2}$, $7\frac{1}{2}$, $8\frac{1}{4}$, 8, $7\frac{1}{2}$, 9, $8\frac{1}{2}$, $7\frac{3}{4}$, $8\frac{1}{2}$, and $8\frac{1}{2}$.

7. At the end of each month, a manufacturer with several sales outlets measures the performance of its 15 credit men by expressing their respective bad debts as a percentage of sales, getting for one month 0.80, 0.94, 1.10, 1.00, 0.99, 0.98, 1.00, 4.20, 0.95, 0.98, 1.01, 0.90, 0.88, 0.75, and 0.80 per cent. Find the mean, the median, and the mode of these percentages, and comment on the suitability of each as a way of expressing average performance of the group.

8. A bridge is designed to carry a maximum load of 55,000 pounds. If at a given moment it is loaded with 15 vehicles having an average (mean) weight of 3,000 pounds, is the bridge overloaded?

9. A passenger elevator has a rated maximum capacity of 4,000 pounds. Is the elevator overloaded if at one time it carries

(a) Twenty-five passengers whose average (mean) weight is 167 pounds?

(b) Ten children whose average weight is 65 pounds and 20 adults whose average weight is 150 pounds?

(c) Twelve women whose average weight is 115 pounds and 15 men whose average weight is 180 pounds?

10. It has been reported that "the typical embezzler is about 34, married and the father of two children, belongs to the middle income group, and works on the average five years before he begins to rob the till." Comment on the statistical aspects of this statement.

11. A bill was introduced in a state legislature to repeal the sales tax on prescription drugs. Comment on the argument of the state finance director that "the average per capita prescription bill for the past three years is a trifling $2.00, which is not really a burden to anyone."

12. (*Weighted mean*) In averaging quantities it is often necessary to account for the fact that not all of them are equally important in the phenomenon being described. For instance, if a person makes three investments which return him 4, 5, and 6 per cent, respectively, his average return is $\dfrac{4+5+6}{3} = 5$ per cent

only if he puts the same amount in each of the three investments. In order to give quantities being averaged their proper degree of importance, it is necessary to assign them (relative importance) *weights*, and then calculate a *weighted mean*. In general, the weighted mean $\bar{x}_w$ of a set of numbers $x_1, x_2, \ldots, x_n$, whose relative importance is expressed numerically by a corresponding set of numbers $w_1, w_2, \ldots, w_n$, is given by

$$\bar{x}_w = \frac{w_1 x_1 + w_2 x_2 + \cdots + w_n x_n}{w_1 + w_2 + \cdots + w_n} = \frac{\Sigma\, w \cdot x}{\Sigma\, w} \qquad \bigstar$$

Note that if all the weights are equal, the formula gives the ordinary (arithmetic) mean.

(a) Use the formula for the weighted mean to show that if a person invests $1,000 at 4 per cent, $2,000 at 5 per cent, and $20,000 at 6 per cent, his average return on these investments is 5.83 per cent.

(b) If an investor bought 50 shares of General Motors stock at $80 a share, 80 shares at $110 a share, and 70 more shares at $90 a share, what was his average cost per share?

(c) If the Federal tax on corporation earnings is 22 per cent of the first $25,000 and 48 per cent of the remainder, what is the average tax rate on earnings of $60,000? Of $125,000?

(d) If in a manufacturer's closeout a person bought 24 LP's at 66 cents, 16 at 88 cents, and 10 at 99 cents, received a 10 per cent quantity discount and paid a 5 per cent sales tax, what was his average cost per record?

(e) A student scored 70, 80, and 90 points, respectively, on his first quiz, midterm, and final exam in a course on money and banking. If the instructor considers the midterm to be three times as important as the first quiz and the final exam three times as important as the midterm in determining the course grade, what was the student's average on this part of the course?

(f) In a study concerned with the impact of rising home construction and replacement costs on the adequacy of home insurance policies, a sample of homes is taken in a certain region. Among them the Type A homes were underinsured on the average by $4,000, the Type B homes were underinsured on the average by $5,200, and the Type C homes were underinsured on the average by $5,900. If in the study there were twice as many Type B homes as Type A homes and twice as many Type C homes as Type B homes, what is the average amount by which all these homes are underinsured?

13. A business consultant carpets the three main rooms in his office as follows: the 15-foot by 20-foot room with carpeting costing $30 per square yard, the 12-foot by 15-foot room with carpeting costing $18 per square yard, and the 15-foot by 15-foot room with carpeting costing $12 per square yard. What is the average price he pays per square yard of carpeting?

14. The following is a special case of the weighted-mean formula of Exercise 12: Given k sets of data having the means $\bar{x}_1, \bar{x}_2, \ldots, \bar{x}_k$ and consisting, respectively, of $n_1, n_2, \ldots, n_k$ observations, the *grand mean* of all the data is given by the formula

$$\frac{n_1\bar{x}_1 + n_2\bar{x}_2 + \cdots + n_k\bar{x}_k}{n_1 + n_2 + \cdots + n_k} = \frac{\Sigma\, n \cdot \bar{x}}{\Sigma\, n} \qquad \bigstar$$

where the numerator represents the total of all the observations and the denominator represents the total number of observations.

(a) If in 3 consecutive years 4.2, 4.0, and 4.5 million spending units spent on the average $260, $275, and $285, respectively, on refrigerators, find the average expenditure for refrigerators per spending unit.

(b) In one year a large national manufacturer made Plan A awards averaging $4,200 to 150 employees and Plan B awards averaging $4,900 to 350 employees. What was the average award made to these 500 employees?

(c) On opening day of the duck season in one state, 600, 620, 580, 170, and 300 hunters on the five state-operated waterfowl management areas where public hunting is allowed folded on the average 3.9, 3.9, 3.5, 3.0, and 2.3 ducks, respectively. What was the average number of ducks folded per hunter?

(d) A class in organizational behavior consists of some Juniors, some Seniors, and some graduate students. If the 8 Juniors averaged 70 on the final exam, the 12 Seniors averaged 75 points, and the 4 graduate students averaged 60 points, find (i) the mean score in the final exam for all the undergraduates in the class, and (ii) the overall mean score for the entire class.

15. (*Geometric mean*) The geometric mean of a set of n positive numbers is the nth root of their product. If the numbers are all the same, the geometric mean equals the arithmetic mean, but otherwise the geometric mean is always less than the arithmetic mean. For example, the geometric mean of the numbers 1, 1, 2, and 8 is $\sqrt[4]{1 \cdot 1 \cdot 2 \cdot 8} = 2$, whereas their (arithmetic) mean is 3. The geometric mean is used mainly to average ratios, rates of change, and index numbers (see Chapter 4), and in practice it is usually calculated by making use of the fact that *the logarithm of the geometric mean of a set of numbers equals the arithmetic mean of their logarithms.*

(a) Find the geometric mean of the numbers 8 and 32.

(b) Find the geometric mean of the numbers 4, 6, and 9.

(c) Find the geometric mean of the numbers 1, 2, 4, 8, and 16.

(d) A company takes a sample of 5 items it buys and finds that they cost 112, 120, 104, 240, and 116 per cent of what they cost a year earlier. Find the geometric mean of these percentages.

16. (*Harmonic mean*) The harmonic mean of n numbers $x_1, x_2, \ldots, x_n$ is defined as n divided by the sum of the *reciprocals* of the n numbers, namely, as $\dfrac{n}{\Sigma\,(1/x)}$.

It has limited usefulness, but it is appropriate in some special situations. For instance, if a commuter drives 10 miles on the freeway at 60 miles per hour and the next 10 miles off the freeway at 30 miles per hour, he will *not* have averaged

$\dfrac{60 + 30}{2} = 45$ miles per hour. He will have driven 20 miles in a total of 30 minutes, so that his average speed is 40 miles per hour.

(a) Verify that the harmonic mean of 60 and 30 is 40, so that it gives the appropriate "average" in the above example.

(b) If a merchant spends \$12 on novelty items costing 40 cents a dozen and another \$12 on other novelty items costing 60 cents a dozen, what is his average cost per dozen? Verify that the harmonic mean of 60 and 40 gives the correct answer.

(c) If a restaurant buys \$72 worth of butter at 60 cents per pound, \$72 worth at 72 cents per pound, and \$72 at 90 cents per pound, calculate the average price per pound and verify that it is the harmonic mean of 60, 72, and 90.

(The following exercises pertain to the material in Technical Note 1.)

17. Write each of the following as a summation:

(a) $y_1^2 + y_2^2 + \cdots + y_{30}^2$;

(b) $x_3 y_3 + x_4 y_4 + \cdots + x_9 y_9$;

(c) $x_1 f_1 + x_2 f_2 + \cdots + x_n f_n$;

(d) $A_5 + A_6 + \cdots + A_{10}$;

(e) $2x_1 + 2x_2 + \cdots + 2x_{20}$;

(f) $(z_1 - y_1) + (z_2 - y_2) + \cdots + (z_n - y_n)$.

18. Write each of the following expressions without summation signs:

(a) $\displaystyle\sum_{i=1}^{9} x_i$;

(b) $\displaystyle\sum_{i=1}^{6} (x_i - k)$;

(c) $\displaystyle\sum_{i=3}^{7} x_i y_i$;

(d) $\displaystyle\sum_{i=4}^{8} y_i^2$;

(e) $\displaystyle\sum_{i=2}^{5} 3x_i^2$;

(f) $\displaystyle\sum_{j=1}^{n} (x_j - z_j)$.

19. Given $x_1 = 2$, $x_2 = 1$, $x_3 = 2$, $x_4 = 1$, and $x_5 = 3$, find

(a) $\sum x$; (c) $\sum (x - 2)$; (e) $\sum (3x - 2)$;

(b) $\sum x^2$; (d) $\sum (x - 2)^2$; (f) $[\sum (3x - 2)]^2$.

20. Given $x_1 = 1$, $x_2 = 2$, $x_3 = -4$, $f_1 = 3$, $f_2 = -1$, $f_3 = 2$, $y_1 = 3$, $y_2 = 1$, and $y_3 = 5$, find

(a) $\sum x \cdot f$; (c) $\sum x \cdot y$; (e) $\sum (x - y)^2$;

(b) $\sum x^2 \cdot f$; (d) $\sum (x - y)$; (f) $\sum x \cdot y \cdot f$.

21. Given that $\displaystyle\sum_{i=1}^{7} x_i = 17$ and $\displaystyle\sum_{i=1}^{7} x_i^2 = 53$, find

(a) $\displaystyle\sum_{i=1}^{7} (x_i - 2)$; (b) $\displaystyle\sum_{i=1}^{7} (2x_i + 1)$; (c) $\displaystyle\sum_{i=1}^{7} (x_i + 3)^2$.

22. Prove that

(a) $\displaystyle\sum_{i=1}^{n} (x_i - k) = \sum_{i=1}^{n} x_i - nk$;

(b) $\sum_{i=1}^{n} (x_i - \bar{x}) = 0$, where $\bar{x}$ is the mean of the x_i;

(c) $\sum_{i=1}^{n} (x_i - k)^2 = \sum_{i=1}^{n} x_i^2 - 2k \cdot \sum_{i=1}^{n} x_i + nk^2$.

23. Is it true in general that $\left[\sum_{i=1}^{n} x_i\right]^2 = \sum_{i=1}^{n} x_i^2$? (*Hint:* Check whether the equation holds for $n = 2$.)

MEASURES OF VARIATION

One of the most important characteristics of a set of data is that the values are generally *not all alike;* indeed, the precise extent to which they are unalike, or vary among themselves, is of basic importance in statistics. The various measures of central location discussed in the preceding section describe one important property of a set of data, their middle or their "average," but they do not tell us anything about this other basic characteristic. Hence, we require ways of measuring the extent to which statistical data are dispersed, spread out, or bunched, and the statistical measures which provide us with this information are called *measures of variation*.

First, let us give a few examples to illustrate the importance of measuring variability. Suppose, for instance, that we are interested in buying a certain stock and we find that its latest closing price on the New York Stock Exchange is $51\frac{1}{8}$ (dollars per share). To decide whether this is a good buy, many things will have to be taken into account; among other things we may want to know how much the price of this stock *fluctuates* and we thus look for the year's High and Low. Supposing that these figures are $68\frac{1}{2}$ and 45, respectively, we now have some information about the *variation* in the price of the stock, and this may be critical in deciding whether or not it should be bought. Of course, an intelligent investor would want to compare the fluctuations of this stock with those of other stocks and, probably, also with those of the market as a whole.

The concept of variability or dispersion is of fundamental importance in statistical inference (estimation, testing hypotheses, making predictions, etc.), where key questions always involve the concept of *chance variation*. To illustrate what this means, suppose that a coin is flipped 100 times and that we obtain 29 heads and 71 tails. In order to decide whether this *supports or contradicts* an assumption that the coin is balanced (that heads and tails are equally likely), let us see for a moment what we might reasonably expect. Although we would, on the average, expect 50 heads and 50 tails, we would not be surprised if we got, say, 48 heads and 52 tails, 54 heads and 46 tails, or 43 heads and 57 tails, attributing the few extra heads or tails entirely to chance. To see whether 29 heads in 100 flips, a discrepancy of 21, might also be attributed to chance, let us investigate this kind of chance effect by repeatedly flipping a supposedly balanced coin 100 times and observing the results. Suppose, then, that we perform 10 such "experiments" and that we obtain 51, 54, 48, 55, 41, 49, 58, 52, 46, and 51 heads. These results provide us with some idea about the magnitude of the fluctuations (variations) produced by chance in the number of heads we may obtain in 100 flips of a coin. Judging from the above data, in which the number of heads varied from 41 to 58, we might feel that anything from 40 to 60 heads is "not unusual," but that 29 heads in 100 tosses is completely "out of line." We thus conclude that there must be something wrong with the coin; perhaps it is worn on one side or bent. This whole argument has been presented on a rather intuitive basis, in order to illustrate the need for measuring chance variation; in later chapters we shall treat the same problem more rigorously and in more detail.

To consider another example where the concept of variability plays an important role in a problem of statistical inference, suppose we want to estimate the true mean (net) weight of all cans of beef hash in one very large production lot put out by a food processor. Suppose, furthermore, that the observed weights in a sample of three cans from the lot were 15.0, 14.8, and 15.2 ounces. The mean of the sample weights is $\bar{x} = 15.0$ ounces, and in the absence of any further information we might use this figure as an estimate of the actual mean weight of all the cans. Now, the "goodness" of this estimate will obviously depend on the *size* of the sample, which was very small, but it will also depend in a very real way on the *variability of the population*, namely, the variability of the weights of *all* of the cans of beef hash in the entire production lot. To illustrate this, let us consider the following two possibilities:

Case 1: The true mean weight is 15.1 ounces, the filling process is *very consistent*, and all the cans in the lot weigh somewhere between 14.8 and 15.4 ounces.

Case 2: The true mean weight is 15.1 ounces, but the filling process is *very inconsistent*, and the weights of the cans in the lot vary widely from 13.0 to 17.2 ounces.

If the population of weights whose mean we are trying to estimate is the (relatively homogeneous) one described in Case 1, we can be *sure* that the sample mean will not differ from the true mean by much, regardless of the size of the sample. In fact, a sample mean cannot possibly be off by more than 0.3 ounces, and off by this much only if the sample weights are all

14.8 or all 15.4 ounces. The population described in Case 2, however, is a much less homogeneous collection of weights, even though its mean is the same as in Case 1. If, by chance, the three sample weights chosen from the second population were all 13.0 or all 17.2 ounces, the sample mean would differ from the population mean μ (which we are trying to estimate) by 2.1 ounces. This should make it very clear that *in order to judge the "closeness" of an estimate or the "goodness" of a generalization based on a sample, we must know something about the variation of the population from which the sample was obtained.*

The purpose of all these examples has been to demonstrate how the concept of variability plays an important role in practically all aspects of statistics. While giving the first example, we actually introduced one way of indicating (or measuring) variability, namely, by giving the two *extreme values* of a sample. More or less the same is accomplished by giving the *difference between the two extremes*, which we refer to as the sample *range*. Thus, for the first example of this section the *range* of the prices paid for a share of the given stock is $68\frac{1}{2} - 45 = 23\frac{1}{2}$ dollars, and for the last example the range of the weights obtained for the three cans of beef hash is $15.2 - 14.8 = 0.4$ ounces.

In spite of the obvious advantages of the range, that it is *easy to calculate* and *easy to understand*, it does not provide a very useful measure of variation in a wide variety of statistical problems. Being based only on the two extremes, its main shortcoming is that it does not tell us anything about the dispersion of the data which fall in between. Each of the following three sets of data

5	17	17	17	17	17	17	17	17	17
5	5	5	5	5	17	17	17	17	17
5	6	8	10	11	14	14	15	16	17

has a range of $17 - 5 = 12$, but the dispersions of the data are by no means the same. Thus, the range is used mainly in situations where it is desired to get a quick, though not necessarily a very accurate, picture of the variability of a set of data. In some cases, where the sample size is very small, the range is quite adequate as a measure of variation; it is thus used widely in *quality control*, where it is important to keep a continuous check on the variability of raw materials, machines, or manufactured products by regularly taking small samples.

To define the *standard deviation*, by far the most useful measure of variation, let us observe that the dispersion of a set of data is *small* if the numbers are bunched closely about their mean, and that it is *large* if the numbers are spread over considerable distances away from their mean. Hence, it would seem reasonable to measure the variation of a set of data in terms of the amounts (distances or deviations) by which the various numbers depart from their mean. If a set of numbers $x_1, x_2, \ldots, x_N$, looked upon as a population, has the mean μ, the differences $x_1 - \mu, x_2 - \mu, \ldots, x_N - \mu$ are called the *deviations from the mean*, and it seems that we might use their average, namely, their mean, as a measure of the variation of the population. This would not be a bad idea, except for the fact that we would always get 0 for an answer, no matter how widely dispersed the data might be. Using the rules of Technical Note 1, it can easily be shown that $\Sigma(x - \mu)$, that is, the sum of the deviations from the mean, is always equal to zero—some of the deviations are positive, some are negative, and their sum and their mean both are always equal to zero (see also Exercise 22 on page 38).

Since we are really interested in the *magnitude* of the deviations and not in their signs, we might simply "ignore" the signs and, thus, define a measure of variation in terms of the *absolute values* of the deviations from the mean. Indeed, adding the values of the deviations from the mean as if they were all positive and dividing by N, we obtain an intuitively appealing measure of variation called the *average deviation*. However, using precisely the same deviations from the mean, there is another way of eliminating their signs, which is preferable on theoretical grounds. The *squares* of the deviations from the mean cannot be negative; in fact, they are positive unless an x happens to coincide with the mean, in which case both $x - \mu$ and $(x - \mu)^2$ are equal to zero. It thus seems reasonable to measure the variability of a set of data in terms of the *squared deviations from the mean*, and this leads us to the following formula, defining what is called the *population variance:*

$$\sigma^2 = \frac{\Sigma (x - \mu)^2}{N}$$

This measure of variation, which is denoted by σ^2 (where σ, *sigma*, is the lowercase Greek letter for s), is simply the mean of the squared deviations from the mean, and it is sometimes called the "mean-square deviation."

The variance of a set of data is an extremely important measure of variation and it is used extensively in statistical work. By reason of squaring the deviations, however, the variance is not in the same units of measurement as the data themselves—if the data are in inches the variance is in inches *squared*, if the data are in pounds the variance is in pounds *squared*, and so on. To compensate for this, we simply take the square root of the population variance, getting the *population standard deviation*

$$\sigma = \sqrt{\frac{\Sigma (x - \mu)^2}{N}}$$

which is sometimes called the "root-mean-square deviation." Indeed, it is the square root of the mean of the squared deviations from the mean, and it is in the same units of measurement as the original data.

Having defined the standard deviation of a population, it may seem logical to use the same formula also for a sample, with n and $\bar{x}$ substituted for N and μ. This is almost, but not quite, what we do. Instead of dividing the sum of the squared deviations from the mean by n, we divide it by $n - 1$, and accordingly we define the *sample standard deviation* as

$$s = \sqrt{\frac{\Sigma\,(x - \bar{x})^2}{n - 1}}$$

and its square, the *sample variance*, as

$$s^2 = \frac{\Sigma\,(x - \bar{x})^2}{n - 1}$$

In using $n - 1$ instead of n in the denominator of these two formulas, we are not just being arbitrary. There is a good reason for it, which is explained in Technical Note 2 on page 62.

To illustrate the calculation of a sample standard deviation, let us find s for the lengths (in inches) of the following sample of six pieces of scrap metal left after a cut-off operation in a mill: 4.3, 3.9, 4.0, 4.2, 3.5, and 4.1. In calculating a sample standard deviation by means of the above formula, we must (1) find $\bar{x}$, (2) determine the n deviations from the mean $x - \bar{x}$, (3) square these deviations, (4) sum the squared deviations, (5) divide this sum by $n - 1$, and (6) take the square root of the quantity arrived at in Step (5).* For the lengths of the six pieces of scrap metal we thus get

x	$x - \bar{x}$	$(x - \bar{x})^2$	
4.3	0.3	0.09	$\bar{x} = \dfrac{24.0}{6} = 4.0$ inches
3.9	−0.1	0.01	
4.0	0.0	0.00	
4.2	0.2	0.04	$s = \sqrt{\dfrac{0.40}{5}} = \sqrt{0.08} = 0.28$ inches
3.5	−0.5	0.25	
4.1	0.1	0.01	
24.0	0.0	0.40	

* Square roots may be obtained from Table XII, whose use is explained in Technical Note 3 on page 63.

On the basis of this sample we can thus estimate the mean μ of the given population of lengths of scrap metal as 4.0 inches, and we can estimate the standard deviation σ of this population as 0.28 inches.

It was easy in this example to calculate s because the measurements were given to one decimal and the mean was a whole number. Had this not been the case, the calculations could have been fairly cumbersome, and it might well have been preferable to use the following *short-cut formula for* s

$$s = \sqrt{\frac{n(\Sigma\, x^2) - (\Sigma\, x)^2}{n(n-1)}} \qquad \bigstar$$

This formula does *not* constitute an approximation and it can be derived from the other formula for s by using the rules for summations given in Technical Note 1 on page 61. The advantage of this short-cut formula is that we do not have to go through the process of actually finding the deviations from the mean; instead we calculate $\Sigma\, x$, the sum of the x's, $\Sigma\, x^2$, the sum of their squares, and substitute directly into the formula.

Referring again to the six scrap metal cut-offs, we now get

x	x^2
4.3	18.49
3.9	15.21
4.0	16.00
4.2	17.64
3.5	12.25
4.1	16.81
24.0	96.40

$$s = \sqrt{\frac{6(96.40) - (24.0)^2}{6 \cdot 5}}$$

$$= \sqrt{\frac{2.40}{30}} = \sqrt{0.08} = 0.28 \text{ inches}$$

In this particular example it may seem that the "short-cut" method is actually more involved; this may be the case, but in actual practice, when we are dealing with realistically complex data (and not merely a blackboard example), the short-cut formula usually provides considerable simplifications. Incidentally, the same short-cut formula can be used to find σ, provided we substitute n for the factor $n-1$ in the denominator before we replace s and n with σ and N. A further simplification in the calculation of s or σ consists of subtracting a suitable constant (the *same* number) from each of the values for which we want to compute the standard deviation. The advantage of doing this is illustrated in Exercise 9 on page 47.

In the argument which led to the definition of the standard deviation, we observed that the dispersion of a set of data is small if the values are bunched closely about their mean and that it is large if the values are spread over considerable distances away from the mean. Correspondingly, we can now say that if the standard deviation of a set of data is small, the values are concentrated near the mean, and if the standard deviation is large, the values are scattered widely about the mean. To present this argument on a less intuitive basis (after all, what is *small* and what is *large?*), let us briefly mention an important theorem, which will be taken up later in Chapter 7. According to this theorem, called *Chebyshev's Theorem*, we can be *sure* for any kind of data (populations as well as samples) that, among other things, *at least 75 per cent of the data must fall within two standard deviations of the mean.* Thus, if a set of data has the mean $\bar{x} = 122$ and $s = 15$, we can be *certain* that at least 75 per cent of the data fall on the interval from 92 to 152. Had the standard deviation of this set of data been smaller, say, $s = 3$, we could have argued that at least 75 per cent of the data must fall on the much smaller interval from 116 to 128. According to Chebyshev's Theorem, we can also assert that at least 88.8 per cent of the data must fall within *three* standard deviations of the mean and that at least 96 per cent must fall within *five* standard deviations of the mean. Thus, if two populations have the same mean of $\mu = 240$ while their standard deviations are $\sigma = 50$ and $\sigma = 20$, respectively, we can assert for the first population that at least 75 per cent of the data must fall between 140 and 340, while for the second population we can make the much stronger statement that at least 96 per cent of the data must fall on the same interval. This illustrates how the magnitude of the standard deviation "controls" the concentration of a set of data about its mean.

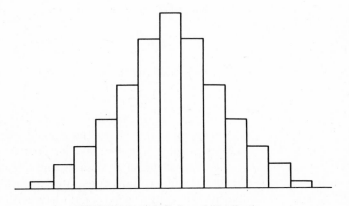

FIGURE 3.1   A bell-shaped distribution

An important feature of Chebyshev's Theorem is that it holds for any kind of data. However, if we *do* have some information about the overall shape of a set of data, that is, the overall shape of their distribution, we can often make much stronger statements. For instance, if a distribution is *bell-shaped* like the one shown in Figure 3.1, we can expect roughly 95 per cent (instead of *at least* 75 per cent) of the data to fall within two standard deviations of the mean and over 99 per cent (instead of *at least* 88.8 per cent) to fall within three standard deviations of the mean. The percentages given here pertain to the so-called *normal distribution*, which will be discussed in detail in Chapter 7.

In the beginning of this section we demonstrated that there are many ways in which knowledge of the variability of a set of data can be important. Another interesting application arises in the comparison of numbers belonging to *different sets of data*. Suppose, for instance, that a large national securities firm administers a battery of tests to all job applicants and that Mr. Hilliard scored 135 points on the General Information (GI) test and 265 points on the Accounting and Finance (AF) test. At first glance it may seem that Mr. Hilliard was much better (nearly twice as good) in accounting and finance than he was in general information. However, if the mean score which thousands of applicants obtained on the GI test was 100 points with a standard deviation of 15 points, and the mean score which all these applicants obtained on the AF test was 250 with a standard deviation of 30 points, we can argue that Mr. Hilliard's score on the GI test was

$$\frac{135 - 100}{15} = 2\tfrac{1}{3} \text{ standard deviations}$$

above the mean of the distribution of all the scores on this test, while his score on the AF test was only

$$\frac{265 - 250}{30} = \tfrac{1}{2} \text{ standard deviation}$$

above the mean of the distribution of all the scores on this test. These two figures can now be meaningfully compared, whereas the original raw scores cannot. Clearly, Mr. Hilliard rates much higher on his command of general information than he does on his knowledge of accounting and finance.

What we did here was to convert the two raw scores into so-called *standard units*. If x is a measurement belonging to a set of data having the mean $\bar{x}$ (or μ) and the standard deviation s (or σ), then its value in *standard units*, denoted by the letter z, is given by

$$z = \frac{x - \bar{x}}{s} \qquad \text{or} \qquad z = \frac{x - \mu}{\sigma} \qquad \bigstar$$

depending on whether the data constitute a sample or a population. In these units, z tells us how many standard deviations a value lies above or below the mean of the set of data to which it belongs. Further examples illustrating the importance of converting into standard units will be discussed in Chapter 7.

EXERCISES

1. Using the basic formula defining s on page 43, calculate the standard deviation of the five percentages of Exercise 3 on page 34.

2. Calculate the sample standard deviations of the cut-out syrup densities of the three grades of dried prunes given in Exercise 4 on page 34. Make the calculations for the fancy grade using *both* the formula defining s and the short-cut formula, but use only the short-cut formula for the other two grades.

3. Use the short-cut formula to calculate s for each of the three samples of food prices of Exercise 5 on page 34.

4. Calculate s for the following five weights (in ounces): 2.2, 1.8, 2.0, 1.6, and 1.4.

5. Calculate σ for the finite population which consists of the integers 1, 2, 3, 4, 5, 6, 7, and 8.

6. Calculate s for the 15 percentages of Exercise 7 on page 35.

7. Find the ranges of the cut-out syrup densities of the three grades of dried prunes given in Exercise 4 on page 34.

8. Find the ranges of the three samples of food prices of Exercise 5 on page 34.

9. In four attempts, it took a man 48, 55, 51, and 50 minutes to do a certain job.
   (a) Find the mean, the range, and the standard deviation of these four sample values.
   (b) Subtract 50 minutes from each of the times, recalculate the mean, the range, and the standard deviation, and compare the results with those obtained in part (a).
   (c) Add 10 minutes to each of the times, recalculate the mean, the range, and the standard deviation, and compare the results with those obtained in part (a).
   (d) Multiply each of the sample values by 2, recalculate the mean, the range, and the standard deviation, and compare the results with those obtained in part (a).
   (e) In general, what effect does (i) adding a constant to each sample value, and (ii) multiplying each sample value by a positive constant, have on the mean, the range, and the standard deviation of a sample?

10. The commissions earned by a sample of four automobile salesmen in a given week were $320, $280, $311, and $290. Calculate the standard deviation of these data making use of what the reader has learned from Exercise 9.

11. In a final examination in accounting, 630 students averaged 69.5 with a standard deviation of 8.2. *At least* how many students must have received grades from 44.9 to 94.1?

12. The manager of a movie theater knows that his average daily attendance is 372 with a standard deviation of 20.5. *At least* what part of the time will the attendance be anywhere from 331 to 413?

13. An investment service reports for each stock that it lists the price at which it is currently selling, its average price over a certain period of time, and a measure of its variability. Stock A, it reports, has a normal (average) price of $56 with a standard deviation of $11, and is currently selling at $74.50; Stock B sells normally for $35, has a standard deviation of $4, and is currently selling at $47. If an investor owns both stocks and wants to dispose of one, which one might he sell and why?

14. Mrs. Jones belongs to an age group for which the average weight is 136 pounds with a standard deviation of 15 pounds, and Mrs. Smith belongs to an age group for which the average weight is 121 pounds with a standard deviation of 18 pounds. If Mrs. Jones weighs 160 pounds and Mrs. Smith weighs 146 pounds, which of the two is more seriously overweight compared to her group?

15. In a large Western city the average cost of an appendectomy is $237.50 (with a standard deviation of $24.60), the average cost of a tonsillectomy is $112.50 (with a standard deviation of $16.25), and the average cost of having a tooth filled is $8.67 (with a standard deviation of $1.80). Among two surgeons and a dentist practicing in this city, Dr. A charges $275.00 for an appendectomy, Dr. B charges $135.00 for a tonsillectomy, and Dr. C charges $10.00 for filling a tooth. Which of the three doctors is relatively most expensive?

16. (*Relative variation*) One disadvantage of the standard deviation as a measure of variation is that it depends on the units of measurement. Thus, if a set of measurements, say, of the weight of bags of peanuts, has a standard deviation of 0.16 ounces, we would look at this variability differently if we were dealing with 8-ounce bags or with 2-pound bags. What we need in a situation like this is a measure of *relative variation*, such as the *coefficient of variation*

$$V = \frac{s}{\bar{x}} \cdot 100 \qquad \text{or} \qquad V = \frac{\sigma}{\mu} \cdot 100 \qquad\qquad ★$$

which expresses the standard deviation as a percentage of the mean.

(a) Use the results of Exercise 2 on page 47 and Exercise 4 on page 34 to calculate the coefficient of variation for the cut-out syrup densities of the three grades of prunes.

(b) Use the results of Exercise 3 on page 47 and Exercise 5 on page 34 to calculate the coefficient of variation for each of the three samples of food prices.

(c) Calculate the coefficient of variation for the four commissions of Exercise 10 above.

(d) Referring to the two stocks of Exercise 13 above, which is, relatively speaking, more variable?

THE DESCRIPTION OF GROUPED DATA

Since the calculation of the various measures which we have discussed is fairly easy and straightforward, there is usually no need to look for further short cuts or simplifications. However, if the numbers are unwieldy, that is, if each number has many digits, or if the sample (or population) size is very large, it may be advantageous to group the data before calculating any statistical descriptions. Another reason why we shall have to study the description of grouped data is that published data are often available only in the form of distributions.

Earlier, in Chapter 2, we observed that the grouping of data entails some loss of information. Each item, so to speak, loses its identity (we only know how many items fall into each class), and the *actual* values of the statistical descriptions discussed so far in this chapter can no longer be calculated. However, good approximations can often be obtained by *assigning the value of the class mark to each item falling into a given class*, and this is how we shall *define* the mean and the standard deviation of a distribution. As we shall see on page 53, the *definition* of the median of a distribution will be based on a different assumption. Thus, to calculate the mean or the standard deviation of the seat-occupancy-time distribution on page 12, the 3 values falling into the first class will be treated as if they all equaled 14.5, the 14 values falling into the second class will be treated as if they all equaled 24.5, . . . , and the 2 values falling into the last class will be treated as if they both equaled 94.5. This procedure is generally very satisfactory, since the errors which are thus introduced will more or less "average out."

To obtain formulas for the mean and the standard deviation of a distribution with k classes, let us designate the successive class marks $x_1, x_2, \ldots, x_k$, and the corresponding class frequencies $f_1, f_2, \ldots, f_k$. The *total* that goes into the numerator of the formula for the mean is thus obtained by adding f_1 times the value x_1, f_2 times the value x_2, . . . , and f_k times the value x_k; in other words, it is equal to $x_1f_1 + x_2f_2 + \cdots + x_kf_k$. Using again the Σ notation introduced on page 30, we can now write the formula for the mean of a distribution as

$$\bar{x} = \frac{\Sigma x \cdot f}{n} \qquad \star$$

where $\Sigma x \cdot f$ represents, in words, the sum of the products obtained by multiplying each class mark by the corresponding class frequency. (When dealing with a population instead of a sample, we have only to substitute μ for $\bar{x}$ in this formula and N for n.) Similarly, the *total* that goes into the numerator of the formula for the standard deviation or the variance of a sample is obtained by adding f_1 times the quantity $(x_1 - \bar{x})^2$, f_2 times the quantity $(x_2 - \bar{x})^2, \ldots$, and f_k times the quantity $(x_k - \bar{x})^2$, and in the corresponding short-cut formulas $\Sigma\, x$ is replaced by $\Sigma\, x \cdot f = x_1 f_1 + x_2 f_2 + \cdots + x_k f_k$, and $\Sigma\, x^2$ is replaced by $\Sigma\, x^2 \cdot f = x_1^2 f_1 + x_2^2 f_2 + \cdots + x_k^2 f_k$. Thus, the corresponding formulas become

and

$$s = \sqrt{\frac{\Sigma\, (x - \bar{x})^2 \cdot f}{n - 1}} \qquad \bigstar$$

$$s = \sqrt{\frac{n(\Sigma\, x^2 f) - (\Sigma\, xf)^2}{n(n - 1)}} \qquad \bigstar$$

where, of course, the x's are the class marks of the distribution and the f's are the corresponding class frequencies. (To obtain corresponding formulas for σ, we have only to substitute μ for $\bar{x}$ and N for the $n - 1$ in the denominator of the first formula, and N for n in the numerator and N^2 for $n(n - 1)$ in the denominator of the second.)

To illustrate the calculation of the mean and the standard deviation of grouped data, let us refer again to the seat-occupancy-time distribution on page 12. All the necessary sums can be obtained as in the following table, where the first and third columns are copied from page 12, the second column contains the class marks (obtained by averaging the respective class limits), and the fourth and fifth columns contain the respective products $x \cdot f$ and $x^2 \cdot f$:

Seat occupancy time (minutes)	Class marks x	Frequencies f	$x \cdot f$	$x^2 \cdot f$
10–19	14.5	3	43.5	630.75
20–29	24.5	14	343.0	8,403.50
30–39	34.5	29	1,000.5	34,517.25
40–49	44.5	22	979.0	43,565.50
50–59	54.5	14	763.0	41,583.50
60–69	64.5	10	645.0	41,602.50
70–79	74.5	4	298.0	22,201.00
80–89	84.5	2	169.0	14,280.50
90–99	94.5	2	189.0	17,860.50
		100	4,430.0	224,645.00

Thus,

$$\bar{x} = \frac{4,430}{100} = 44.3 \text{ minutes}$$

and (using the short-cut formula)

$$s = \sqrt{\frac{100(224,645) - (4,430)^2}{100 \cdot 99}} = \sqrt{286.8} = 16.9 \text{ minutes}$$

As is apparent from this example, the arithmetic required to find the mean and/or the standard deviation of a distribution can be quite involved. This is not serious, however, because there exists a simplification which is based on *coding* the class marks so that we will have smaller numbers to work with. Provided the class intervals are all equal, this process of coding consists of assigning the value 0 to one of the class marks, preferably (in manual calculations) near the center of the distribution, and representing all of the class marks by means of *consecutive integers*. For instance, if a distribution has seven classes and the class mark of the middle class is assigned the value 0, the successive class marks of the distribution are assigned the values -3, -2, -1, 0, 1, 2, and 3.

Of course, when we code the class marks like this, we also have to account for it in the formulas we use to calculate the mean and the standard deviation. Referring to the new (coded) class marks as u's, the formula for the mean of a distribution becomes

$$\bar{x} = x_0 + \frac{\Sigma u \cdot f}{n} \cdot c \qquad \bigstar$$

where x_0 is the class mark (in the original scale) to which we assign 0 in the new scale, c is the class interval, n is the number of items grouped, and $\Sigma u \cdot f$ is the sum of the products obtained by multiplying each of the new class marks by the corresponding class frequency. Similarly, the formula for the standard deviation becomes

$$s = c\sqrt{\frac{n(\Sigma u^2 \cdot f) - (\Sigma u \cdot f)^2}{n(n-1)}} \qquad \bigstar$$

where $\Sigma u^2 \cdot f$ is the sum of the products obtained by multiplying the *squares* of the new class marks by the corresponding class frequencies.

To illustrate the simplifications thus introduced, let us use this kind of coding to recalculate the mean and the standard deviation of the seat-occupancy-time distribution. Rearranging the work, as before, in a table, we now get

x	u	f	$u \cdot f$	$u^2 \cdot f$
14.5	-3	3	-9	27
24.5	-2	14	-28	56
34.5	-1	29	-29	29
44.5	0	22	0	0
54.5	1	14	14	14
64.5	2	10	20	40
74.5	3	4	12	36
84.5	4	2	8	32
94.5	5	2	10	50
		100	-2	284

where the class mark 44.5 was taken to be 0 in the u-scale (shown in the second column of the table). Of course, we could have used any class mark as the zero of the u-scale, for instance 54.5, but as the reader will be asked to verify in Exercise 6 on page 56, this would make the arithmetic a little harder.

Substituting $c = 10$, $x_0 = 44.5$, $n = 100$, $\Sigma\, u \cdot f = -2$, and $\Sigma\, u^2 \cdot f = 284$ into the above formulas for $\bar{x}$ and s, we obtain

$$\bar{x} = 44.5 + \frac{(-2)}{100} \cdot 10 = 44.5 - 0.2 = 44.3 \text{ minutes}$$

$$s = 10\sqrt{\frac{100(284) - (-2)^2}{100 \cdot 99}} = 10\sqrt{2.868} = 16.9 \text{ minutes}$$

These results are, as they should be, identical with the ones obtained earlier —the use of this kind of coding is *not* a method of approximation, and the results should always be identical with those given by the other formulas for $\bar{x}$ and s. To use this kind of coding when dealing with a population, we have only to substitute N for n in the formula for the mean, and N for n in the numerator and N^2 for $n(n-1)$ in the denominator of the formula for the standard deviation.

If we want to determine the *median* of grouped data, we find ourselves in a position similar to the one in which we found ourselves on page 49; that is, its *actual* value can no longer be found. If we proceeded as in the case of the mean and the standard deviation (and assigned the value of the class mark to each item within any given class), the median would simply be the class mark of the class into which the middle item falls. It is easy to see why this would be unreasonable—if the median happened to be the 59th largest of the 60 values which fall in a class, it stands to reason that the median should be very close to the upper class boundary rather than the class mark. For this reason, the median of a distribution is defined as *a number, a point, which is such that half the total area of the rectangles of the histogram of the distribution lies to its left and half lies to its right (see Figure 3.2).* This means that the sum of the areas of the rectangles to the left of

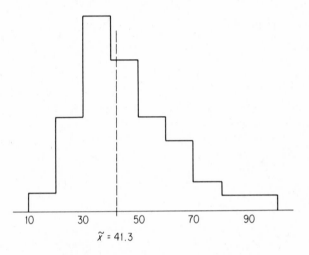

$$\tilde{x} = 41.3$$

FIGURE 3.2   The median of a distribution

the dashed line of Figure 3.2 equals the sum of the areas of the rectangles to its right. Note that this definition amounts to the assumption that, within each class, the items are spread (or distributed) evenly throughout the class interval.

To find the median of a distribution containing n items, we first determine the class into which the median must fall by counting $n/2$ items starting at either end. For the seat-occupancy-time distribution we have $n = 100$, and hence we must count $n/2 = 50$ items starting at either end. Counting from the bottom (that is, beginning with the smallest values) of this distribution, we find that $3 + 14 + 29 = 46$ items are less than 40, and $3 + 14 + 29 + 22 = 68$ items are less than 50. Thus, we must count 4 more items in addition to the 46 which are less than 40, and on the assumption that the 22 values of the 40–49 class are evenly spread throughout its interval, this can be accomplished by adding $\frac{4}{22}$ of the class interval of 10 to 39.5, the lower boundary of the class. Accordingly, we have

$$\bar{x} = 39.5 + \frac{4}{22} \cdot 10 = 41.3 \text{ minutes}$$

for the median of the seat-occupancy-time distribution. In general, if L is the lower boundary of the class containing the median, f is its frequency, c is the class interval, and j the number of items we still lack when reaching L, then the *median of the distribution* is given by the formula

$$\bar{x} = L + \frac{j}{f} \cdot c \qquad\qquad \star$$

It is possible, of course, to arrive at the median of a distribution by starting at the other end and *subtracting* an appropriate fraction of the class interval from the upper boundary U of the class into which the median must fall. For the seat-occupancy-time distribution we thus get

$$\bar{x} = 49.5 - \frac{18}{22} \cdot 10 = 41.3 \text{ minutes}$$

and the two answers are identical, as they should be. A general formula for the case where we start counting at the top (beginning with the largest values) of a distribution is given by

$$\bar{x} = U - \frac{j'}{f} \cdot c \qquad\qquad \star$$

where j' is the number of items we still lack when reaching U.

The method we have just described can also be used to determine more general *positional measures* of a distribution called *fractiles* or *quantiles*. By definition, *a fractile (or quantile) is a value at or below which we will find a given fraction of the data*. There are, for instance, the three *quartiles* Q_1, Q_2, Q_3, which are such that 25 per cent of the data are less than or equal to Q_1, 50 per cent are less than or equal to Q_2, and 75 per cent are less than or equal to Q_3. Also, there are the nine *deciles* D_1, D_2, . . . , and D_9 which are such that 10 per cent of the data are less than or equal to D_1, 20 per cent are less than or equal to D_2, . . . ; and there are the 99 percentiles P_1, P_2,

..., and P_{99} which are such that 1 per cent of the data are less than or equal to P_1, 2 per cent are less than or equal to P_2, ... From this it should be clear that Q_2, D_5, and P_{50} are all equal to the median, and that P_{25} and P_{75} equal, respectively, Q_1 and Q_3. As the reader will be asked to verify in Exercise 10 on page 56, for the seat-occupancy-time distribution we have

$$Q_1 = 29.5 + \frac{8}{29} \cdot 10 = 32.3 \text{ minutes}$$

$$D_9 = 69.5 - \frac{2}{10} \cdot 10 = 67.5 \text{ minutes}$$

and

$$P_{15} = 19.5 + \frac{12}{14} \cdot 10 = 28.1 \text{ minutes}$$

where we used the median formulas, counting in each case the appropriate percentage of the items in the distribution.

To conclude this discussion of positional measures of grouped data, let us point out that there exist fairly elaborate ways of defining the mode of a distribution, but in most situations a reference to the *modal class*, the class with the highest frequency, is all one needs.

EXERCISES

1. Calculate the mean and the standard deviation of the distribution of sack weights of Exercise 2 on page 14 (a) without coding and (b) with coding.

2. The following is a distribution of the final examination grades obtained by 150 students in a course on finance:

Grades	Frequency
0–19	17
20–39	45
40–59	53
60–79	27
80–99	8

Find the mean and the standard deviation of these grades, looked upon as a sample, (a) without coding and (b) with coding.

3. Find the mean and the standard deviation of whichever data you grouped among those of Exercises 9, 11, and 12 on pages 16 and 17. (Treat the data as a sample.)

4. Find the mean and the variance of whichever data you grouped among those of Exercises 13 and 14 on page 17. (Treat the data as a population.)

5. Find the mean and the standard deviation of whichever data you grouped among those of Exercises 15 and 16 on pages 17 and 18. (Treat the data as a sample.)

6. Recalculate the mean and the standard deviation of the seat occupancy distribution using a coding which assigns 0 to 54.5 in the original scale, and compare the results with those obtained on page 52.

7. Find the median of the weekly earnings distribution of the 1,216 secretaries in the Phoenix area as given on page 8.

8. Referring to the same distribution as in Exercise 1, find (a) the median, (b) the two quartiles Q_1 and Q_3, (c) the three deciles D_2, D_4, and D_8, and (d) the two percentiles P_5 and P_{95}.

9. With reference to the distribution of Exercise 2, find (a) the median, (b) the two quartiles Q_1 and Q_3, (c) the two deciles D_1 and D_9, and (d) the two percentiles P_{35} and P_{65}.

10. Verify the values given on page 55 for Q_1, D_9, and P_{15} of the seat-occupancy-time distribution, and also determine (a) the third quartile Q_3, (b) the deciles D_1, D_4, and D_6, and (c) the percentiles P_5, P_{85}, and P_{95}.

11. Given the quartiles of a distribution, we sometimes use the *midquartile* $\dfrac{Q_1 + Q_3}{2}$ as a measure of central location instead of using the median or the mean, the *interquartile range* $Q_3 - Q_1$ or the *semi-interquartile range* $\dfrac{Q_3 - Q_1}{2}$ as a measure of variation, and the *coefficient of quartile variation* $\dfrac{Q_3 - Q_1}{Q_1 + Q_3} \cdot 100$ as a measure of *relative variation*.

(a) Calculate all these statistical descriptions for the seat-occupancy-time distribution.

(b) With reference to the distribution of Exercise 8, find the midquartile and the interquartile range.

(c) With reference to the distribution of Exercise 9, find the midquartile, the semi-interquartile range, and the coefficient of quartile variation.

FURTHER DESCRIPTIONS

So far we have discussed statistical descriptions coming under the general headings of "measures of location" and "measures of variation." Actually, there is no limit to the number of ways in which statistical data can be described, and statisticians are continually developing new methods of describing characteristics of numerical data that are of interest in particular problems. In this section we shall briefly study the problem of describing the *overall shape* of a distribution.

Although frequency distributions can assume almost any shape or form, there are certain standard types which fit most distributions we meet in actual practice. Foremost among these is the aptly described *bell-shaped* distribution, which is illustrated by the histogram of Figure 3.1. One often runs into this kind of distribution when dealing with actual data, and there are certain theoretical reasons why, in many problems, one can actually *expect* to get bell-shaped distributions. Although the distribution of Figure 3.3 is also more or less bell-shaped, it differs from the one of Figure 3.1 in-

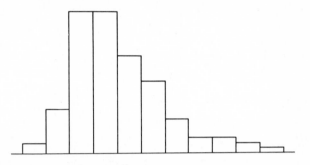

FIGURE 3.3   A skewed distribution

asmuch as the latter is *symmetrical* while the former is *skewed*. Generally speaking, a distribution is said to be *symmetrical* if we can picture the histogram folded (say, along the dashed line of Figure 3.2) so that the two halves will more or less coincide. If a distribution has a more pronounced "tail" on one side, such as the distribution of Figure 3.3, we say that the distribution is *skewed*. Note that the weekly earnings distribution on page 8 is skewed with a "tail" on the right—we say that it has *positive skewness* or that it is *positively skewed*. Correspondingly, if a distribution has a pronounced "tail" on the left, we say that the distribution has *negative skewness* or that it is *negatively skewed*.

There are several ways of measuring the extent to which a distribution is skewed. A relatively easy one is based on the fact, illustrated in Figure 3.4, that if a distribution has a "tail" on the right, its median will generally be exceeded by its mean. (If the "tail" is on the left, this order will be reversed, and the median will generally exceed the mean.) Based on this difference, the so-called *Pearsonian coefficient of skewness* measures the skewness of a distribution by means of the formula

$$\frac{3(\text{mean} - \text{median})}{\text{standard deviation}} \qquad \bigstar$$

Here three times the difference between the mean and the median is divided by the standard deviation to make this description of the shape of the distribution independent of the units of measurement of the data.

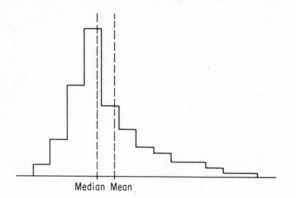

FIGURE 3.4   The median and the mean of a positively skewed distribution

If we substitute into this formula the values obtained for the mean, median, and standard deviation of the seat-occupancy-time distribution, we find that the Pearsonian coefficient of skewness equals

$$\frac{3(44.3 - 41.3)}{16.9} = 0.53$$

This value is positive, but small enough to permit us to say that the distribution is only moderately skewed.

Besides bell-shaped distributions, two other kinds—*J-shaped* and *U-shaped* distributions—are sometimes, though less frequently, met in actual practice. As is illustrated by means of the histograms of Figure 3.5, the

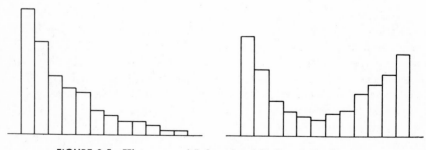

FIGURE 3.5   Histograms of J-shaped and U-shaped distributions

names of these distributions literally describe their shapes; examples of *J*-shaped and *U*-shaped distributions may be found in Exercises 4 and 5 below.

EXERCISES

1. Use the results of Exercise 1 on page 55 and Exercise 8 on page 56 to determine the Pearsonian coefficient of skewness for the distribution of sack weights.

2. Use the results of Exercise 2 on page 55 and Exercise 9 on page 56 to find the Pearsonian coefficient of skewness for the grade distribution.

3. Give one example each of actual data whose distribution might reasonably be expected to be (a) bell-shaped and symmetrical, (b) bell-shaped but not symmetrical, (c) negatively skewed, (d) positively skewed, and (e) *J*-shaped.

4. If we roll a pair of dice, the number of sixes we obtain is either 0, 1, or 2. Roll a pair of dice 100 times, and construct a distribution showing how often 0, 1, and 2 sixes were obtained, and draw a histogram of this distribution, which should be *J*-shaped with its tail on the right.

5. If a coin is flipped five times, the result may be represented by means of a sequence of H's and T's (for example, HHTTH), where H stands for *heads* and T for *tails*. Having obtained such a sequence of H's and T's, we can then check after each successive flip whether the number of heads exceeds the number of tails. For example, for the sequence HHTTH, heads is ahead after the first flip, after the second flip, after the third flip, *not* after the fourth flip, but again after the fifth flip; altogether, it is ahead *four times*. Repeat this experiment 50 times, and construct a histogram showing in how many cases heads was ahead 0 times, *1 time, 2 times, . . . , and 5 times. The resulting distribution should be *U*-shaped; can you explain why?

A WORD OF CAUTION

The fact that there is a certain amount of arbitrariness in the selection of statistical descriptions has led some persons to believe that they can take a set of data, apply the magic of statistics, and prove almost anything they want. To put it more bluntly, a nineteenth-century British statesman once said that there are three kinds of lies: lies, damned lies, and statistics. To give an example where such a criticism might be justified, suppose that a paint manufacturer asks his research department to "prove" that on the average a gallon of his paint covers more square feet than that of his two principal competitors. Suppose, furthermore, that the research department tests five cans of each brand, getting the following results (in square feet per gallon can):

> *Brand A:*   505,   516,   478,   513,   503
>
> *Brand B:*   512,   486,   511,   486,   510
>
> *Brand C:*   496,   485,   490,   520,   484

If the manufacturer's own brand is Brand A, the person who is doing this analysis will find to his delight that the means of the three samples are, respectively, 503, 501, and 495. He can, thus, claim that in actual tests a can of his employer's product covered on the average more square feet than those of his competitors.

Now suppose, for the sake of argument, that the manufacturer's own brand is Brand B. Clearly, the analyst can no longer base the comparison on the sample means; *this would not prove his point.* Trying instead the sample medians, he finds that they are, respectively, 505, 510, and 490, and this provides him with exactly the kind of ammunition he wants. The median is a perfectly respectable measure of the "average" or "center" of a set of data, and using the medians he can claim that his employer's product came out best in the test.

Finally, suppose that the manufacturer's own brand is Brand C. After going down the list of various measures of central location, the analyst comes upon one that does the trick. The *midrange* is defined as the mean of the smallest and largest values in a sample, and for the given data, the midranges of the three samples are, respectively, 497, 499, and 502. Gleefully, he thus points out that Brand C, his employer's product, scored on the average highest in the test. The moral of this example is that *if data are to be compared, the method of comparison should be decided on beforehand, or at least without looking at the actual data.* All this is aside from the fact that comparisons based on samples are often far from conclusive. It is quite possible that whatever differences there may be among the three means (or three other descriptions) can be attributed entirely to chance.

Another point which must be remembered is that a statistical measure always describes a particular characteristic of a set of data and that it, furthermore, describes this characteristic in a special way. Whether this "special way" is appropriate for a given situation is something which will have to be examined individually in each case. Suppose, for instance, that we want to buy a house and that we are shown a house in a neighborhood where, according to the broker, average family income is in excess of $24,000. This gives the impression of a relatively prosperous neighborhood, but it could well be a neighborhood where most families have incomes of less than $4,000 while one very wealthy family has an income of several hundred thousand dollars a year. This is the kind of situation we warned about on page 32, namely, a situation where the mean is greatly affected by one extreme value. It would be much more informative to say in this case that the median family income is less than $4,000, mentioning, perhaps, the special situation created by the one wealthy family. Further examples of this kind may be found in *How to Lie with Statistics*, the book by D. Huff referred to in the Bibliography at the end of this book.

TECHNICAL NOTE 1 (Summations)

In the abbreviated notation introduced on page 30, $\Sigma\, x$ does not make it clear which, or how many, values of x are to be added. This is taken care of by the more explicit notation

$$\sum_{i=1}^{n} x_i = x_1 + x_2 + \cdots + x_n$$

where it is made clear that we are adding the x's whose subscripts i are $1, 2, \ldots,$ and n. We did not use this notation in the text, in order to simplify the overall appearance of the various formulas, assuming that it is clear in each case what x's we are referring to and how many there are.

Using the Σ notation, we shall also have occasion to write such expressions as $\Sigma\, x^2,\, \Sigma\, xy,\, \Sigma\, x^2 f, \ldots$, which (more explicitly) represent the sums

$$\sum_{i=1}^{n} x_i^2 = x_1^2 + x_2^2 + x_3^2 + \cdots + x_n^2$$

$$\sum_{j=1}^{m} x_j y_j = x_1 y_1 + x_2 y_2 + \cdots + x_m y_m$$

$$\sum_{i=1}^{n} x_i^2 f_i = x_1^2 f_1 + x_2^2 f_2 + \cdots + x_n^2 f_n$$

Working with two subscripts (as we shall do in Chapter 11), we may have to evaluate a *double summation* such as

$$\sum_{j=1}^{3} \sum_{i=1}^{4} x_{ij} = \sum_{j=1}^{3} (x_{1j} + x_{2j} + x_{3j} + x_{4j})$$

$$= x_{11} + x_{21} + x_{31} + x_{41} + x_{12} + x_{22}$$
$$+ x_{32} + x_{42} + x_{13} + x_{23} + x_{33} + x_{43}$$

To verify some of the formulas involving summations that are stated but not proved in the text, the reader will find it convenient to use the following rules:

$$\text{Rule A:} \quad \sum_{i=1}^{n} (x_i \pm y_i) = \sum_{i=1}^{n} x_i \pm \sum_{i=1}^{n} y_i$$

$$\text{Rule B:} \quad \sum_{i=1}^{n} k \cdot x_i = k \cdot \sum_{i=1}^{n} x_i$$

$$\text{Rule C:} \quad \sum_{i=1}^{n} k = n \cdot k$$

The first of these rules states that the summation of the sum (or difference) of two terms equals the sum (or difference) of the individual summations, and it can be extended to the sum or difference of more than two terms. The second rule states that we can, so to speak, factor a constant out of a summation, and the third rule states that the summation of a constant is simply n times that constant. All of these rules can be proved by actually writing out in full what each of the summations represents.

TECHNICAL NOTE 2 (Unbiased Estimators)

Ordinarily, the purpose of calculating a sample statistic (such as the mean, the standard deviation, or the variance) is to *estimate* the corresponding population parameter. If we actually took many samples (of size greater than one) from a population which has the mean μ, calculated the sample means $\bar{x}$, and then averaged all these estimates of μ, we would find that their average is very close to μ. On the other hand, if we calculated the variance of each sample by means of the formula $\dfrac{\Sigma (x - \bar{x})^2}{n}$, and then averaged all these estimates of σ^2, we would find that this average is very close to $\dfrac{n-1}{n}$ times as large as σ^2. It is for this reason that we multiply $\dfrac{\Sigma (x - \bar{x})^2}{n}$ by $\dfrac{n}{n-1}$, and canceling the n's get the formula $s^2 = \dfrac{\Sigma (x - \bar{x})^2}{n-1}$. Thus, the values of s^2 calculated for many samples from the same population should on the average equal the population variance σ^2. Estimators which have the desirable property that *the average of their values is equal to the quantity they are supposed to estimate* are said to be *unbiased;* otherwise, they are said to be *biased*. The sample mean $\bar{x}$ is an *unbiased estimator* of the population mean μ, and s^2 is an *unbiased estimator* of the population variance σ^2, but we add that it does *not* follow from this that s is an unbiased estimator of the population standard deviation σ.

TECHNICAL NOTE 3 (The Use of Square Root Tables)

Although square root tables are relatively easy to use, most beginners seem to have some difficulty in choosing the right column and in placing the decimal point correctly in the answer. Table XII, in addition to containing the *squares* of the numbers from 1.00 to 9.99 spaced at intervals of 0.01, gives the *square roots* of these numbers rounded to 6 decimals. To find the square root of any positive number rounded to 3 significant digits, we have only to use the following rule in deciding whether to take the entry of the $\sqrt{n}$ or the $\sqrt{10n}$ column:

> *Move the decimal point an even number of places to the right or to the left until a number greater than or equal to 1 but less than 100 is reached. If the resulting number is less than 10 go to the $\sqrt{n}$ column; if it is 10 or more go to the $\sqrt{10n}$ column.*

Thus, to find the square roots of 12,800, 379, and 0.0812, we go to the $\sqrt{n}$ column since the decimal point has to be moved, respectively, 4 places to the left, 2 places to the left, and 2 places to the right, to give 1.28, 3.79, and 8.12. Similarly, to find the square roots of 5248, 0.281, and 0.0000259 we go to the $\sqrt{10n}$ column since the decimal point has to be moved, respectively, 2 places to the left, 2 places to the right, and 6 places to the right, to give 52.48, 28.1, and 25.9.

Having found the entry in the appropriate column of Table XII, the only thing that remains to be done is to put the decimal point in the right position in the answer. Here it will help to use the following rule:

> *Having previously moved the decimal point an even number of places to the left or right to get a number greater than or equal to 1 but less than 100, the decimal point of the entry of the appropriate column in Table XII is moved half as many places in the opposite direction.*

For example, to determine the square root of 12,800, we first note that the decimal point has to be moved *four places to the left* to give 1.28. We thus take the entry of the $\sqrt{n}$ column corresponding to 1.28, move its decimal point *two places to the right*, and get $\sqrt{12,800} = 113.137$. Similarly, to find the square root of 0.0000259, we note that the decimal point has to be moved *six places to the right* to give 25.9. We thus take the entry of the $\sqrt{10n}$ column corresponding to 2.59, move the decimal point *three places to the left*, and get $\sqrt{0.0000259} = 0.00508920$. In actual practice, if a number whose square root we want to find is rounded, the square root will have to be rounded to as many significant digits as the original number.

4

SUMMARIZING DATA: INDEX NUMBERS

BASIC PROBLEMS

As their name implies, index numbers are measures which *indicate* something, usually how much things have changed or how they compare with one another. For instance, if we compare the 14,340 tons of California raisins shipped to all markets in January 1971 with the 19,042 tons shipped in January 1970, we find that the shipments for the more recent month were 75.3 per cent of the earlier shipments. Similarly, a comparison of the 840,000 proof gallons of U.S.-produced beverage brandy entering distribution channels in the United States in March 1970 with the 234,000 proof gallons of imported beverage brandy entering these channels in the same month shows that the U.S.-produced total was 359.0 per cent of the imported total. Further comparisons show that the wholesale prices of processed foods and feeds in the United States in January 1970 were 125.1 per cent of the average prices of these commodities in the years 1957–1959; and also that San Francisco Bay Area workers earned 106.8 per cent as much in March 1969 as they earned in the same month a year earlier, but their wages in the latter month bought only 99.5 per cent as much as they bought in March 1968.

These percentages, each of which compares two things, are simple examples of index numbers; we call them "simple" because there are also index numbers which are intended to indicate changes in such complicated phenomena as wholesale prices, consumer prices, total industrial production, business cycles, and stock prices. In fact, there may be several, sometimes many, different indicators for the same phenomenon. There are, for instance, at least 16 indicators of the condition of stocks on the New York Stock Exchange, or of the direction the market is heading, including a confidence index, a breadth-of-the-market index, a short-interest-trend indicator, a disparity indicator, and a quarterly turn indicator.

Although index numbers are commonly associated with business and economics, they are also widely used in other fields. Psychologists measure intelligence quotients, which are essentially index numbers comparing a person's intelligence with that of an average for his or her age. Health authorities prepare indexes to display changes in the adequacy of hospital facilities, educational research organizations devise indexes to measure the effectiveness of school systems, sociologists construct indexes measuring population changes, the Weather Bureau has devised a so-called "Discomfort Index" to measure the combined effect of heat and humidity on individuals, and so on. The principal use of index numbers in business is to make comparisons between two different time periods, as, for instance, to compare the price of tin in June 1971 to its price one year earlier. However, index numbers can also be used to make other kinds of comparisons, such as a comparison of the average price of grains and feeds with the average price of fats and oils in May 1971, or a comparison of the 1971 production of cotton in Texas with the 1971 production of cotton in California.

In recent years, the use of index numbers has extended to so many fields of human activity that some knowledge of these important measures really belongs under the heading of "general education." Index numbers showing changes in consumer prices are of vital importance to all consumers, and the Bureau of Labor Statistics' *Consumer Price Index* is closely watched by many, including millions of workers whose wages automatically go up or down as the level of this index rises or falls by a specified amount. Index numbers are also of great concern to farmers whose subsidies depend on the *Parity Index* of the Federal government, and they are no less important to business firms and individuals for whom they provide actual insurance against changing prices. Index numbers have also found their way into alimony agreements, trust fund payments, and legacies, which can be made to vary with the index of the purchasing power of the dollar.

Like other statistical measures, index numbers are usually constructed to serve definite purposes. Sometimes, the stated purpose of an index is such that the only problem which might arise in its construction is that of locating the necessary data. For instance, in a comprehensive study of the role of mutual funds in the American economy, it may be necessary to construct an index showing the change in mutual fund assets from 1940 to 1968. Comparable figures from industry records show that these assets were $500 million in 1940 and $5.2 billion in 1968, so that

$$\text{index} = \frac{5,200,000,000}{500,000,000} = 10.4 \text{ (or 1,040 per cent)}$$

and the problem is solved. In contrast, there are many situations in which some very complex problems arise as soon as the purpose of an index has been stated. The most critical among these are (1) the availability and comparability of data, (2) the selection of items to be included in the comparison(s), (3) the choice of time periods (localities, etc.) that are to be compared, (4) the selection of appropriate weights measuring the relative importance of various items which are to be compared, and (5) the choice of a suitable index number formula by means of which the data are to be combined. Some of these basic problems will be discussed below.

The availability and comparability of data. It is hardly necessary to point out that comparisons cannot be made and index numbers cannot be constructed unless the required data can be obtained. Many research workers have been frustrated by the fact that essential information needed by townships was tabulated by counties, sales data needed by brand were available only by type of merchandise, insurance losses were given per risk and not per claim, and so on. The question of availability also arises if we want to make a comparison, say, of the cost of living in 1972 with that of the year 1914. Nowadays, television sets, major appliances, contact lenses, frozen foods, and many transistorized devices are widely used, but none of them was sold commercially in 1914, and we may be forced to invent fictitious prices for what such items might have cost had they been available in 1914.

The question of comparability can also be quite troublesome. In congressional hearings some labor organizations have complained that, to some extent, the Consumer Price Index reflects deterioration in quality rather than an actual change in prices. It does not matter to us here whether this criticism is valid, but it serves to indicate that it can be very difficult to make sure that prices are actually comparable, namely, that they refer to goods and services which are identical in quality.

The comparability of statistical data may also be questioned if parts of the data are obtained from different sources. It is very confusing, for example, to note that the Bureau of the Census reports that the 1939 production of nonferrous minerals in the United States had a total value of $350 million, while the Bureau of Mines reports a corresponding figure of $434 million. (This discrepancy arises from the fact that the Bureau of the Census figures are those given by producers while the Bureau of Mines figures are those given by purchasers and transportation companies.) The situation with regard to import-export data supplied by the partners in trade is sometimes so bad that one wonders whether any confidence at all can be placed in international trade statistics. For example, the following table, based on figures published by the United Nations in *Commodity Trade Statistics*, shows the importer's and exporter's versions of how many thousands of U.S. dollars' worth of certain commodities the second country received from the first during the first 6 months of 1969:

	Exporter's data		Importer's data	
Woven textiles (noncotton)	United Kingdom	3,818	New Zealand	4,817
Wool and animal hair	New Zealand	33,151	United Kingdom	35,465
Lumber, shaped, conifer	Canada	284,301	United States	313,461
Medicinal products	United States	16,609	Canada	21,784

These figures were deliberately selected to illustrate our point; similar difficulties arise, say, with employment and production figures quoted by different sources, or sickness and accident data supplied by different agencies. It is hard to say without further investigation whether such instances are the exception or the rule; in any case, they stress the seriousness of the problem of obtaining relevant and comparable data.

The selection of items to be included in the comparison. If an index is designed for the special purpose of comparing the price of a commodity at two different times, there is no question as to what figures should be included. However, the situation is entirely different in the construction of so-called *general-purpose indexes*, for instance, those designed to measure general changes in wholesale or consumer prices. It must be clear that it is physically impossible, or at least highly impractical, to include in such a comparison all commodities from alum to zircon and to include, furthermore, all prices at which these commodities are traded in every single transaction throughout the entire country. The only thing anyone can do is to take samples in such a way that, in the professional judgment of the persons responsible for constructing the index, the items and transactions included adequately reflect the overall situation. For example, the Consumer Price Index is based on about 400 items (goods and services) playing a significant role in the average budget of persons belonging to a certain population group. The prices included in this (judgment) index are samples both of the goods and services that are included and also of the stores and cities from which the data are gathered.

The choice of time periods that are to be compared. If an index number is designed for the specific purpose of comparing 1971 figures with those of, say, 1965, it is customary to refer to 1971 as the *given year*, and to 1965 as the *base year*, indicating the latter by writing "1965 = 100." In general, the year or period which we want to compare is called the *given year* or *given period*, while the year or period relative to which the comparison is made is called the *base year* or *base period*.

The choice of a base year or base period does not present any problems if an index is constructed for a specific comparison. So far as general-purpose indexes describing complex phenomena are concerned, it is desirable to base the comparison on a period of relative economic stability that is not too distant in the past. The reason for the first stipulation is that during times of abnormal economic conditions (for example, during a war) there may be no free trading of some commodities, there may be black markets, and the buying habits of the public may be irregular due to shortages of products that would otherwise be readily available.

One reason for choosing a relatively recent base period is that some newly developed items must usually be incorporated in an index; the farther back the base period, the more difficult (if not actually impossible) it becomes to find comparative data relating both to the present and to the earlier period. Base periods that are too far in the past also raise problems not unlike those faced by an art critic who instead of judging two paintings by holding them next to one another is forced to compare them individually with a third, and then relate the individual comparisons.

The Federal government has made it a practice to establish a standard reference base for use by Federal agencies. About every 10 years, this base period has been brought forward. In 1940, for example, a reference base of 1935–1939 was established, in 1951 a new base of 1947–1949 was designated, and in 1960 a new 3-year base period 1957–1959 was established by the Bureau of the Budget for all government general-purpose indexes. Where feasible, conversion to the 1957–1959 base began with the January 1962 indexes. It is worth noting that the 1957–1959 period is probably as stable a period as can be found in the post-World War II era. But, as the government has pointed out, the selection of a base period does not imply "normality" in any real sense—a base period is merely a convenient and necessary reference point if comparisons are to be made. At the time this is written, most government indexes refer to the 1957–1959 base, but some (the indexes of prices received and prices paid by farmers, for example) are still tied by law to the pre-World War I period 1910–1914. Also, before another revision of this book appears, the base period will most likely have been brought forward to the more recent past.

The choice of appropriate weights. As was illustrated in connection with the weighted mean in Chapter 3, there are many situations in which figures cannot be meaningfully averaged without paying due attention to the relative importance of each item; this applies particularly to index numbers. Suppose, for example, that a manufacturer wants to construct an index comparing the 1971 and 1965 prices of replacement parts required for machines of a certain type, and that he arbitrarily chooses parts AX345 and AX765 for study. Suppose, further, that for part AX345 the index is $\dfrac{\$3.60}{\$2.40} \cdot 100 = 150$ per cent and that for part AX765 it is $\dfrac{\$4.80}{\$1.20} \cdot 100 = 400$ per cent. On the basis of these figures alone, we might assert that the price of replacement parts in 1971 is $\dfrac{150 + 400}{2} = 275$ per cent of what it was in 1965. This result would be valid if equal numbers of each part wore out and were replaced, but if in 1971 the company actually replaced, say, 100 times as many of the first part as of the second, then the two price changes are clearly not equally important. In general, valid comparisons of price changes require that prices be weighted in some way so as to account for their relative importance in practice.

The problem of choosing suitable weights in the construction of an index number is often a hard one. It depends on whether we want to average prices, quantities, or (as in the illustration above) indexes of individual commodities. As it is virtually impossible to treat this problem without referring to specific index number formulas, we shall defer discussion of this matter until we study weighted index number formulas in the section beginning on page 73.

The choice of a suitable formula. In the same way in which the average of a set of data can be described by using the mean, the median, the mode, or some other measure of central location, relative changes can be described by means of any one of several index number formulas. In the next two sections we shall treat some of the simpler of these formulas, and after reading these sections, it will be clear that any choice among the formulas will ultimately have to depend on practical considerations as well as on some of their mathematical "niceties." The symbolism we shall use in the remainder of this chapter consists of referring to index numbers as I, base-year prices as p_0, given-year prices as p_n, base-year quantities as q_0, and given-year quantities as q_n.

UNWEIGHTED INDEX NUMBERS

To illustrate some of the simplest methods used in index number construction, let us compare the May 1969 prices received by farmers for four kinds of farm animals with the prices received for these animals in

May 1968. The following are the prices given in dollars per hundredweight (cwt):

	May 1968	May 1969
Hogs	18.30	22.30
Beef cattle	23.70	28.60
Sheep	6.62	8.02
Lambs	25.80	28.00

If we divide the sum of the May 1969 prices by the sum of the May 1968 prices, we get

$$I = \frac{22.30 + 28.60 + 8.02 + 28.00}{18.30 + 23.70 + 6.62 + 25.80} = \frac{86.92}{74.42} = 1.168 \text{ (or 116.8 per cent)}$$

and this tells us that the combined May 1969 prices are 116.8 per cent of those of May 1968. Thus, we might say that with May 1968 = 100, this hog-cattle price index stood at 116.8 in May 1969. The method illustrated here is called the *simple aggregative method*, and the index it leads to is called a *simple aggregative index*. The general formula for a simple aggregative index is

$$I = \frac{\Sigma\,p_n}{\Sigma\,p_0} \cdot 100 \qquad\qquad \star$$

where $\Sigma\,p_n$ is the sum of the given-year prices, $\Sigma\,p_0$ is the sum of the base-year prices, and the ratio of the two is multiplied by 100 to express the index as a percentage.

A simple aggregative index is easy to compute and easy to understand, but it does not meet a criterion of adequacy called the *units test*—that is, depending on the units for which the prices of the various index items are quoted, the index can yield substantially different results. Had we combined, say, the prices of 1 pound of hog, 1 pound of beef cattle, 5 pounds of sheep, and 20 pounds of lamb, we would have found that the index of the prices of these items stood at 110.1 instead of 116.8 (see Exercise 6 on page 78). Largely for this reason, simple aggregative indexes are now rarely used, and very few examples of them appear in publications.

Another way to compare the two sets of prices is to calculate first a separate index for each of the four kinds of animals, and then average these so-called *price relatives* using some measure of central location. Writing the individual price relatives as percentages, we obtain

Price relatives

Hogs	$\dfrac{22.30}{18.30} \cdot 100 = 122$ per cent
Beef cattle	$\dfrac{28.60}{23.70} \cdot 100 = 121$ per cent
Sheep	$\dfrac{8.02}{6.62} \cdot 100 = 121$ per cent
Lambs	$\dfrac{28.00}{25.80} \cdot 100 = 109$ per cent

To construct an overall index comparing the prices of these animals at the two different times, we can now calculate the arithmetic mean, the geometric mean, or the median of the price relatives. Using the (arithmetic) mean, we get

$$I = \frac{122 + 121 + 121 + 109}{4} = 118.2 \text{ per cent}$$

and this index is called an *arithmetic mean of price relatives*. Symbolically, the formula for this kind of index is

$$I = \frac{\Sigma \dfrac{p_n}{p_0} \cdot 100}{k} \qquad\qquad ★$$

where k is the number of index items whose price relatives are thus being combined.

If we had calculated the geometric mean of the price relatives in this example, we would have obtained $I = 118.1$, and if we had calculated the median, we would have obtained $I = 121$ (see Exercise 7 on page 78). In principle, price relatives can be averaged with any measure of central location, but in practice, the arithmetic mean and the geometric mean are most widely used.

It is a matter of historical interest that the earliest index number on record is an arithmetic mean of price relatives. In the middle of the eighteenth century, G. R. Carli, an Italian, calculated the effect of the import of silver on the value of money, using a formula like the one given above to compare the 1750 prices of oil, grain, and wine with those of the year 1500.

Today the need for employing weights has been almost universally accepted, and very few indexes are actually computed in this simple way. Prior to 1914, the *Wholesale Price Index* of the Bureau of Labor Statistics was computed as an arithmetic mean of the price relatives of about 250 commodities. As a result of an important study by W. C. Mitchell in 1915, which has affected index number construction since that time, the index was changed to a weighted index. Among the important government indexes only the daily *Index of Spot Market Prices* is still calculated as a simple (unweighted) geometric mean of price relatives.

The formulas we have given in this section are all price index formulas. However, if we want to compare quantities rather than prices, we have only to replace the p's with q's and use the same formulas for *quantity indexes*.

WEIGHTED INDEX NUMBERS

To show how an index number can be made to account for differences in importance, let us consider the problem of measuring the change from 1963 to 1968 in the prices of grapes for all uses produced in California. The required data, as well as 1964 data which will be used later, are given in the following table, where the prices of the grapes produced for wine, table, and raisin uses are in dollars per ton, and the production figures are in thousands of tons:

| | *Prices* | | | *Quantities* | | |
	1963	*1964*	*1968*	*1963*	*1964*	*1968*
Wine varieties	54.90	67.90	71.10	624	610	650
Table varieties	41.30	54.30	55.80	622	525	470
Raisin varieties	44.70	52.20	57.20	2,191	2,028	2,135

The importance of the price of a commodity in trade or use is best determined by the quantity which is bought or sold, or consumed or produced. Thus, we can arrive at an index measuring the grape price change from 1963 to 1968 by using as weights the respective quantities produced in the base year (1963), the given year (1968), or some other year (such as 1964). Choosing for illustration the base-year production figures as weights, we shall thus first calculate the weighted means of the 1968 prices and the 1963 prices, and then define an index in terms of the ratio of the two.

Ignoring the denominators of the two weighted means, which cancel, and multiplying the ratio by 100, we get

$$I = \frac{71.10(624) + 55.80(622) + 57.20(2,191)}{54.90(624) + 41.30(622) + 44.70(2,191)} \cdot 100 = 129.5 \text{ per cent}$$

An index constructed in this way is called a *weighted aggregative index with base-year weights*. It is also known as a *Laspeyres Index*, named after the statistician who first suggested its use. In general, the formula for this index is

$$I = \frac{\Sigma \; p_n q_0}{\Sigma \; p_0 q_0} \cdot 100 \qquad\qquad \star$$

where the numerator contains the sum of the products of the respective given-year prices and base-year quantities, whereas the denominator contains the sum of the products of the respective base-year prices and base-year quantities. Since we are using the same quantities in each case, the index clearly reflects changes in the prices of the index items.

We cannot construct a price index by weighting the given-year prices with given-year quantities and the base-year prices with base-year quantities. Since $\Sigma \; p_n q_n$ is the *total value* of the goods in the given year and $\Sigma \; p_0 q_0$ is the *total value* of the goods in the base year, the ratio of $\Sigma \; p_n q_n$ to $\Sigma \; p_0 q_0$ reflects changes in value rather than changes in price; it is, in fact, a *value index*. We can, however, use given-year quantities to weight the base-year prices as well as the given-year prices, and the resulting *weighted aggregative index with given-year weights*, sometimes called a *Paasche Index*, is given by the formula

$$I = \frac{\Sigma \; p_n q_n}{\Sigma \; p_0 q_n} \cdot 100$$

Thus using the 1968 production totals as weights in our numerical example, it is easy to verify that the formula leads to $I = 129.3$ (see also Exercise 8 on page 78).

Most of the important index numbers constructed by the Federal government are published *in series*, that is, regularly every day, every week, every month, or every year. For these it would be highly impractical to use the Paasche formula, since this formula would continually require new quantity weights. An index that is currently in great favor is the *fixed-weight aggregative index*, whose formula is

$$I = \frac{\Sigma \, p_n q_a}{\Sigma \, p_0 q_a} \cdot 100 \qquad \qquad \bigstar$$

where the weights are quantities referring to some period other than the base year 0 or the given year n. Although it is actually calculated somewhat differently, one of the most important fixed-weight aggregative indexes is the Wholesale Price Index of the Bureau of Labor Statistics. Its current base period is the 3-year period 1957–1959 and the q_a are quantities marketed in 1958 (see also Exercise 15 on page 79).

In addition to weighted aggregative indexes of the sort we have discussed above, we can also obtain weighted indexes by weighting the individual price relatives. For example, we can write the formula for a *weighted arithmetic mean of price relatives* as

$$I = \frac{\Sigma \, \dfrac{p_n}{p_0} \cdot w}{\Sigma \, w} \cdot 100 \qquad \qquad \bigstar$$

where the w's are suitable weights assigned to the individual price relatives, which are now written as proportions.

Since the importance of relative change in the price of a commodity is most adequately reflected by the *total amount of money* that is spent on it, it is customary to use *value weights* for the w's of this last formula. This raises the question whether to use the values (prices times quantities) of the base year, those of the given year, or perhaps some other fixed-value weights. It will be left to the reader to show in Exercise 9 on page 78 that base-year value weights $p_0 q_0$ would not yield a new index; with these weights, the last formula reduces to that of the Laspeyres Index given on page 74. To give an example in which we use value weights pertaining to some year other than the base year or the given year, let us refer again to the example on page 73 and compute an index showing the change in grape prices from 1963 to 1968 using as weights the corresponding values for 1964. Calculating first the price relatives $\dfrac{p_n}{p_0} = \dfrac{p_{68}}{p_{63}}$ and the value weights $p_a q_a = p_{64} q_{64}$ (in thousands of dollars), we get

	Price relatives p_{68}/p_{63}	Values $p_{64} q_{64}$
Wine	1.295	41,419
Table	1.351	28,508
Raisin	1.280	105,862

Then, substituting into the index number formula, we obtain

$$I = \frac{1.295(41,419) + 1.351(28,508) + 1.280(105,862)}{41,419 + 28,508 + 105,862} \cdot 100$$

$$= 129.5 \text{ per cent}$$

Having determined indexes measuring the change from 1963 to 1968 in the prices of California grapes for all uses in three different ways, we obtained 129.3, 129.5, and 129.5 per cent. Although these values are quite similar, different methods applied to the measurement of the same phenomenon can sometimes lead to substantially different results. If millions of dollars ride on an increase or decrease of one point (as in some labor-management agreements containing escalator clauses), the question of choosing an appropriate index is a very serious matter.

EXERCISES

1. The following are the retail prices, in cents, of selected dairy products for 1966, 1967, 1968, and 1969:

	1966	1967	1968	1969
Milk, fresh (grocery), $\frac{1}{2}$ gallon	49.8	51.7	53.7	54.6
Ice cream, $\frac{1}{2}$ gallon	80.6	80.9	80.7	80.8
Butter, pound	82.2	83.0	83.6	84.1
Cheese, American process, $\frac{1}{2}$ pound	42.2	43.6	44.4	45.6

   (a) Construct a simple aggregative price index for the given 4-year period with 1966 = 100; that is, calculate the 1966, 1967, 1968, and 1969 values of such an index.

   (b) Comment on the applicability of these results to the changing dairy food costs of an elderly couple, a family with several very small children, and a teenage social club.

   (c) Find the arithmetic mean of the relatives comparing the 1969 prices with those for 1966.

2. The total 1967 and 1968 output of selected electrical appliances (in thousands of units) was

	1967	*1968*
Refrigerators	4,713	5,151
Freezers	1,100	1,124
Ranges	1,910	2,307
Water heaters	1,168	1,410
Dishwashers	1,585	1,961
Disposers	1,357	1,741
Dehumidifiers	280	478
Air conditioners	4,129	4,026

(a) Calculate a simple aggregative index comparing the 1968 production of these goods with that of 1967.

(b) Find the arithmetic mean of the relatives comparing the 1968 production of these goods with that of 1967.

3. The following are the annual average prices (in dollars per pound) of three selected imported metals:

	1960	*1965*	*1966*	*1967*	*1968*
Copper	0.305	0.348	0.394	0.492	0.550
Nickel	0.731	0.736	0.757	0.858	0.933
Tin	0.980	1.733	1.636	1.481	1.423

(a) Find the 1965, 1966, 1967, and 1968 values of a simple aggregative index with 1960 = 100.

(b) Find a simple aggregative index comparing the 1968 prices with those of 1966.

(c) Find for 1967 and 1968 the arithmetic mean of the price relatives using 1960 = 100.

(d) Find the arithmetic mean of the relatives comparing the 1968 prices with those of 1966.

(e) Calculate the geometric mean of the relatives comparing the 1968 prices of copper and nickel with those of 1966.

4. The number of immigrants admitted to the United States during the years 1964–1968 were (in thousands) 292, 297, 323, 362, and 454. Construct an index number *series* for the number of immigrants admitted to the United States during the years 1964–1968 with 1966 = 100.

5. The following figures, in millions of dollars, are the cash receipts of farms in the East North Central United States and the state of Kansas in 1968:

	Crops	Livestock	Government payments
Ohio	519	704	95
Indiana	607	745	120
Illinois	1,367	1,224	161
Michigan	396	454	65
Wisconsin	215	1,243	52
Kansas	534	1,002	228

(a) Construct a simple aggregative index for each of the East North Central states, comparing its total cash receipts of farms with that of Kansas.

(b) Find the arithmetic mean of the relatives comparing the cash receipts of farms for Indiana with those for Illinois.

6. Verify the value of 110.1 obtained on page 71 for an index based on the prices of 1 pound of hog, 1 pound of beef cattle, 5 pounds of sheep, and 20 pounds of lamb.

7. Verify the values given on page 72 for the geometric mean and the median of the price relatives.

8. Verify that the Paasche Index for the data on page 74 equals 129.3.

9. Show that if we substitute base-year value weights into the formula for a weighted arithmetic mean of price relatives, we obtain the formula for the Laspeyres Index.

10. It is interesting to note that the Laspeyres formula can generally be expected to *overestimate* price changes, while the Paasche formula will generally do just the opposite. Explain why this is so.

11. In the so-called *Ideal Index* (which has never been widely used for practical reasons) the biases referred to in Exercise 10 are compensated for by taking the geometric mean of the Laspeyres Index and the Paasche Index computed from the same data. Use the results obtained in the text to compute the Ideal Index for the data on page 73. Also write a general formula for the Ideal Index.

12. The following table contains wholesale prices and production totals of three petroleum products, prices being given in cents per gallon and production figures in millions of barrels:

	Prices			Quantities		
	1966	1967	1968	1966	1967	1968
Gasoline	11.7	11.4	11.3	1,793	1,846	1,940
Kerosene	10.4	11.0	11.3	102	100	101
Distillate fuel oil	9.4	10.0	10.3	786	805	841

(a) Using 1966 quantities as weights and 1966 = 100, find weighted aggregative indexes for the 1967 and 1968 prices of these petroleum products.

(b) Calculate a weighted aggregative index comparing the 1968 prices of the three petroleum products with those of 1966, using the 1968 quantities as weights.

(c) Calculate a weighted aggregative index comparing the 1968 prices of the three petroleum products with those of 1966, using the *averages* of the 1967 and 1968 quantities as weights.

(d) With 1966 = 100, calculate for 1968 the weighted arithmetic mean of price relatives, using base-year values as weights.

(e) With 1966 = 100, calculate for 1968 the weighted arithmetic mean of price relatives, using given-year values as weights. Comment on the practical aspects of this index number.

(f) Interchanging the p's and q's in the formula used in part (e), construct an index comparing the 1968 production of the three petroleum products with that of 1966.

13. The following are the 1966, 1967, and 1968 prices (in cents per pound) and production totals (in millions of pounds) of three food items:

	Prices			Quantities		
	1966	1967	1968	1966	1967	1968
Cheese	52.7	52.1	54.8	1,220	1,276	1,282
Butter	67.2	67.5	67.8	1,112	1,223	1,772
Margarine	26.6	25.7	25.6	2,110	2,114	2,141

(a) Using the 1966 quantities as weights, construct aggregative indexes comparing the 1967 and 1968 prices, respectively, with those of 1966.

(b) Using the 1967 quantities as weights, construct aggregative indexes comparing the 1967 and 1968 prices, respectively, with those of 1966.

(c) Using the 1968 quantities as weights, construct an aggregative index comparing the 1968 prices with those of 1966.

(d) Using the means of the 1966 and 1968 quantities as weights, construct an aggregative index comparing the 1968 prices with those of 1966.

(e) With 1966 = 100, calculate for 1968 the weighted arithmetic mean of the price relatives, using base-year values as weights. Compare this result with the 1968 index number calculated in part (a).

(f) With 1966 = 100, calculate for 1968 the weighted arithmetic mean of the price relatives, using given-year values as weights.

14. Compare the 1968 prices of the three foods in Exercise 13 with those of 1966 by means of the Ideal Index (see Exercise 11).

15. Show that the formula for the weighted arithmetic mean of price relatives with value weights of the form $w = p_0 q_a$ reduces to the formula for a fixed-weight aggregative index.

THREE IMPORTANT INDEXES

Among the many important indexes intended to describe assorted phenomena, some are prepared by private organizations. Financial institutions and utility companies, for example, often prepare indexes of such things as employment, factory hours and wages, and retail sales for the regions they serve; trade associations prepare indexes of price and quantity changes vital to their particular interests; and so on. Many of these indexes are widely used and highly regarded indicators of the phenomena they aim to describe. The most widely circulated and the most widely used indexes, however, are those prepared by the Federal government. Of the many important government indexes, we shall briefly describe three, two price indexes and one quantity index; they are the Consumer Price Index, the Wholesale Price Index, and the Index of Industrial Production.

The Consumer Price Index, constructed by the Bureau of Labor Statistics, is designed to measure the effect of price changes of a collection of goods and services called a "market basket," on the living costs of urban clerical workers and wage earners, both families and single persons living alone. This market basket consists of fixed quantities of some 400 consumption items, described by detailed specifications to assure comparability in successive periods. Included among these items are meats, dairy products, residential rents, clothing, appliances, new and used cars, gasoline and oil and parking fees, physicians' and dentists' services, drugs, haircuts, toothpaste, TV sets and replacement tubes, newspapers, cigarettes, wine, and the like, which are of major importance in consumer purchases and which, in fact, account for the greater part of consumer spending.

The items which comprise the market basket are periodically priced in a sample of 56 Standard Metropolitan Statistical Areas (SMSA's) and cities chosen to represent all urban places in the United States whose populations are more than 2,500. The prices that enter the index are for items classified into six major groups and subgroups: (1) food (at home and away from home), (2) housing (shelter, fuel and utilities, household furnishings and operation), (3) apparel and upkeep (men's and boys', women's and girls', footwear, miscellaneous apparel, and apparel services), (4) transportation (private and public), (5) health and recreation (medical care, personal care, reading and recreation), and (6) other goods and services. Group indexes are regularly calculated for 26 cities (or areas), and they are then combined into overall city indexes, group indexes for all cities combined, and one index covering all groups and all cities. When the data for the individual groups are combined into a city index, each group is assigned a weight (differing from city to city) which is intended to represent the "relative importance" of the group in the average budget, or family expenditure, of families covered by the index. The combined index for the whole country, the Consumer Price Index, is calculated from the data obtained for the various cities, giving those for each city a weight in proportion to the part of the total wage-earner and clerical-worker population it represents in the overall index. The current base period for the index is 1957–59.

In calculating the Consumer Price Index and its various group and sub-indexes, the Bureau of Labor Statistics uses two formulas which differ in the mechanics of calculation, but which are mathematically the same. Both formulas represent fixed-weight aggregative indexes of the sort now commonly in use. Looking at the index as a weighted average of price relatives, the value (price times quantity) weights used vary from time to time, but this is due entirely to changes in prices, since the quantities making up the market basket are held fixed. In this way, the index measures the change in the prices of fixed quantities of goods and services.

A crucial factor in an index such as this is establishing as part of the value weights the "relative importance" of the different groups of goods and services in the expenditures of the index families and individuals. Weighting factors for the "new" (current) series which began January 1964 were calculated on the basis of a sample of wage-earner and clerical-worker families and individuals carefully selected in 1960–1961, with the data of that year being adjusted to 1963 prices and consumption levels. Under normal conditions, the relative importance of the various categories of goods and services in a person's expenditures (food 22.94 per cent, housing 32.89 per cent, etc.) changes fairly slowly, so that the weights need only infrequent revision. Nevertheless, the relative importance weights are only approximations to the actual expenditure distribution at any time other than at the point when revised weights are introduced into the index.

The new series starting in January 1964 was intended to account for the major social and economic developments which had taken place in this country in the several years prior to that time. The major revision of the older index (which began in January 1953) leading to this new series included, among other things, increasing the number of sample cities; increasing the number of prices collected by one-fifth; increasing the sample of discount houses, suburban stores, and physicians; and increasing the market basket from 325 to 400 items, giving broader coverage to such categories as apparel, home furnishings, automobiles, and restaurant meals, and pricing a number of items including precooked foods and the so-called "miracle" fabrics for the first time.

The basic and most time-consuming factor of any major revision of the Consumer Price Index is the necessary survey of consumer expenditures, incomes, and savings. Based upon such a survey, the Bureau of Labor Statistics analyzed about 10,000 usable schedules from 70 sample cities to determine both the content of the market basket for the new series and the weights for the revised index. These weights continue to be used at the present time, and it is believed that the index continues to reflect, as nearly as possible, changes in the prices paid by urban families headed by a wage earner or clerical worker, and by individual wage earners and clerical workers.

Also constructed by the Bureau of Labor Statistics, the Wholesale Price Index is intended to measure changes in the prices of commodities at their first important commercial transaction. Thus, the word "wholesale" does not refer to prices received by wholesalers, jobbers, and distributors, but to prices of large lots in primary markets. Most of the prices used are those quoted on organized exchanges or markets, or received by manufacturers and other producers.

Like other comprehensive indexes, the Wholesale Price Index is based on a sample. Because of the importance of wholesale price movements in the many subdivisions of the economy, more than 6,300 price quotations are periodically collected on about 2,300 commodities. These index commodities are not randomly selected, but instead are carefully chosen by professionals in whose judgment they appear to be the most important or representative ones in their categories.

At present, separate price indexes are calculated for commodities classified in various ways. One major classification is by stages of processing (crude materials for further handling, intermediate materials, supplies and components, and finished goods). In another classification, all farm products are subclassified as fresh and dried fruits and vegetables, grains, live poultry, and livestock; industrial commodities are grouped in 13 subclassifications including chemical and allied products, furniture and household durables, lumber and wood products, machinery and equipment, rubber and rubber products, textile products and apparel, and transportation equipment. Some of these in turn are further subdivided; the furniture and household durables group, for instance, includes household appliances, household furniture, and home electronic equipment, all of which are themselves groupings of other commodities.

Basically, though not strictly, the Wholesale Price Index is calculated as a fixed-weight aggregative index. It can easily be shown, however, that this type of index is identical with one in which price relatives, p_n/p_0, are weighted with value weights $p_0 q_a$ (see Exercise 15 on page 79). Commencing with the January 1967 figures, the weights used in calculating the index represent the total net selling value of commodities produced in, processed in, or imported into the United States and flowing into primary markets in the year 1963. For technical reasons, the formula actually used to calculate the index is a variation of the one on page 75, with $w = p_0 q_a$, and currently 1957–59 = 100.

Unlike indexes measuring changes in prices, the Federal Reserve Board's Index of Industrial Production measures changes in volume or quantity. Since its origin in the 1920's with data based on 60 series relating to manufacturing and mining, the index has been periodically revised in an effort to keep pace with the output of a rapidly expanding economy. At present, the index is intended to measure changes in the physical volume or quantity of output of the nation's factories, mines, and electric and gas utilities. (The output of the latter was added to the index in a 1959 revision aimed at giving broader coverage and also permitting better comparisons with production indexes of other countries.) Directly, the index thus accounts for about 35 per cent of the value of the total output of goods and services in this country, but indirectly it reflects an additional 25 per cent or so of total activity, namely, the distribution of industrial products and their use in construction. Production in the construction industry itself, however, on farms, in transportation, and in various trade and service industries is not covered by the index.

The monthly composite index showing changes in the nation's total industrial production is arrived at by combining 207 individual monthly series into a number of different groups and then combining the corresponding group indicators into an overall figure. One such grouping of individual series is the so-called "industry grouping," which has the three main components: manufacturing, mining, and utilities. In this classification, manufacturing, for instance, is separated into durable and nondurable manufactures. Making up the durable manufactures group are primary metals; fabricated metal products; machinery; transportation equipment; instruments and related products; clay, glass, and stone products; lumber and related products; furniture and fixtures; and miscellaneous manufactures. For its part, the transportation group, for instance, includes motor vehicles and parts, as well as aircraft and other equipment. The nondurable manufactures group consists of 10 subgroups including textile mill products, apparel products, petroleum products, rubber and plastics products, and foods and beverages, some of which are themselves groups of further subdivisions. For their part, mining and utilities are also groupings (utilities of electricity and gas, for instance).

In addition to the industry groupings, the 1959 revision added another combination of production series called "market groupings." In this arrangement, the 207 component series of the total industrial production index are grouped broadly according to three major market sectors (consumer goods, equipment including defense, and materials), with further divisions and subdivisions as in the industry groupings. One major advantage of the new market groupings is that they make possible careful studies of the relationships between changes in production and dollar expenditures. Somewhat more broadly, it is hoped that the improved physical volume measures in the revised index will permit penetrating analyses of economic developments in the complex and rapidly changing American economy.

We shall not describe in detail the fairly complicated way in which the individual series are combined into group indexes and eventually into the overall production index. Let us point out, however, that for the *total index* the calculations actually performed lead to a fixed-weight aggregative index which is the ratio of total value added in a given month to the corresponding value added in 1957–1959 (using in both cases 1957 unit prices); that is

$$I = \frac{\Sigma \ q_n p_{57}}{\Sigma \ q_{57-59} p_{57}} \cdot 100$$

Currently, the published index relates to the base 1957–59.

Unlike the Consumer and Wholesale Price Indexes, the Index of Industrial Production is intended to be a measure of current business conditions. Along with such other indicators of business activity as personal income, the average workweek of production workers, new orders in durable goods industries, new plant and equipment expenditures, new business formations and failures, and business sales and inventories—called "business indicators"—the production index is carefully watched by all those whose current activities and future plans are in some way affected by the ever-changing overall business conditions.

SOME SPECIAL APPLICATIONS

One of the most widely discussed of all economic phenomena is what is often called the "shrinking value of the dollar." Of course, there are all sorts of "dollars"—money spent for construction, money spent for food or housing, money spent for medical care, and so on—and what happens to a particular "dollar" is not of equal concern to everyone. Referring to the different dollars, we often hear such statements as "compared to 1913 the construction dollar is worth only 10 cents," "compared to 1950 the food dollar is worth only 70 cents," "compared to 1957–59 the housing dollar is worth only 80 cents," and so forth. This is another way of saying that what bought a given amount of construction in 1913 now buys only 10 per cent of it, the amount of money necessary to feed a family for 100 days in 1950 now feeds it for only 70 days, and what in 1957–59 paid for a year's housing now pays for about $9\frac{1}{2}$ months' housing. In each of these examples, a change in price is expressed in terms of a so-called change in the value of a dollar.

It is easy to illustrate how the value of some single-commodity dollar is obtained. If an ice cream maker bought a certain amount of vanilla extract for \$4.10 in 1950 and had to pay \$8.20 for the same amount of the identical vanilla in 1970, the value of the "vanilla dollar" in 1970 was only 50 cents compared to 1950. This simply means that the price has doubled and that the value or the *purchasing power* of the "vanilla dollar" has been cut in half. Generally speaking, the purchasing power of a dollar is simply the *reciprocal* of an appropriate price index, written as a proportion. If prices have doubled from some reference period, the price index is 2.00 (200 per cent), and what a dollar will buy is only $\frac{1}{2.00} = \frac{1}{2}$ of what it would buy in the earlier period. In other words, the purchasing power is $\frac{1}{2}$ of what it was, or 50 cents. Similarly, if the price had risen by 50 per cent, the price index would stand at 1.50 and the purchasing power of the dollar would be $\frac{1}{1.50} = \frac{2}{3}$ of what it was, or about 67 cents.

The same argument applies also when we speak of the purchasing power of, say, a construction dollar, a food dollar, a rent dollar, or a medical care dollar, none of which refers to a single commodity. Presuming that comprehensive indexes of construction, food, rent, and medical care prices are available, we have only to take their reciprocals to arrive at the purchasing powers of the respective dollars. Thus in the beginning of this section, we arrived at a purchasing power of 70 cents for the food dollar by taking the reciprocal of 1.43, the current value of an appropriate food price index (written as a proportion). To obtain an estimate of the purchasing power of what is the nearest thing to an omnibus dollar used to buy the full range of goods and services available in the economy, called *"the* dollar," we might take the reciprocal of such comprehensive indexes as the Consumer Price Index or the Wholesale Price Index. For instance, for the month of March 1969, the purchasing power of the dollar was $\frac{1}{1.256} = 0.796$ (or 79.6 cents) based on the index of consumer prices, and it was $\frac{1}{1.117} = 0.895$ (or 89.5 cents) based on the index of wholesale prices, where in both cases 1957–59 = 100.

Another important application of price indexes is in the calculation of so-called "real wages." Since money is not an end in itself, wage earners are usually interested in what their wages will buy, and this depends on two things: How much they earn and the prices of the things they wish to buy. Clearly, a person would be worse off than before if his wages *doubled* while, at the same time, the prices of the things he wants to buy *tripled.*

To calculate "real wages," we can either multiply actual money wages by a quantity measuring the purchasing power of the dollar, or better, divide actual wages by an appropriate price index. This process is referred to as *deflating*, and the price index used as a divisor is called a *deflator*. As an illustration, let us investigate the change in real wages from 1950 to 1970 of production workers in nondurable goods industries in a large metropolitan area. The first column of the following table shows the average weekly gross earnings in these industries for the two years, while the second column shows corresponding values of an index of consumer prices in that area, with its base *shifted* from 1957–59 = 100 to 1950 = 100 (see Exercise 10 on page 88). The values in the third column are the real wages, obtained by dividing the index numbers (expressed as proportions) into the corresponding actual earnings.

	Average weekly wages	Consumer prices 1950 = 100	Real wages
1950	$ 54.35	100.0	$54.35
1970	$106.51	141.0	$75.54

As can easily be verified, the actual wages increased by 96.0 per cent from 1950 to 1970, and if prices had remained unchanged, real wages would have increased by precisely the same amount. However, prices have increased by 41.0 per cent, and there has been a decrease in what the dollar will buy. Consequently, real wages have increased by only 39.0 per cent, since $\frac{75.54}{54.35} = 1.39$. Expressed in another way, a market basket of goods and services which could have been bought for $75.54 in 1950 cost $106.51 in 1970.

The method we have illustrated here is frequently used to deflate individual values, value series, or value indexes. It is applied in problems dealing with such diverse things as dollar sales or dollar inventories of manufacturers, wholesalers, and retailers; total values of construction contracts or construction put in place; money incomes; and money wages. The only real problem in deflating value series such as these is that of finding appropriate indexes, or as we have called them here, appropriate deflators.

EXERCISES

1. Explain why an economist might have been justified in saying that he was not surprised that the Consumer Price Index reached a new high of 134.6 in May 1970 when prices on the stock market were declining and unemployment had risen to a high of 5 per cent and was increasing.

2. An elderly couple, planning to move from Baltimore to either San Francisco or Detroit, finds from a government publication that the most recent value of the Consumer Price Index for San Francisco stands at 115.6, while that for Detroit is only 111.1. Comment on their "deduction" from these figures that it would cost them 4.5 per cent more to live in San Francisco than in Detroit.

3. Comment on the following statements, both wrong, which appeared in a "popular" article on index numbers:

(a) "Probably the most important use of the Wholesale Price Index is in forecasting later movements in the Consumer Price Index."

(b) "A direct comparison of the Wholesale Price Index and the Consumer Price Index gives a very close estimate of the profit margins between primary markets and other distributive levels."

4. A business magazine reported that in May of a certain year the Index of Industrial Production was up 3 per cent and that the Federal Reserve Board had hinted that for June the index would likely return to its "pre-recession" level. Commenting on this, the magazine said, "In weighing these figures it is worth noting that, in a dynamic economy, such indexes always understate production." Why is a bias of this sort inherent in production indexes?

5. For March 1970 relative to March 1969, department store sales were down 5 per cent in San Francisco and up 4 per cent in Oakland, California. Is it reasonable to conclude that for the given month department store sales in San Francisco were 9 per cent lower than in Oakland?

6. For January through May 1970 the gross average weekly earnings (in dollars) in manufacturing were 131.93, 130.94, 132.40, 131.80, and 133.67, respectively, and the corresponding values of the Consumer Price Index (1957–59 = 100) were 131.8, 132.5, 133.2, 134.0, and 134.6.

(a) Use the values of the Consumer Price Index to express the weekly earnings in constant 1957–59 dollars; that is, use them to deflate the given weekly earnings.

(b) Use the results of part (a) to compare the actual percentage increases in earnings from January to May 1970 with the corresponding percentage change in the real earnings.

(c) Using the data of this exercise, calculate the January through May 1970 values of an index of the purchasing power of the dollar with 1957–59 = 100.

7. Given that the spendable weekly earnings of workers with three dependents for the same months as in Exercise 6 were 114.48, 113.69, 114.85, 114.37, and 115.36 dollars, deflate these 1970 figures to constant 1957–59 dollars. Comment on the use of the Consumer Price Index in connection with this problem.

8. The following table contains the average monthly earnings of wage earners in manufacturing industries in Mexico and values of an index of consumer prices for Mexico with 1963 = 100:

	Average monthly earnings (pesos)	Index of consumer prices (1963 = 100)
1962	964	99
1963	1,135	100
1964	1,239	102
1965	1,324	106
1966	1,385	110
1967	1,468	114
1968	1,544	116
1969	1,578	119

Deflate this series of average monthly earnings to constant 1963 pesos. Also, calculate the actual percentage increase in these earnings from 1963 to 1969 and compare it with the corresponding percentage change in the real earnings.

9. When we deflate the 1971 value of a single commodity to, say, 1969 prices, we divide its value by an index expressing the 1971 price of the commodity as a relative of the 1969 price. Show symbolically that this process leads to the value of the commodity in 1971 at 1969 prices. (Although this argument does not apply strictly when we deflate an aggregate of the values of several commodities, we are in a sense estimating the total of the *same* goods at base year prices.)

10. It is often desirable (or necessary) to change the point of reference, or *shift the base*, of an index number series from one period to another. Ordinarily, this is done simply by dividing each value by the original index number for the period which is to be the new base (and multiplying by 100). For instance, if we wanted to shift the base of the Consumer Price Index for the years 1960 through 1969 from 1957–59 = 100 to 1960 = 100, we would divide each of the 10 yearly values by 103.1, the original value of the index for 1960. Given the following values of an index of man-hours worked in mining for the months of January through May 1970 with 1957–59 = 100, shift the base to January 1970: 80.9, 82.2, 81.6, 81.1, and 80.6.

11. Shift the base of the index of consumer prices in Mexico of Exercise 8 to 1966 = 100; that is, calculate the values of the index for the given years, so that the base year is 1966 instead of 1963.

12. In 1970 the average weekly earnings of laborers in a certain region were $122.50 and in 1966 the corresponding earnings were $103.00. A regional "cost-of-living" index stood at 130 for 1970 and at 116 for 1966 with 1947 = 100. Express the 1970 dollar earnings of these laborers in terms of constant 1966 dollars.

13. Since index numbers are designed to compare two sets of figures, it seems reasonable that if an index for 1971 with the base year 1967 stands at 200, the same index for 1967 with the base year 1971 should be equal to 50. (If one thing is twice as big as another, the second should be half as big as the first.) To test whether an index meets this criterion, called the *time reversal test*, we need only interchange the subscripts 0 and n wherever they appear in the formula, and then see whether the resulting index (written as a proportion) is the reciprocal of the first. Determine which indexes among the simple aggregative index, the weighted aggregative index, the arithmetic mean of price relatives, the geometric mean of price relatives, and the Ideal Index (see Exercise 11 on page 78) satisfy this criterion.

14. As has been suggested in the text, price index formulas can be changed into quantity index formulas simply by replacing the p's with q's and the q's with p's. Using this relationship between the formula for a price index and the corresponding formula for a quantity index, the *factor reversal test* requires that the product of the two (written as proportions) equal the value index $\dfrac{\Sigma\, p_n q_n}{\Sigma\, p_0 q_0}$. Show that this criterion is satisfied if we compare the prices, quantities, and values of a single commodity and for the Ideal Index (see Exercise 11 on page 78), but not for any of the other index number formulas given in this chapter.

A WORD OF CAUTION

Having already discussed some of the problems encountered in index number construction, let us now add a word of caution about their use and their interpretation. Difficulties always arise when attempts are made to generalize beyond the stated purpose of an index to phenomena it was never intended to describe. The word "general" serves well enough to distinguish more or less comprehensive general-purpose indexes from those that are deliberately limited, or "special," in scope, but it is quite misleading in another sense: *Most "general-purpose" indexes are themselves strictly limited in purpose and in scope.*

Perhaps the most widely misunderstood index of all is the Consumer Price Index of the Bureau of Labor Statistics. The index is widely thought to measure not only the "cost of living" for everybody everywhere, but also to measure current business conditions—neither of which it does. In view of the government's many careful explanations of just what the Consumer Price Index is and is not intended to measure, this is hard to understand. Whatever remote or indirect connection there may exist between the limited phenomenon described by the index and business conditions in general, is unintended. Thus, there is really nothing surprising about the fact that in February 1958, when the country was in a strong recession and unemployment was growing, the Consumer Price Index reached a (then) record high of 122.5 (1947–49 = 100). Moreover, there is little basis for thinking of the Consumer Price Index as a measure of everyone's cost of living, even though it was once officially called a cost-of-living index. Actually, as the government is now careful to point out, the index measures the effect of *price changes* of a (fixed) market basket of goods and services on the cost of living of the families and individuals to which it applies. Clearly, a person's cost of living depends to some extent on his *level of living*, and changes in the level of living are not reflected in the index because purchases are held constant. Also, the index does not take into account Federal and state income taxes, among other things, nor does it (usually) account for price reductions on sale and discount items in regular trade. Unfortunately, no true cost-of-living index—one which, for example, would measure price changes while holding satisfaction or utility, rather than purchases, constant—exists for this country. Nevertheless, some professionals have asserted that, under normal conditions, the Consumer Price Index can be considered to be a good approximation to changes in the cost of living, but no knowledgeable person seems willing to claim more than that.

There are some persons who would like to see the government develop bigger, better, and more general indexes, say, a truly "general" consumer price index (or even a cost-of-living index) covering *all* families and *all* goods and services. Others feel that the value of an index decreases more or less in proportion to the increase in its scope. From this latter point of view, such phenomena as changing retail prices, wholesale prices, industrial production, and so on are far too broad ever to be described in terms of a single number. No matter how one feels about this problem, it is difficult to get away from the fact that the reduction of a large set of data to a single number generally entails the loss of such a tremendous amount of information that the whole procedure may have little practical value, if any. There are, indeed, some formidable difficulties connected both with the construction of index numbers by the professional and their interpretation by the layman. As one economist has pointed out:

> "*It ought to be conceded that index numbers are essentially arbitrary. Being at best rearrangements of data wrenched out of original market and technological contexts, they strictly have no economic meaning. Changes in tastes, technology, population composition, etc., over time increase their arbitrariness. But, of course, there is no bar to the use of indexes 'as if' they did have some unequivocal meaning provided that users remember that they themselves made up the game and do not threaten to 'kill the umpire' when the figures contradict expectations.*" *

---

* I. H. Siegel, in a letter to the editor, *The American Statistician*, February 1952.

5

PROBABILITY

INTRODUCTION

We can hardly predict what records will be among the "top 10" unless we know at least which ones are on the market, and we cannot very well predict the outcome of a presidential election unless we know what candidates are running for office. More generally, we cannot make intelligent predictions or decisions unless we know at least what is possible: To put it somewhat differently, *we must know what is possible before we can judge what is probable*. Thus, the next two sections will be devoted to "what is possible" in a given situation, after which the remainder of the chapter will be devoted to "what is probable."

MATHEMATICAL PRELIMINARY: SETS, OUTCOMES, AND EVENTS

In statistics, it is customary to refer to any process of observation as an experiment. Thus, using this term in a very wide sense, the simple process of determining the number of mistakes a typist makes on a page and the much more complicated process of obtaining and evaluating data to predict gross national product are both regarded as experiments. The results which one obtains from an experiment, whether they are instrument readings, counts, "yes" or "no" answers, or other kinds of measurements, are called the *outcomes* of the experiment, and the set (totality) of all possible outcomes of an experiment is called the *sample space*. (Other terms used instead of "sample space" are "possibilities space," "universal set," and "universe of discourse.")

In most cases it is convenient to think of the outcomes of an experiment, the elements of the sample space, as a set of points. Thus, if an experiment consists of determining whether on a certain day the price of a given stock goes down, remains unchanged, or goes up, the sample space may be thought of as consisting of three points, say, the three points of Figure 5.1,

FIGURE 5.1 A sample space

to which we arbitrarily assigned the numbers 1, 2, and 3. (Instead, we could have used any other configuration of three points and we could have assigned to them any arbitrary numbers.) Had we been interested in the behavior of two stocks, the sample space could have been given as in Figure 5.2, where 1, 2, and 3 again stand for a stock's price going down, remaining

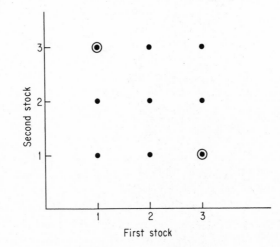

FIGURE 5.2 A sample space

unchanged, and going up, while the two coordinates refer to the respective stocks. Thus, the point (2, 3) represents the outcome that the price of the first stock remains unchanged while that of the second stock goes up. Note that if we had been interested in n stocks in this experiment, the corresponding sample space would have consisted of 3^n points. For instance, for five stocks there would have been $3^5 = 3 \cdot 3 \cdot 3 \cdot 3 \cdot 3 = 243$ points, that is, 243 different outcomes, among which (1, 2, 1, 1, 3) represents the case where the prices of the first, third, and fourth stocks go down, that of the second stock remains unchanged, while that of the fifth stock goes up. (Since this sample space has more than three dimensions, that is, each point has more than three coordinates, it cannot readily be pictured geometrically like the sample spaces of Figures 5.1 and 5.2.)

What kind of a sample space is actually appropriate for a given experiment will have to depend on what we look upon as an individual outcome. If in the two-stock example we had been interested only in the total number of stocks whose prices go down, remain unchanged, or go up, we could have used the three-dimensional sample space of Figure 5.3 instead of the

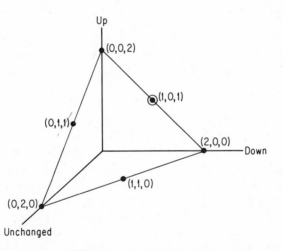

FIGURE 5.3 A sample space

one of Figure 5.2. Here the first coordinate gives the number of stocks whose prices go down, the second coordinate gives the number of stocks whose prices remain unchanged, while the third coordinate gives the number of stocks whose prices go up. Thus (0, 2, 0) represents the outcome that the prices of both stocks remain unchanged, while (0, 1, 1) represents the outcome that the price of one stock remains unchanged while the price of the other goes up. Note that the two points circled in Figure 5.2 correspond to the single point circled in Figure 5.3. It will be left to the reader to check what point or points of Figure 5.2 correspond to the other points of Figure 5.3.

Generally speaking, it is desirable to use sample spaces whose elements cannot be further subdivided; that is, the individual points should not represent two or more outcomes which are distinguishable in some fashion. Thus, in the example dealing with the two stocks, it would in most instances be preferable to use the sample space of Figure 5.2 rather than that of Figure 5.3.

Having defined a sample space as the set of all possible outcomes of an experiment, let us now explain what we mean by an *event*. This is important because probabilities invariably refer to the occurrence or nonoccurrence of some event. We assign a probability to the *event* that a given executive will be promoted, to the *event* that a shipment will be delayed, to the *event* that 60 of the 82 employees of a company belong to a union, and so forth. Hence, let us state formally that *when we speak of an event, we are speaking of an appropriate subset of a sample space* (by *subset* we mean any part of a set including the set as a whole and, trivially, the empty set which has no elements at all). Thus, in Figure 5.2 the subset which consists only of the point (1, 1) represents the *event* that the prices of both stocks go down; the subset which consists of the two points circled in Figure 5.2 represents the *event* that the price of one stock goes up while that of the other goes down; and the subset of the sample space of Figure 5.2 which consists of the five points (3, 1), (1, 3), (3, 2), (2, 3), and (3, 3) represents the event that the price of *at least one* of the stocks goes up.

Still referring to the example of the two stocks, let us now suppose that X stands for the event that the price of the first stock goes down regardless of what happens to the price of the second, Y stands for the event that the price of the second stock remains unchanged regardless of what happens to the price of the first, and that Z stands for the event that the prices of both stocks go up. Referring to Figure 5.4, giving the same sample space as Figure 5.2, it can be seen that X consists of the three points inside the dotted line, Y consists of the three points inside the dashed line, while Z consists of the point (3, 3). Note that the events X and Z have no points in common; they are referred to as *mutually exclusive events*, which means that they cannot both occur in the same experiment. On the other hand, X and Y are *not* mutually exclusive, since these two events both occur when the price of the first stock goes down while that of the second stock remains unchanged.

There are many problems of probability in which we are interested in events that can be expressed in terms of two or more other events. For instance, in our example we might be interested in the event that *either the price of the first stock goes down or the prices of both stocks go up*, or we might be interested in the event that *the price of the first stock goes down and the price of the second stock remains unchanged.* In the first case we are interested in the event that *either X or Z* occurs and in the second case we are interested in the event that *X and Y both* occur.

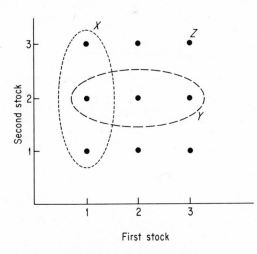

FIGURE 5.4 A sample space

In general, if A and B are any two events we define their *union* $A \cup B$ as the event which consists of all the individual outcomes contained either in A, in B, or in both. It is customary to read $A \cup B$ as "A cup B" or simply as "A or B." With reference to the above example, we find that $X \cup Y$ consists of the five points $(1, 1)$, $(1, 2)$, $(1, 3)$, $(2, 2)$, and $(3, 2)$, while $X \cup Z$ consists of the four points $(1, 1)$, $(1, 2)$, $(1, 3)$, and $(3, 3)$. Also, if A and B are any two events we define their *intersection* $A \cap B$ as the event which consists of all the individual outcomes contained in both A and B. We read $A \cap B$ as "A cap B" or simply as "A and B." With reference to the above example we find that $X \cap Y$ consists of the point $(1, 2)$ and that $X \cap Z$ has no elements at all, it is empty. Denoting the empty set by the symbol ϕ we write $X \cap Z = \phi$.

To complete our notation, let us define the *complement* A' of an event A with respect to a sample space S as the event which consists of all the individual outcomes of S that are not contained in A. With reference to our example, X' stands for the event that the price of the first stock does not go down, and it consists of the six points $(2, 1)$, $(2, 2)$, $(2, 3)$, $(3, 1)$, $(3, 2)$, and $(3, 3)$. Also, $(X \cup Y)'$, the complement of the event $X \cup Y$, consists of the points $(2, 1)$, $(3, 1)$, $(2, 3)$, and $(3, 3)$; it represents the event that the price of the first stock will not go down while that of the second stock will not remain unchanged.

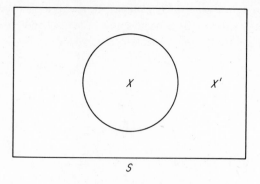

FIGURE 5.5 Venn diagram

When dealing with sample spaces, subsets, and events, it is often helpful to represent the entire sample space S by means of a rectangle and events by means of regions within the rectangle (usually circles or pieces of circles). Such representations are called *Venn diagrams* and some examples are shown in Figures 5.5, 5.6, and 5.7. Looking upon the rectangle of Figure 5.5 as representing the sample space for the experiment with the two stocks, event X could be represented by the circle, in which case its complement X' would be represented by the remainder of S. Similarly, the two circles of Figure 5.6 could be used to represent the events X and Y, in which case the region common to the two circles would represent the event $X \cap Y$, while the entire shaded region would represent the event $X \cup Y$. When dealing with three events, it is customary to draw the respective circles as in Figure 5.7. Note that the shaded region represents the event $X \cap (Y \cup Z)$,

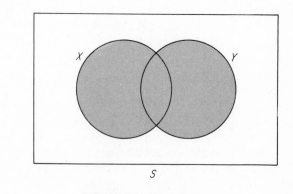

FIGURE 5.6 Venn diagram

namely, the event which consists of all outcomes belonging to X and also to either Y or Z. As indicated in the exercises below, Venn diagrams are often used to verify various kinds of relationships among sets, subsets, or events, without requiring rigorous proofs based on a formal Algebra of Sets.

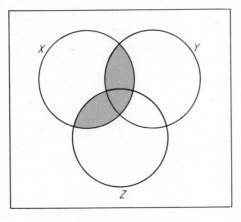

FIGURE 5.7 Venn diagram

EXERCISES

1. Having moved to a new city, a doctor is looking for a new one-, two-, or three-bedroom house, and he finds that none of the houses advertised for sale have fewer bedrooms than baths, but each has, of course, at least one bath.

   (a) Draw a diagram similar to that of Figure 5.2 (with the x-coordinate denoting the number of bedrooms and the y-coordinate the number of baths) which shows the six different ways in which his choice can be made. For instance, $(3, 1)$ represents the event that he chooses a three-bedroom house with one bath, and $(2, 2)$ represents the event that he chooses a two-bedroom house with two baths.

   (b) Describe *in words* the event which is represented by each of the following sets of points: event E which consists of the points $(2, 2)$ and $(3, 1)$, event F which consists of the points $(2, 1)$, $(3, 1)$, and $(3, 2)$, and event G which consists of the points $(1, 1)$, $(2, 1)$, and $(2, 2)$. Also indicate these three sets on the diagram of part (a) by enclosing the corresponding points by means of solid, dotted, or dashed lines.

   (c) Referring to part (b), describe each of the following events *in words* and list the points which it contains:

   (i) F';   (iii) $E \cap F$;   (v) $E \cap G$;
   (ii) G';   (iv) $F \cup G$;   (vi) $E \cup G'$.

(d) Referring to part (b), which of the following pairs of subsets represent *mutually exclusive events:*

(i) E and G;
(iii) F' and G;
(ii) F and G;
(iv) E and $F \cap G$?

2. Among six applicants for an executive job Mr. A is a college graduate, foreign born, and single; Mr. B is not a college graduate, foreign born, and married; Mr. C is a college graduate, native born, and married; Mr. D is not a college graduate, native born, and single; Mr. E is a college graduate, native born, and married; and Mr. F is not a college graduate, native born, and married. One of these applicants is to get the job, and the event that the job is given to a college graduate, for example, is denoted $\{A, C, E\}$. Indicate the following in a similar manner:

(a) The event that the job is given to a married person.

(b) The event that the job is given to a native-born college graduate.

(c) The event that the job is given to a single person who is foreign born.

(d) The event that the job is given to a person who is either a college graduate or married and foreign born.

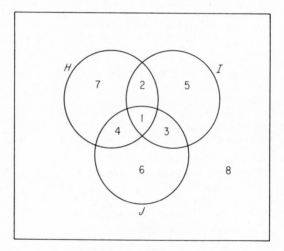

FIGURE 5.8   Venn diagram

3. Suppose that a graduate of a two-year program in business administration is looking for a job and that H is the event that he will get a job paying a high salary, I is the event that he will get a job with a good future, and J is the event that he will get a job with a good pension plan. With reference to the Venn diagram of Figure 5.8 list the regions or combinations of regions which represent the following events:

(a) The event that he will get a job paying a high salary, which has a good future but does not have a good pension plan.

(b) The event that he will get a job paying a high salary, which has neither a good future nor a good pension plan.

(c) The event that he will get a job paying a high salary, which has a good pension plan.

(d) The event that he will get a job which does not pay a high salary, but has a good pension plan.

(e) The event that he will get a job which does not pay a high salary, does not have a good future, and does not have a good pension plan.

4. With reference to Exercise 3 and the Venn diagram of Figure 5.8, explain *in words* what events are represented by the following regions:

(a) Region 1;   (d) Regions 1 and 3 together;

(b) Region 4;   (e) Regions 4 and 7 together;

(c) Region 7;   (f) Regions 5 and 8 together.

5. A coin is tossed three times in succession and the result is recorded as HHT if there are first two heads and then a tail, THT if there is first a tail, then a head, and then another tail, and so on. List the eight possible outcomes of this "experiment."

6. Referring to Exercise 5, let E represent the event that the first two tosses are heads, and let F represent the event that at least one of the tosses is tails. List the outcomes belonging to

(a) E;   (e) $E' \cap F'$;

(b) F;   (f) $(E \cap F') \cup (E \cap F)$;

(c) $E \cap F$;   (g) $E \cap (E \cup F)$;

(d) $E \cup F'$;   (h) $E \cup (E \cap F)$.

7. If in Exercise 5 we let 0 stand for "tail" and 1 for "head," we can let the point $(1, 1, 0)$ represent HHT, $(0, 1, 0)$ represent THT, and so on. Draw a figure showing the eight possible outcomes in a three-dimensional sample space.

8. A small taxicab company has two cabs, of which the larger can carry five passengers and the smaller can carry four passengers. Using the point $(0, 2)$ to indicate that at a given moment the larger cab is empty while the smaller one has two passengers, the point $(4, 3)$ to indicate that the larger cab has four passengers while the smaller one has three passengers, and so on, draw a figure showing the 30 points which correspond to the different ways in which the cabs can be occupied. Also, if D stands for the event that at least one of the cabs is empty, E stands for the event that together they carry two, four, or six passengers, and F stands for the event that there is the same number of passengers in each cab, list the points of the sample space belonging to

(a) D;   (d) $D \cup E$;   (g) $D \cup F$;

(b) E;   (e) $D \cap E$;   (h) $D \cap F'$;

(c) F;   (f) $E \cup F$;   (i) $E' \cap D'$.

9. It was pointed out in the text that the points $(1, 3)$ and $(3, 1)$ of Figure 5.2 together represent the same event as the point $(1, 0, 1)$ of Figure 5.3. Indicate

what point or points of Figure 5.2 correspond to each of the other points of Figure 5.3.

10. A company plans to build a new plant in one of Connecticut's eight counties. T is the event that it will build in Hartford County or in New Haven County, U is the event that it will build in Hartford County or in Litchfield County, V is the event that it will build in New Haven County or in New London County, and W is the event that it will build in Fairfield County or in New Haven County. The other three counties are Middlesex, Tolland, and Windham counties. List the outcomes which belong to each of the following events:

(a) T';

(b) W';

(c) $V \cap W$;

(d) $U \cap V$;

(e) $U \cup V$;

(f) $T \cup U$;

(g) $V \cup W$;

(h) $(U \cup V)'$.

11. Which of the following pairs of events are mutually exclusive:
    (a) Having rain and having sunshine on a given day;
    (b) Wearing green sox and wearing black shoes (at the same time);
    (c) Being intoxicated and being sober (at the same time);
    (d) Obtaining a king and obtaining a queen when drawing one card from an ordinary deck of 52 playing cards;
    (e) Obtaining a king and obtaining a black card when drawing one card from an ordinary deck of 52 playing cards;
    (f) Owning a Chevrolet and owning a Ford (at the same time);
    (g) Driving a Chevrolet and driving a Ford (at the same moment);
    (h) Being under 25 years of age and being President of the United States?

12. Use Venn diagrams to verify that $(A \cap B)' = A' \cup B'$, and also that $(A \cup B)' = A' \cap B'$.

13. Use Venn diagrams to verify that $A \cup (A \cap B) = A$.

14. Use Venn diagrams to verify that $A \cup (B \cap C) = (A \cup B) \cap (A \cup C)$.

15. Use Venn diagrams to verify that $(A \cap B) \cup (A \cap B') = A$.

16. Use Venn diagrams to verify that $A \cup B = (A \cap B) \cup (A \cap B') \cup (A' \cap B)$.

MATHEMATICAL PRELIMINARY: COUNTING

In contrast to the complexity of many of the modern methods used in business and economics, the simple process of counting still plays an important role. One still has to count 1, 2, 3, 4, . . . , say, when taking inventory, when counting cars passing a given spot, or when preparing a report showing how many rainy days there were during a given month. Sometimes the actual process of counting can be simplified by the use of mechanical devices, by performing the count indirectly (e.g., when determining the total number of sales from the serial numbers of invoices), or by using some of the mathematical theory which we shall discuss below.

To perform an actual count, it is always necessary to list, align, or otherwise arrange the objects to be counted in some way. Although this may sound easy, the following example will illustrate that sometimes it may not be. Suppose, for instance, that two used-car salesmen make a bet each week as to who will be the first to make three sales. If we let X denote a sale made by the first salesman and Y a sale made by the second, $XXYX$ is one sequence which leads to the first salesman winning the bet, and so are the sequences $YXXX$ and $XYXYX$. On the other hand, YYY, $YXXYY$, and $XYYY$ are sequences which lead to the second salesman winning the bet. Continuing this way very carefully, we may be able to list all possible sequences of X's and Y's which lead to one of the two salesmen winning the bet and come up with the correct answer that there are altogether 20 possibilities.

To handle problems like this systematically, it is helpful to refer to a *tree* diagram like that of Figure 5.9. This diagram shows that for the first letter there are two possibilities (two branches) corresponding to which salesman makes the first sale; for the second letter there are two branches emanating from each of the two branches; similarly, for the third letter there are two branches emanating from each of the four branches. After that, things begin to change: For the fourth letter there are two branches emanating from each of the six middle branches, but the first and last branches terminate for they represent XXX and YYY, namely, situations in which one of the salesmen wins. For the fifth letter there are two branches emanating from 6 of the remaining 12 branches while the others terminate, and it can thus be seen that there are altogether 20 different paths along the "branches" of the tree diagram of Figure 5.9. In other words, there are 20 possible outcomes to this "experiment." It can also be seen from the tree diagram that in *two* of the cases the issue was decided after three sales, in *six* of the cases the issue was decided after four sales, and that in *twelve* of the cases the issue was not decided until the fifth sale. Note that this kind of scheme applies also to a play-off in sports where a team must win three games out of five.

To consider another example where a tree diagram can be of some aid, suppose that a dispatcher has four drivers and three trucks that he can assign to a job. If we label the four drivers A, B, C, and D, and the three trucks I, II, and III, we find from the tree diagram of Figure 5.10 that there are 12 different ways in which this can be done. The first path along the branches of the tree corresponds to the choice of Driver A and Truck I, the second path corresponds to the choice of Driver A and Truck II, . . . , and the twelfth path corresponds to the choice of Driver D and Truck III.

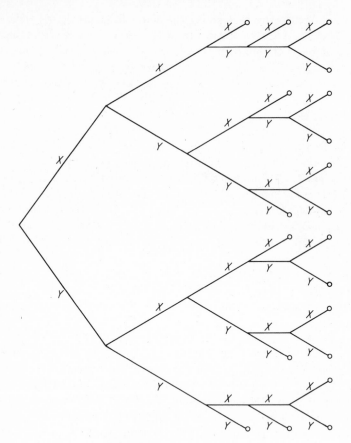

FIGURE 5.9 Tree diagram

Note that the answer which we obtained for this example is the *product* of 4 and 3, namely, the number of ways in which the dispatcher can first select one of the drivers and then one of the trucks. Generalizing from this example, let us thus state the following rule:

> *If a selection consists of two separate steps, the first of which can be made in m ways and the second in n ways, then the whole selection can be made in m · n ways.*

Thus, if a drive-in restaurant offers the choice of eight kinds of hamburgers and six kinds of soft drinks, there are altogether $8 \cdot 6 = 48$ ways in which one can order a hamburger and a soft drink.

Using appropriate tree diagrams, it is easy to generalize the above rule so that it will apply to selections involving more than two steps. If there are k steps we have the following:

> *If a selection consists of k separate steps, the first of which can be made in n_1 ways, the second in n_2 ways, . . . , and the kth in n_k ways, then the whole selection can be made in $n_1 \cdot n_2 \cdot \cdots \cdot n_k$ ways.*

Thus, if the dispatcher in the example on page 101 also had to choose one of five helpers, he could have chosen a driver, a truck, and a helper in $4 \cdot 3 \cdot 5 = 60$ ways, and if the drive-in restaurant mentioned above also offered four kinds of pies, there would be $8 \cdot 6 \cdot 4 = 192$ ways in which one could order a hamburger, a soft drink, and a piece of pie.

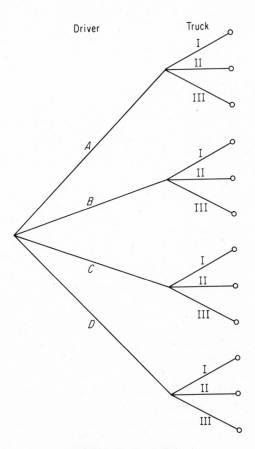

FIGURE 5.10 Tree diagram

The rule of the preceding paragraph is often applied when *several selections are made from one and the same set, and the order in which they are made is of importance.* To illustrate, suppose that the five finalists of the Miss Arizona contest are Miss Cochise, Miss Maricopa, Miss Navajo, Miss Pima, and Miss Yuma, and that the judges have to choose a winner and a first runner-up. To determine the number of ways in which this can be done, suppose that they first choose the winner and then the first runner-up. Clearly, the winner can be chosen in five ways, and since this leaves only four of the finalists for the choice of the runner-up, the whole selection can be made in $5 \cdot 4 = 20$ different ways. Had the judges been asked to pick the winner, a first runner-up, a second runner-up, a third runner-up, and a fourth runner-up from among the five finalists, they could have done so in $5 \cdot 4 \cdot 3 \cdot 2 \cdot 1 = 120$ different ways; clearly, *after each choice there is one less to choose from for the next selection.* To consider another example, if the 35 members of a fraternal organization wanted to select a president, a vice-president, and a secretary-treasurer, they could do this in $35 \cdot 34 \cdot 33 = 39{,}270$ different ways; again, there is one less to choose from after each selection.

In general, if r objects are selected from a set of n objects, any particular arrangement of these r objects is referred to as a *permutation*. For instance, 45132, 13524, and 25341 are three different permutations of the first five positive integers; *Idaho, New Mexico, and Utah* is a permutation (a particular ordered arrangement) of three of the eight Mountain States; *DCGE, CFGA,* and *BCDF* are three of many possible permutations of four of the first seven letters of the alphabet; and if we were asked to list *all possible* permutations of two of the five vowels a, e, i, o, and u, our answer would be

ae	*ai*	*ao*	*au*	*ei*	*eo*	*eu*	*io*	*iu*	*ou*
ea	*ia*	*oa*	*ua*	*ie*	*oe*	*ue*	*oi*	*ui*	*uo*

So far as the counting of possible permutations is concerned, direct application of the rule on page 103 leads to the following result:

The number of permutations of r objects selected from a set of n objects is $n(n - 1)(n - 2) \cdots (n - r + 1)$, which we shall denote $_nP_r$. [*]

To prove this formula, we have only to observe that the first selection is made from the whole set of n objects, the second selection is made from the $n - 1$ objects which remain after the first selection has been made, the

---

[*] The following are some alternate symbols used to denote the number of permutations of r objects selected from a set of n objects: $P(n, r)$, P_r^n, $P_{n,r}$, and $(n)_r$.

third selection is made from the $n - 2$ objects which remain after the first two selections have been made, and the rth and final selection is made from the

$$n - (r - 1) = n - r + 1$$

objects which remain after the first $r - 1$ selections have been made. Thus, the number of ways in which four new territories can be assigned (one each) to four of a company's seven salesmen is $7 \cdot 6 \cdot 5 \cdot 4 = 840$, and the number of ways in which a college senior can pick a first, second, and third choice among 45 potential employers is $45 \cdot 44 \cdot 43 = 85,140$. Incidentally, it has been assumed in our discussion that the objects with which we are dealing are *distinguishable* in some way; otherwise, there are complications, which we shall discuss in Exercises 20 and 21 on pages 109 and 110.

Since products of consecutive integers arise in many problems dealing with permutations and other kinds of special arrangements, we write $5 \cdot 4 \cdot 3 \cdot 2 \cdot 1$, for instance, as 5! and refer to it as "5 *factorial*." More generally, $n!$, which we call "*n factorial*," denotes the product

$$n(n - 1)(n - 2) \cdots 3 \cdot 2 \cdot 1$$

for any positive integer n. Thus, $1! = 1, 2! = 2 \cdot 1 = 2, 3! = 3 \cdot 2 \cdot 1 = 6,$ $4! = 4 \cdot 3 \cdot 2 \cdot 1 = 24,\ 5! = 5 \cdot 4 \cdot 3 \quad 2 \cdot 1 = 120,\ \ 6! = 6 \cdot 5 \cdot 4 \cdot 3 \cdot$ $2 \cdot 1 = 720, \ldots,$

and in this notation it is also customary to let $0! = 1$ *by definition*. Using this notation we can now say that

The number of permutations of n objects taken all together is n!

This is simply a special case of the rule on page 104 with $r = n$, so that the last factor becomes $n - r + 1 = n - n + 1 = 1$. Thus, the number of ways in which seven instructors can be assigned to seven sections of a course is $7! = 5,040$, and the number of ways in which the 10 football teams in a conference can finish the season is $10! = 3,628,800$ (not counting possible ties in the final standings).

There are many problems in which we are interested in the number of ways in which r objects can be selected from a set of n objects, but where we do not care about the order in which the selection is made. We may thus want to know in how many ways a committee of 5 can be selected from among the 100 employees of a company, or the number of ways in which

10 stocks can be selected from a list of 24. To obtain an appropriate formula, let us first consider the following 24 permutations of three of the first four letters of the alphabet:

$$
\begin{array}{cccccc}
abc & acb & bac & bca & cab & cba \\
abd & adb & bad & bda & dab & dba \\
acd & adc & cad & cda & dac & dca \\
bcd & bdc & cbd & cdb & dbc & dcb
\end{array}
$$

If we are interested only in the number of ways in which three of these four letters can be selected and not in the order in which they are selected, it should be noted that each row in the above table contains $3 \cdot 2 \cdot 1 = 6$ permutations of the same three letters selected from among the first four letters of the alphabet. In fact, there are only four ways in which three letters can be selected from among the first four letters of the alphabet if we are not interested in the order in which the selection is made; they are a, b, and c; a, b, and d; a, c, and d; and b, c, and d. In general, there are $r!$ permutations of r specific objects selected from a set of n objects and, hence, the $n(n-1) \cdots (n-r+1)$ permutations of r objects selected from a set of n objects contain each set of r objects $r!$ times. (In our example, the 24 permutations of 3 letters selected from among the first 4 letters of the alphabet contained each set of 3 letters $3! = 6$ times.) Dividing $n(n-1) \cdots (n-r+1)$ by $r!$, we have the following rule:

The number of ways in which r objects can be selected from a set of n objects is

$$
\frac{n(n-1) \cdots (n-r+1)}{r!} \qquad \bigstar
$$

Symbolically, we write the number of ways in which r objects can be selected from a set of n objects as $\binom{n}{r}$, and we refer to it as "the number of *combinations* of n objects taken r at a time." * An easy way of determining these quantities is indicated in Exercise 12 on page 108. Otherwise, the values of $\binom{n}{r}$ for $n = 2$ through $n = 20$ can be obtained directly from Table X at the end of the book, where these quantities are referred to as *binomial coefficients;* the reason for this is explained in Exercise 22 on page 110.

---

* The following are some alternate symbols used to denote the number of combinations of r objects selected from a set of n objects: $_nC_r$, $C(n, r)$, and C_r^n.

Applying the above rule for the number of combinations, we find, for example, that a person can select 5 of the 8 stocks suggested to him by his broker in $\dfrac{8 \cdot 7 \cdot 6 \cdot 5 \cdot 4}{5 \cdot 4 \cdot 3 \cdot 2 \cdot 1} = 56$ ways, an employer can fire 3 of his 10 salesmen in $\dfrac{10 \cdot 9 \cdot 8}{3!} = 120$ ways, and a secretary can address 2 of 5 letters incorrectly in $\dfrac{5 \cdot 4}{2!} = 10$ ways.

EXERCISES

1. With reference to the illustration on page 104, draw a tree diagram which shows the 20 ways in which the judges can choose Miss Arizona and the first runner-up from among five finalists.

2. A theatrical promoter plans to put $20,000 into *one* Broadway play each year, so long as the number of flops in which he invests does not exceed the number of hits. (It will be assumed here that each play can be classified as being either a hit or a flop.)

   (a) Draw a tree diagram which shows the eight possible situations that can arise during the first four years the plan is in operation. (Note that if the first year's play is a flop, that branch of the tree diagram ends right then and there.)

   (b) In how many of the situations described in part (a) will the promoter continue to invest in the fifth year?

   (c) If the promoter loses his total investment in a flop and doubles his money in a hit, in how many of the situations described in part (a) will he be exactly even after four years?

3. The manager of a TV-stereo shop stocks two color television sets of the same kind, reordering at the end of each day (for delivery early the next morning) if and only if both sets have been sold. Construct a tree diagram which shows that if he starts on a Monday with two of the color television sets in stock, there are altogether eight different ways in which he can make sales on Monday and Tuesday of that week.

4. A customer can buy either the standard or deluxe model of an item, in any one of four colors and in any one of four sizes. How many distinct items are available to him?

5. If the NCAA has applications from Los Angeles, Dallas, Miami, and Boston for hosting its intercollegiate tennis championships in 1974 and 1975, in how many ways can they select the tournament sites (assuming that they are not to be the same)?

6. The Standard and Poor's Corporation regularly rates common stocks, assigning them the ratings A+, A, A−, B+, B, B−, and C.

(a) In how many ways can they assign ratings to three different stocks?

(b) In how many ways can they assign at least B+ ratings to two different stocks?

(c) In how many ways can they assign ratings to two different stocks, if one of the stocks (but not both) is to have a rating of A or A+?

7. If there are five motor routes from Town A to Town B, in how many different ways can a motorist travel from A to B and back, if

(a) he must not travel both ways by the same route;

(b) he must travel both ways by the same route;

(c) he can go and return by any route he chooses?

8. A hearse manufacturer makes 9 different types of hearses, each of which is available in 3 different roof finishes and 18 different body colors. How many distinct vehicles does the manufacturer provide for conveying a person to his eternal rest?

9. If a test consists of 20 true-false questions, in how many different ways can a student mark his test?

10. Trailer license plates in a given state consist of three digits, the first of which cannot be 0, followed by two letters of the alphabet, the first of which cannot be I, O, or Q. How many different plates are possible with this scheme?

11. The number of ways in which n distinct objects can be arranged in a circle is $(n - 1)!$.

(a) Present an argument to justify this formula.

(b) In how many ways can six persons be seated at a round table?

(c) In how many ways can a window dresser display four shirts in a circular arrangement?

12. (*Pascal's triangle*) The number of combinations of r objects selected from a set of n objects, namely, the quantities $\binom{n}{r}$, can be determined by means of the following arrangement, called Pascal's triangle,

$$
\begin{array}{ccccccccccc}
 & & & & 1 & & 1 & & & & \\
 & & & 1 & & 2 & & 1 & & & \\
 & & 1 & & 3 & & 3 & & 1 & & \\
 & 1 & & 4 & & 6 & & 4 & & 1 & \\
1 & & 5 & & 10 & & 10 & & 5 & & 1 \\
\end{array}
$$

....

where each row begins with a 1, ends with a 1, and each other entry is given by the sum of the nearest two entries in the row immediately above.

(a) Use Table X to verify that the *third* row of the triangle contains the values of $\binom{3}{r}$ for $r = 0, 1, 2,$ and 3, and that the *fourth* row contains the values of $\binom{4}{r}$ for $r = 0, 1, 2, 3,$ and 4.

(b) Construct the next four rows of the table and verify the results with the use of Table X.

13. To use Table X it is sometimes necessary to make use of the fact that

$$\binom{n}{r} = \binom{n}{n-r}$$

(a) Justify this formula informally.

(b) Prove the formula by using the formula for $\binom{n}{r}$ given in Exercise 14 below.

14. Verify symbolically that the number of permutations of n objects taken r at a time can be written $\dfrac{n!}{(n-r)!}$ and, hence, that the number of combinations of n objects taken r at a time can be written $\dfrac{n!}{(n-r)!r!}$.

15. Among the eight nominees for the Board of Directors of a grocery cooperative are four men and four women. In how many ways can the members elect as directors

(a) any two of the nominees;

(b) two of the male nominees;

(c) one of the male nominees and one of the female nominees;

(d) two of the female nominees?

Explain why the sum of the results of parts (b), (c), and (d) should equal the result of part (a).

16. How many four-man bowling teams can be chosen from 12 candidates in the shipping department of a large company?

17. In how many ways can a subcommittee of 4 persons be chosen from a committee of 10 persons if

(a) the chairman of the full committee is required to be on the subcommittee;

(b) neither the chairman nor the vice-chairman of the full committee is allowed to be on the subcommittee?

18. In how many ways can nine distinct cases be assigned to three case workers, A, B, and C, so that A gets four cases, B gets three cases, and C gets two cases?

19. The personnel manager of a store wants to fill five openings in its training program with three college graduates and two persons who are not college graduates. In how many ways can these openings be filled if among 26 applicants 15 are college graduates?

20. If among n objects r are alike, while the others are all distinct, the number of permutations of these n objects taken all together is $\dfrac{n!}{r!}$.

(a) How many permutations are there of the letters in the word "jeep"?

(b) How many distinct six-digit numbers can be formed with the digits 5, 5, 5, 6, 7, and 8?

(c) Justify the formula given in this exercise.

21. If among n objects r_1 are identical, another r_2 are identical, and the rest are all distinct, the number of permutations of these n objects taken all together is $\dfrac{n!}{r_1!r_2!}$.

(a) How many permutations are there of the letters in the word "greater"?

(b) How many distinct seven-digit numbers can be formed with the digits 3, 3, 3, 5, 5, 5, 7?

(c) How many distinct eight-character codes can be formed with the following eight characters *, *, *, *, ¢, ¢, ¢, and +?

(d) Justify this extension of the formula of Exercise 20, and, if possible, make it more general. How many permutations are there of the letters in the word "reiterate"?

22. The quantity $\binom{n}{r}$ defined on page 106 is referred to as a *binomial coefficient* because it is, in fact, the coefficient of a^r in the binomial expansion of $(a + b)^n$. Verify that this is true for $n = 2, 3,$ and 4, by expanding $(a + b)^2$, $(a + b)^3$, and $(a + b)^4$ and comparing the coefficients with the corresponding values of $\binom{n}{r}$ given in Table X.

PROBABILITIES AND ODDS

So far we have studied only what is *possible* in a given situation. In some instances we listed all possibilities and in others we merely determined how many different possibilities there are. *Now we shall go one step further and judge also what is probable and what is improbable.*

The most common way of measuring the uncertainties connected with events (e.g., the success of a new product, the outcome of a baseball game, the effectiveness of an advertising campaign, the returns of an investment) is to assign them *probabilities,* or to specify the *odds* at which it would be fair to bet that the events will occur. Among the different theories of probability—and there are many—the most widely held is the *frequency concept of probability,* according to which the probability of an event is interpreted as *the proportion of the time that events of the same kind will occur in the long run.* Thus, if we say that there is a probability of 0.82 that a jet from New York to Miami, Florida, arrives on time, this means that such flights will arrive on time about 82 per cent of the time. More generally, we say that an event has a probability of, say, 0.90, in the same sense in which we might say that our car will start in cold weather about 90 per cent of the time. *We cannot guarantee what will happen at any particular time— the car may start and then it may not—but it would be reasonable to bet $9 against $1 or 90 cents against 10 cents (namely, at odds of 9 to 1) that the car will start at any given try.* This would be "fair," "reasonable," or "equitable," for we would win $1 (or 10 cents) about 90 per cent of the time, lose $9 (or 90 cents) about 10 per cent of the time, and we can therefore expect to break even in the long run.

In accordance with the frequency concept of probability, we *estimate* the probability of an event by observing how often (what part of the time) similar events have occurred in the past. For instance, if data kept by a government agency show that (over a period of time) 492 of 600 jets from New York to Miami, Florida, arrived on time, we *estimate* the probability that any one flight from New York to Miami (perhaps, the next one) will arrive on time as $\frac{492}{600} = 0.82$. Similarly, if 687 of 1,854 freshmen who have attended a given college (over a number of years) dropped out before the end of their freshmen year, we *estimate* the probability that any freshman attending this college will drop out before the end of his freshman year as $\frac{687}{1,854} = 0.37$.

Having defined probabilities in terms of what happens to similar events in the long run, let us consider for a moment whether it is at all meaningful to talk about the probability of an event which *cannot occur more than once.* Can we ask for the probability that Mrs. Barbara Smith's broken right arm will heal within a month, or the probability that a certain major-party candidate will win an upcoming presidential election? If we put ourselves in the position of Mrs. Smith's doctor, we could check medical records, discover that such fractures have healed within a month in (say) 39 per cent of the thousands of reported cases, and apply this figure to Mrs. Smith's arm. This may not be of much comfort to Mrs. Smith, but it does provide a *meaning* for a probability statement concerning her arm—the probability that it will heal within a month is 0.39. Thus, *when we make a probability statement about a specific (nonrepeatable) event, the frequency concept of probability leaves us no choice but to refer to a set of similar events.* This can lead to complications, for the choice of "similar" events is often neither obvious nor easy. With reference to Mrs. Smith's arm, for example, we might consider as "similar" only those cases where the fracture was in the same arm, or we might consider only those cases in which the patients were just as old, or we might consider only those cases in which the patients were also of the same height and the same weight as Mrs. Smith. Ultimately, this is a matter of choice, and it is by no means contradictory that we can arrive at different probabilities concerning Mrs. Smith's arm; it should be observed, however, that the more we narrow things down, the less information we have to estimate the corresponding probability.

As for the second example, the one concerning the presidential election, suppose we ask some persons who have conducted a poll "how sure" they are that the given candidate will actually win. If their answer is "99 per cent sure," that is, if they assign the candidate's election a probability of 0.99, they are not implying that he would win 99 per cent of the time if he ran for office a great many times. *No, it means that the persons who conducted the poll based their conclusion (judgment, or decision) on methods which (in the long run) will "work" 99 per cent of the time.* In this sense, many of the probabilities which we use to express our faith in predictions or decisions are simply "success ratios" that apply to the methods we have employed.

An alternate point of view, which is currently gaining favor among statisticians, is to interpret probabilities as *personal* or *subjective.* To illustrate, suppose that a businessman feels that the *odds* for the success of a new venture, say, a new shoe store, are 3 to 2. This means that he would be willing to bet (or consider it fair to bet) $300 against $200, or perhaps $3,000 against $2,000, that the venture will succeed. In this way he expresses the *strength of his belief* regarding the uncertainties connected with the success of the new store. This method of dealing with uncertainties works well (and is certainly justifiable) in situations where there is very little direct evidence; in that case one may have no choice but to consider pertinent collateral information, "educated" guesses, and perhaps intuition and other subjective factors. Thus, the businessman's odds concerning the success of a new shoe store may well be based on his ideas about business conditions in general, the opinion of an expert, and his own subjective evaluation of the whole situation, including, perhaps, a small dose of optimism.

Regardless of how we interpret probabilities and odds, subjectively, or objectively (in terms of frequencies or proportions), the mathematical relationship between the two is always the same. It is given by the following rule:

> *If somebody considers it fair or equitable to bet a dollars against b dollars that a given event will occur, he is, in fact, assigning the event the probability* $\dfrac{a}{a+b}$.

Thus, the businessman who is willing to give odds of 3 to 2 that the new shoe store will succeed is actually assigning its success a probability of $\dfrac{3}{3+2} = 0.60$. Also, if the odds are 7 to 2 that a student *will not* get an A in a certain history course, then the probability that he will not get an A is $\frac{7}{7+2} = \frac{7}{9}$; correspondingly, the odds that he *will* get an A are 2 to 7, the probability that he will get an A is $\frac{2}{2+7} = \frac{2}{9}$, and it should be observed that the *sum* of these two probabilities is $\frac{7}{9} + \frac{2}{9} = 1$.

To illustrate how probabilities are converted into odds, let us refer back to the example on page 110, where we dealt with the question of whether or not we could start our car. As we pointed out at the time, the probability of 0.90 implies that we should win (namely, get the car started in cold weather) about 90 per cent of the time, lose about 10 per cent of the time, and, hence, that the proper odds are 9 to 1. In general:

> *If the probability of an event is p, and p does not equal 0 or 1, then the odds for its occurrence are p to 1 − p and the odds against its occurrence are 1 − p to p.*

For instance, if the probability that an item lost in a department store will never be claimed is 0.15, then the odds are 0.15 to 0.85, or 3 to 17, that a lost item will never be claimed (and they are 17 to 3 that it will be claimed). Also, if the probability that we shall have to wait for a table at our favorite restaurant is 0.25, then the odds are 0.25 to 0.75, or 1 to 3 that we shall have to wait for a table (and they are 3 to 1 that we shall not have to wait). Note that in both of these illustrations we followed the common practice of quoting odds as *ratios of positive integers* (having no common factors).

EXERCISES

1. If statistics compiled by the management of a department store show that 703 of 925 women who entered the store on a Saturday morning made at least one purchase, estimate the probability that a woman entering this department store on a Saturday morning will make at least one purchase.

2. If 252 of 400 housewives interviewed in a supermarket said that they preferred the "new and improved" detergent over the old kind, estimate the probability that any one housewife living in that area will prefer the "new and improved" detergent over the old kind.

3. A study made by a traffic engineer showed that 2,014 of 5,300 cars which approached a certain intersection from the North made a left turn. Estimate the probability that any one car approaching this intersection from the North will make a left turn.

4. Statistics compiled by a mountain resort in New Hampshire show that it has snowed there on Christmas Day 56 times in the last 70 years.

   (a) Estimate the probability that it will snow there on Christmas Day.

   (b) What are the odds that it will snow there on Christmas Day?

   (c) Would it be wise to bet $3 against $10 that it will *not* snow there on Christmas Day?

5. Among the 375 times that Mr. G has gone fishing, he has come back empty-handed (that is, without a single catch) 125 times.

   (a) Estimate the probability that he will come back empty-handed from his next fishing trip.

   (b) What are the odds that he will catch a fish?

   (c) If someone offered Mr. G *even money* (that is, odds of 1 to 1) that he will *not* make a catch, who would be favored by this bet?

6. In a sample of 128 cans of mixed nuts (taken from a very large shipment), 96 were found to contain mostly peanuts.

   (a) Estimate the probability that one of these cans will contain mostly peanuts.

   (b) What are the odds that any one of these cans will contain mostly peanuts?

   (c) What are the odds that any one of these cans will not contain mostly peanuts?

   (d) If we offered to bet the manager of the store $5 against his $2 that the next can he opens will contain mostly peanuts, would this be a smart thing to do?

7. If a stockbroker feels that 5 to 3 are fair odds that the price of a given stock will go up within a week, what is the probability (his personal probability) that this will be the case?

8. If the owner of a race horse feels that the odds are 7 to 2 against his horse coming in first, what is his personal probability that the horse will win?

9. If the odds are 13 to 7 that a newly hired secretary will get married before she has been with the company for two years, what is the probability that this will be the case?

10. If somebody claims that the odds are 11 to 5 that a certain shipment will arrive on time, what probability does he assign to the shipment's arriving on time?

11. A friend going on an ocean cruise is anxious to bet us $22.50 against our $2.50 that he will not get seasick. What does this tell us about the probability (i.e., his personal probability) that he will not get seasick? (*Hint:* The answer should read "at least. . . .")

12. A businessman refuses to bet $1 against $4 that he will not get stuck in freeway traffic while driving home from work. What does this tell us about the personal probability he assigns to his getting stuck? (*Hint:* The answer should read "greater than. . . .")

RULES OF PROBABILITY

In the study of probability there are basically *three kinds of questions*. First, there is the question of what we *mean*, for example, when we say that the probability for rain is 0.80, when we say that the probability for the success of a new venture is 0.35, or when we say that the probability for a candidate's election is 0.63; then there is the question of how probabilities are *measured* (namely, how their values are determined in actual practice); and finally there is the question of *how probabilities "behave"* (i.e., what mathematical rules they have to obey).

The first and second kinds of questions have already been discussed to some extent in the preceding section. As we have pointed out, there is the *frequency concept* of probability in which a probability is interpreted as a proportion, or percentage, in the long run, and its value is obtained (estimated) by observing what proportion of the time similar events have occurred in the past. Such probabilities are also referred to as *objective*, in contrast to *subjective*, or *personal*, probabilities, which are meant to express the strength of a person's belief. Subjective probabilities could be evaluated by simply asking a person what he considers "fair odds" that an event will occur, and then converting these odds into a probability as on page 112. More realistic, perhaps, would be to make a person "put up or shut up," namely, to see how he would react if there were really something at stake.

In the remainder of this chapter we shall be concerned mainly with the third kind of question mentioned in the beginning of this section, namely, that of determining the probabilities of relatively complex events in terms of known (estimated or assumed) values of the probabilities of simpler kinds of events.

To formulate the *postulates of probability* and some of their immediate consequences, we shall continue the practice of denoting events by means of capital letters, and we shall write the probability of event A as $P(A)$, the probability of event B as $P(B)$, and so forth. Furthermore, we shall follow the common practice of denoting the set of all possible outcomes, the *sample space*, by the letter S. As we shall formulate them here, the three postulates of probability apply only when the sample space S is finite; an example where this assumption of finiteness is dropped will be given in Exercise 13 on page 132.

POSTULATE 1: *The probability of any event is a positive real number or zero; symbolically, $P(A) \geq 0$ for any event A.*

POSTULATE 2: *The probability of any sample space is equal to 1; symbolically, $P(S) = 1$ for any sample space S.*

It is important to note that both of these postulates are satisfied by the frequency interpretation as well as by the subjective concept of probability. So far as the first postulate is concerned, proportions are always positive or zero, and so long as a and b (the amounts bet for and against the occurrence of an event) are positive, the probability $\dfrac{a}{a+b}$ cannot be negative.

The second postulate states indirectly that *certainty* is identified with a probability of 1; after all, it is always assumed that *one of the possibilities included in S must occur*, and it is to this certain event that we assign a probability of 1. So far as the frequency interpretation is concerned, a probability of 1 implies that the event will occur 100 per cent of the time, or in other words, that it is certain to occur. So far as subjective probabilities are concerned, the surer we are that an event will occur, the "better" odds we should be willing to give—say, 100 to 1, 1,000 to 1, or perhaps even 1,000,000 to 1. The corresponding probabilities are $\dfrac{100}{100+1}$, $\dfrac{1,000}{1,000+1}$, and $\dfrac{1,000,000}{1,000,000+1}$ (or approximately 0.99, 0.999, and 0.999999), and it can be seen that *the surer we are that an event will occur, the closer its probability will be to 1.*

In actual practice, we also assign a probability of 1 to events which we are "practically certain" will occur. For instance, we would assign a probability of 1 to the event that at least one person will vote in the next presidential election, and we would assign a probability of 1 to the event that among all the new cars sold during any one model year at least one will be involved in an accident before it has been driven 10,000 miles.

The third postulate of probability is especially important, and it is not quite so "obvious" as the other two:

> **POSTULATE 3:** *If two events are mutually exclusive, the probability that one or the other will occur equals the sum of their respective probabilities; symbolically,* $P(A \cup B) = P(A) + P(B)$ *for any two mutually exclusive events A and B.*

For instance, if the probability that the price of a stock will go up is 0.62 and the probability that its price will remain unchanged is 0.16, then the probability that the price of the stock will either go up or remain unchanged is $0.62 + 0.16 = 0.78$. Similarly, if the probability that an applicant for a certain job with the California Highway Department was born in California is 0.41 and the probability that he was born in Nevada is 0.06, then the probability that he was born in either state is $0.41 + 0.06 = 0.47$. All this agrees with the frequency interpretation of probability: If one event occurs, say, 37 per cent of the time, another event occurs 58 per cent of the time, *and they cannot both occur at the same time* (i.e., they are mutually exclusive), then one or the other will occur $37 + 58 = 95$ per cent of the time. So far as subjective probabilities are concerned, the third postulate does not follow from our discussion on page 112, but proponents of the subjective point of view impose it as what they call the "consistency criterion" (see also Exercises 3, 4, and 5 on page 123).

By using the three postulates of probability, we can derive many further rules according to which probabilities must "behave"—some of them are easy to prove and some are not, but they all have important applications. Among the immediate consequences of the three postulates, we find that *probabilities can never be greater than 1*, that *an event which cannot occur has the probability 0*, and that *the respective probabilities that an event will occur and that it will not occur always add up to 1*. Symbolically,

$$P(A) \leq 1 \qquad \text{for any event } A$$
$$P(\phi) = 0$$

and

$$P(A) + P(A') = 1 \qquad \text{or} \qquad P(A') = 1 - P(A).$$

The first of these results (which the reader will be asked to *prove* in Exercise 16 on page 125) simply expresses the fact that an event cannot occur more than 100 per cent of the time, or that the probability $\dfrac{a}{a + b}$ cannot exceed 1 when a and b are *positive amounts* bet for and against the occurrence of an event. So far as the second result is concerned, it expresses the fact that an impossible event happens 0 per cent of the time; actually, we also assign zero probabilities to events which are *so unlikely* that we are "practically certain" they will not occur. Thus, we would assign a probability of 0 to the event that a monkey set loose on a typewriter will by chance type Plato's *Republic* word for word without a single mistake.

The third result can be derived formally from the three postulates of probability, but it also follows immediately from the frequency interpretation as well as the subjective concept of probability. Clearly, if Miss Jones arrives at work late 18 per cent of the time, then she does *not* arrive late 82 per cent of the time; the respective probabilities are 0.18 and 0.82, and they add up to 1. Subjectively speaking, if a person considers it *fair*, or equitable, to bet a dollars against b dollars that a given event will occur, he is actually assigning the event the probability $\dfrac{a}{a + b}$ and its nonoccurrence the probability $\dfrac{b}{b + a}$; evidently, these two probabilities also add up to 1.

The third postulate of probability applies only to *two* mutually exclusive events, but it can easily be generalized; repeatedly using this postulate, it can be shown that:

If k events are mutually exclusive, the probability that one of them will occur equals the sum of their respective probabilities; symbolically,

$$P(A_1 \cup A_2 \cup \cdots \cup A_k) = P(A_1) + P(A_2) + \cdots + P(A_k)$$

for any mutually exclusive events $A_1, A_2, \ldots,$ and A_k.

For instance, if the probabilities that a person who immigrated to the United States in 1955 came from England, Germany, or Italy are, respectively, 0.06, 0.12, and 0.13, then the probability that he came from one of these countries is $0.06 + 0.12 + 0.13 = 0.31$. Also, if the probabilities that a person ordering one entree in a given restaurant will order steak, pork chops, chicken, or lobster are, respectively, 0.28, 0.09, 0.32, and 0.16, then the probability that he will order one or another of these dishes is $0.28 + 0.09 + 0.32 + 0.16 = 0.85$; the probability that he will order something else instead is $1 - 0.85 = 0.15$.

Since a sample space with five individual outcomes has $2^5 = 32$ subsets and a sample space with 20 individual outcomes has $2^{20} = 1,048,576$ subsets, the problem of determining the probabilities of any particular events in which we may be interested in a given "experiment" is greatly simplified by the following rule:

The probability of any event A is given by the sum of the probabilities of the individual outcomes comprising A.

This rule is illustrated in Figure 5.11, where the dots represent the individual (mutually exclusive) outcomes; the fact that the probability of A is given by the sum of the probabilities of the individual points in A follows immediately from the "addition" rule on this page.

To illustrate the use of this rule, let us refer again to the example of the two stocks on page 92, and let us suppose that the nine possible outcomes have the probabilities shown in Figure 5.12. Then, the probability that the price of the first stock will go down is 4/16, namely, the sum of the probabilities assigned to the points (1, 1), (1, 2), and (1, 3); similarly, the probability that the price of at least one of the stocks will go up is 7/16, namely, the sum of the probabilities assigned to the points (3, 1), (3, 2), (3, 3), (2, 3), and (1, 3).

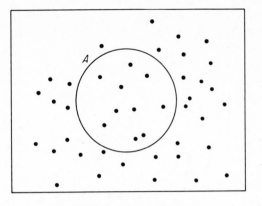

FIGURE 5.11 A sample space

To consider another example, suppose we are interested in the roll of a pair of dice, one red and one green, and that each of the 36 possible outcomes (see Figure 5.13) is assigned a probability of 1/36. Then, the probability of rolling a total of 7 is the *sum* of the probabilities of the points inside the dotted line and it is equal to 1/6; also, the probability of rolling a total of 2, 3, or 12 is the sum of the probabilities assigned to the four points which are circled in Figure 5.13 and it is equal to 1/9. Note that in this example, where the 36 possible outcomes have *equal* probabilities, the probability of any event is given by 1/36 *times* the number of individual outcomes comprising the event. More generally, when dealing with "experi-

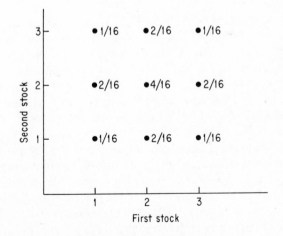

FIGURE 5.12 A sample space

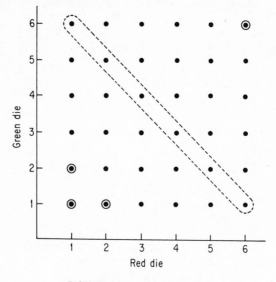

FIGURE 5.13 A sample space

ments" where the individual outcomes are *equiprobable*, we have the following rule:

> *If a sample space consists of n equiprobable outcomes of which s constitute event A, then* $P(A) = \dfrac{s}{n}.$

The ratio s/n of the number of "successes" to the total number of outcomes is sometimes used as a *definition* of probability, but aside from the logical difficulty of defining "probability" in terms of "equiprobable events," it has the shortcoming that it applies only when the individual outcomes all have the same probability. Nevertheless, this is often the case in games of chance, where we can assume that each card in a deck has the same chance of being drawn, that each face of a coin has the same probability of coming up, and likewise for each of the six sides of a balanced die. Thus, the probability of drawing a king from an ordinary deck of playing cards is 4/52 (there are four kings among the 52 cards), the probability of getting heads with a balanced coin is 1/2, and the probability of rolling a 5 or a 6 with a balanced die is 2/6. Note that the special rule for equiprobable events applies also, say, when each farm in a country has the same chance of being included in a survey, when each guinea pig has the same chance of being chosen for an experiment, when each teacher has the same chance for a promotion, or when each tax return has the same chance of being audited by the Internal Revenue Service (which would be the case for incomes within a certain range).

Since the third postulate applies only to mutually exclusive events, it cannot be used, for example, to find the probability that at least one of two roommates will pass a final exam in economics, the probability that a person will break an arm or a rib in an automobile accident, or the probability that a customer will buy a shirt or a tie while shopping at Macy's. In the first case, both roommates can pass the exam; in the second case, the person can break an arm as well as a rib; and in the third case, the customer can buy a shirt as well as a tie. To obtain a formula for $P(A \cup B)$ which holds regardless of whether the events A and B are mutually exclusive, let us consider the situation illustrated by means of the Venn diagram of Figure 5.14; it concerns an insurance salesman's luck with a potential customer.

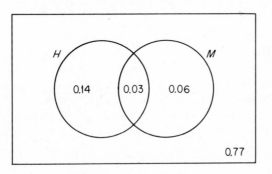

FIGURE 5.14   Venn diagram

The letter H stands for the event that he will sell him a homeowner's policy, M stands for the event that he will sell him a major medical policy, and it can be seen that

$$P(H) = 0.14 + 0.03 = 0.17$$
$$P(M) = 0.06 + 0.03 = 0.09$$

and

$$P(H \cup M) = 0.14 + 0.03 + 0.06 = 0.23$$

We were able to add the respective probabilities since they referred to mutually exclusive events (i.e., regions of the Venn diagram which have no points in common), but had we *erroneously* used the third postulate of probability to calculate $P(H \cup M)$, we would have obtained

$$P(H) + P(M) = 0.17 + 0.09 = 0.26$$

which exceeds the *correct value* by 0.03. What happened is that $P(H \cap M) = 0.03$ was added in *twice*, once in $P(H) = 0.17$ and once in $P(M) = 0.09$, and we could correct for this by *subtracting* $P(H \cap M) = 0.03$ from the final result, writing

$$P(H \cup M) = P(H) + P(M) - P(H \cap M)$$
$$= 0.17 + 0.09 - 0.03$$
$$= 0.23$$

Since this kind of argument holds for any two events A and B, we can now state the following *general addition rule*, which applies regardless of whether A and B are mutually exclusive events:

$$P(A \cup B) = P(A) + P(B) - P(A \cap B) \qquad \bigstar$$

Note that when A and B *are* mutually exclusive, then $P(A \cap B) = 0$ (since *by definition* the two events cannot both occur at the same time), and the new formula reduces to that of the third postulate of probability on page 116. To illustrate this rule, let us refer again to the example of the two stocks and Figure 5.12. If, as before, X is the event that the price of the first stock will go down while Y is the event that the price of the second stock will remain unchanged, it can be seen from Figure 5.12 that $P(X) = \dfrac{1}{16} + \dfrac{2}{16} + \dfrac{1}{16} = \dfrac{1}{4}$, that $P(Y) = \dfrac{2}{16} + \dfrac{4}{16} + \dfrac{2}{16} = \dfrac{1}{2}$, and that $P(X \cap Y) = \dfrac{2}{16} = \dfrac{1}{8}$. Hence, the probability that at least one of these two events will occur is

$$P(X \cup Y) = P(X) + P(Y) - P(X \cap Y) = \frac{1}{4} + \frac{1}{2} - \frac{1}{8} = \frac{5}{8}$$

It is interesting to note that this result can be verified by calculating $P(X \cup Y)$ directly; since $X \cup Y$ consists of the five points (1, 1), (1, 2), (1, 3), (2, 2), and (3, 2), we obtain $\dfrac{1}{16} + \dfrac{2}{16} + \dfrac{1}{16} + \dfrac{4}{16} + \dfrac{2}{16} = \dfrac{10}{16} = \dfrac{5}{8}$.

To consider another example, suppose we want to determine the probability of getting either a red card (event A) or a queen (event B) when drawing a card from an ordinary deck of 52 playing cards. Assuming that each card has a probability of 1/52 of being drawn, we obtain

$$P(A \cup B) = \frac{26}{52} + \frac{4}{52} - \frac{2}{52} = \frac{28}{52}$$

since there are 26 red cards, 4 queens, and 2 red queens. This result can also be verified by observing that the number of "successes" in this case is 28, there being 26 red cards and 2 black queens.

EXERCISES

1. Three production managers are asked to assign probabilities to the events that a particular job will be completed early, on schedule, or late. Mr. A assigns respective probabilities of 0.22, 0.55, and 0.35 to the three events; Mr. B assigns them probabilities of 0.15, 0.65, and 0.20; and Mr. C assigns them probabilities of 0.15, 0.60, and 0.20. Comment on these assignments.

2. Discuss the logic (or the lack of it) of the following assertions:
    (a) The probabilities that there will be 0, 1, 2, or at least 3 wildcat strikes in a company's plants this year are, respectively, 0.01, 0.15, 0.50, and 0.20.
    (b) The probability that a municipal bus will have one serious breakdown this year is 0.50, and the probability that it will have at least one such breakdown is 0.25.
    (c) The probability that Mr. Jones will be called for jury duty this month is 0.25, and the probability that Mr. Jones and his partner will both be called is 0.35.
    (d) The probabilities that at most 1, 2, 3, 4, or at least 5 members of a club will rise to speak at a club meeting are, respectively, 0.10, 0.15, 0.18, 0.20, and 0.37.

3. Asked about his chances of getting an A or a B in a course in business statistics, a student replies that the odds are 10 to 1 against his getting an A and 8 to 1 against his getting a B; furthermore, he feels that the odds are 6 to 1 against his getting either an A or a B. Discuss the *consistency* of this student's probabilities concerning the various events. (*Hint:* Convert the odds into the probabilities that he will get an A, that he will get a B, and that he will get an A or a B.)

4. Asked about his chances of getting a raise of $400 in his annual salary or a raise of $800, a university instructor replies that the odds are 5 to 1 against his getting an $800 raise and 2 to 1 against his getting a $400 raise; furthermore, he feels that it is an even bet (the odds are 1 to 1) that he will get either a $400 raise or an $800 raise. Discuss the *consistency* of the corresponding probabilities.

5. Asked about the chances that business conditions will improve or remain the same, an economist replies that the odds are 2 to 1 that business conditions will improve and 3 to 1 that they will *not* remain the same; furthermore, he feels that the odds are 5 to 1 that business conditions will either improve or remain the same. Discuss the *consistency* of the corresponding probabilities.

6. Given the mutually exclusive events A and B for which $P(A) = 0.28$ and $P(B) = 0.54$, find (a) $P(A')$, (b) $P(B')$, (c) $P(A \cap B)$, (d) $P(A \cup B)$, and (e) $P(A' \cap B')$.

7. Given two events A and B for which $P(A) = 0.24$, $P(B) = 0.52$, and $P(A \cap B) = 0.12$, find (a) $P(A')$, (b) $P(B')$, (c) $P(A \cup B)$, (d) $P(A' \cap B)$, (e) $P(A \cap B')$, and (f) $P(A' \cap B')$. (*Hint:* Draw a Venn diagram and fill in the probabilities associated with the various regions.)

8. If G is the event that a firm offers for public sale an unregistered new stock and H is the event that the SEC issues a stop order, state *in words* what events are referred to by each of the following probabilities: (a) $P(G')$, (b) $P(G \cap H)$, (c) $P(G \cap H')$, and (d) $P(G' \cap H')$.

9. Referring to Exercise 1 on page 97, suppose that each choice where the house has as many bedrooms as baths is assigned a probability of $1/9$, while each other choice is assigned the probability $2/9$. If events E, F, and G are as defined in part (b) of that exercise, find (a) $P(E)$, (b) $P(F)$, (c) $P(G)$, (d) $P(F')$, (e) $P(G')$, (f) $P(E \cap F)$, (g) $P(F \cup G)$, (h) $P(E \cap G)$, and (i) $P(E \cup G')$.

10. If each point of the sample space of Exercise 8 on page 99 is assigned the probability $1/30$ and the events D, E, and F are as defined in that exercise, find (a) $P(D)$, (b) $P(E')$, (c) $P(F')$, (d) $P(D \cup E)$, (e) $P(D \cap E')$, (f) $P(E \cap F)$, (g) $P(E' \cup F')$, (h) $P(D \cup F')$, and (i) $P(D \cap F)$.

11. Among the 80 executives of a corporation 48 are married men, 35 are college graduates, and 22 of the 48 married men are also college graduates. If one of these executives is chosen by lot to attend a convention (i.e., if each of the executives has a probability of $1/80$ of being selected), what is the probability that the person selected is neither married nor a college graduate?

12. Assuming that each of the 52 cards in a standard deck of playing cards is equally likely to be drawn, what is the probability that a single card drawn is (a) a red ace; (b) a red card; (c) an ace, a 10, or a red jack; (d) a black jack or a red queen; (e) either a club or an ace; (f) neither a club nor an ace?

13. The probability that a patient visiting his dentist will have a tooth extracted is 0.06, the probability that he will have a cavity filled is 0.23, and the probability that he will have a tooth extracted as well as a cavity filled is 0.02.

   (a) What is the probability that a patient visiting his dentist will have a tooth extracted but no cavity filled?

   (b) What is the probability that a patient visiting his dentist will have a tooth extracted or a cavity filled, possibly both?

   (c) What is the probability that a patient visiting his dentist will have a tooth extracted or a cavity filled, but not both?

   (d) What is the probability that a patient visiting his dentist will have neither a tooth extracted nor a cavity filled?

14. The probabilities that the public stenographer in a large metropolitan hotel will receive 0, 1, 2, . . . , or 10 *or more* calls for service in a morning are, respectively, 0.01, 0.03, 0.08, 0.14, 0.18, 0.18, 0.15, 0.11, 0.06, 0.04, and 0.02. What is the probability that in one morning the stenographer will receive (a) at most 3 calls, (b) at least 2 calls, (c) 5, 6, or 7 calls, (d) at most 7 calls, (e) either 0 or 10 or more calls?

15. The probabilities that 0, 1, 2, 3, or 4 *or more* customers will ask for a particular shelf item in a paint store on a given day are, respectively, 0.10, 0.40, 0.30, 0.15, and 0.05. What is the probability that

    (a) none or one customer will ask for the item;

    (b) at least two customers will ask for the item;

    (c) at most three customers will ask for the item;

    (d) at least one customer will ask for the item?

16. Making use of the fact that $A \cup A' = S$ and A and A' are mutually exclusive, prove that $P(A) \leq 1$.

17. Making use of the second and third postulates of probability and the fact that ϕ and S are mutually exclusive, prove that $P(\phi) = 0$.

18. Making use of the fact that $A \cup A' = S$ and A and A' are mutually exclusive, prove that $P(A') = 1 - P(A)$.

CONDITIONAL PROBABILITY

Very often, it is meaningless (or at least very confusing) to speak of the probability of an event without specifying the sample space with which we are concerned. For instance, if we ask for the probability that a lawyer makes more than $15,000 a year, we may well get many different answers, *and they can all be correct.* One of these might apply to all lawyers in the United States, another might apply to lawyers handling only divorce cases, a third might apply only to lawyers employed by corporations, another might apply to lawyers handling only tax cases, and so on. Since the choice of the sample space (i.e., the set of all possibilities under consideration) is by no means always self-evident, it is helpful to use the symbol $P(A \mid S)$ to denote the *conditional probability* of event A relative to the sample space S, or as we often call it "the probability of A given S." The symbol $P(A \mid S)$ makes it explicit that we are referring to the sample space S (that is, a *particular* sample space S), and it is generally preferable to the abbreviated notation $P(A)$ unless the tacit choice of S is clearly understood. It is also preferable when we have to refer to *different* sample spaces in one and the same problem, as in the examples which follow.

To elaborate on the idea of a *conditional probability*, let us consider the following problem: There are 200 applicants for a minor position in the personnel department of a large concern, and since there is practically no time to screen the applicants, we shall assume that each one has the same probability of 1/200 of getting the job. It is known, though, that among the 200 applicants some have had previous experience in personnel work and some have had formal training in personnel administration, with the actual breakdown being as follows:

	Formal Training	*No Formal Training*
Previous Experience	16	32
No Previous Experience	24	128

As can be seen from this table, the chance of selecting a person with previous experience *and* formal training is rather slim—the probability is $\frac{16}{200} = 0.08$, to be exact. Letting E denote the selection of an applicant with previous experience, and T the selection of an applicant with formal training, we can write this probability as

$$P(E \cap T) = 0.08$$

Furthermore, it can be seen that the probability of selecting someone with previous experience is

$$P(E) = \frac{16 + 32}{200} = 0.24$$

and the probability of selecting someone with formal training is

$$P(T) = \frac{16 + 24}{200} = 0.20$$

where each of these probabilities was obtained by means of the special formula for equiprobable events on page 120.

Suppose now that the management of the concern decides to limit the selection to applicants who have had some formal training. As a result of this decision, the number of applicants is reduced to 40, and if we assume that each of them still has an equal chance, we find that

$$P(E \mid T) = \frac{16}{40} = 0.40$$

This is the conditional probability of selecting someone with previous experience *given that he has had some formal training*.

Note that this conditional probability can also be written as

$$P(E \mid T) = \frac{16/200}{40/200} = \frac{P(E \cap T)}{P(T)}$$

namely, as the *ratio* of the probability of selecting a person *with previous experience and formal training* to the probability of selecting someone *with formal training*. Generalizing from this example, let us now make the following definition which applies to any two events A and B belonging to a given sample space S:

If $P(B)$ is not equal to zero, then the conditional probability of A relative to B, namely, the "probability of A given B," is given by

$$P(A \mid B) = \frac{P(A \cap B)}{P(B)} \qquad \bigstar$$

Had the selection been limited to applicants *with previous experience* in our example, we could argue that the probability of selecting someone with *formal training* is given by

$$P(T \mid E) = \frac{P(T \cap E)}{P(E)} = \frac{0.08}{0.24} = \frac{1}{3}$$

Of course, this result could easily have been obtained directly by observing that among the 48 applicants with previous experience, 16 (or 1/3) have also had some formal training.

Although we justified the formula for $P(A \mid B)$ by means of an example in which all outcomes were *equiprobable*, this is *not* a requirement for its use. To consider an example in which we are not dealing with equiprobable events, let us refer again to the two-stock example and Figures 5.4 and 5.12. If, as before, X denotes the event that the price of the first stock will go down while Y denotes the event that the price of the second stock will remain unchanged, we find that

$$P(X) = \frac{1}{16} + \frac{2}{16} + \frac{1}{16} = \frac{1}{4}$$

$$P(Y) = \frac{2}{16} + \frac{4}{16} + \frac{2}{16} = \frac{1}{2}$$

and $P(X \cap Y) = \frac{2}{16}$. Then, if we substitute the last two values into the formula for $P(A \mid B)$ with X and Y replacing A and B, we get

$$P(X \mid Y) = \frac{P(X \cap Y)}{P(Y)} = \frac{2/16}{1/2} = \frac{1}{4}$$

and what is *special* (and interesting) about this result is that

$$P(X \mid Y) = \frac{1}{4} = P(X)$$

This means that the probability of event X is the same regardless of whether event Y has occurred (occurs, or will occur), and we say that *event X is independent of event Y*. Intuitively speaking, this means that *the occurrence of X is in no way affected by the occurrence or nonoccurrence of Y*. Note that in the first example of this section, event E was *not* independent of event T; whereas $P(E)$ equaled 0.24, $P(E \mid T)$ equaled 0.40, and this is indicative of the fact that the probability of selecting someone with previous experience *becomes larger* when the selection is restricted to applicants with some formal training.

As it can be shown that *event B is independent of event A whenever event A is independent of event B, that is,* $P(B) = P(B \mid A)$ *whenever* $P(A) = P(A \mid B)$, it is customary to say simply that A *and B are independent* whenever one is independent of the other. Also, when two events A and B are *not independent*, they are said to be *dependent*.

So far we have used the formula $P(A \mid B) = \dfrac{P(A \cap B)}{P(B)}$ only to calculate conditional probabilities. However, if we multiply the expressions on both sides of this equation by $P(B)$, we get

$$P(A \cap B) = P(B) \cdot P(A \mid B) \qquad \bigstar$$

and this provides us with a formula, sometimes referred to as a *multiplication rule,* which enables us to calculate the probability that two events will both occur. In words, the formula states that *the probability that two events will both occur is the product of the probability that one of the events will occur and the conditional probability that the other event will occur given that the first event has occurred (occurs, or will occur).* As it does not matter which event is referred to as A and which event is referred to as B, the above formula can also be written as

$$P(A \cap B) = P(A) \cdot P(B \mid A) \qquad \qquad \star$$

To illustrate the use of these formulas, suppose we want to determine the probability of randomly picking 2 defective television sets from a shipment of 15 sets among which 3 are defective. Assuming equal probabilities for each selection (which is what we mean by "randomly picking" the two sets), we find that the probability that the first one is defective is 3/15, and that the probability that the second one is defective *given that the first set was defective* is 2/14. Clearly, there are only 2 defectives among the 14 sets which remain after one defective set has been picked. Hence, the probability of choosing two sets which are *both defective* is

$$\frac{3}{15} \cdot \frac{2}{14} = \frac{1}{35}$$

A similar argument leads to the result that the probability of choosing two sets which are *not defective* is

$$\frac{12}{15} \cdot \frac{11}{14} = \frac{22}{35}$$

and it follows, by subtraction, that the probability of getting one good set and one defective set is $1 - \dfrac{1}{35} - \dfrac{22}{35} = \dfrac{12}{35}.$ An alternate way of handling problems of this kind will be discussed in Chapter 7.

When A and B are *independent events*, we can substitute $P(A)$ for $P(A \mid B)$ into the first form of the multiplication rule on page 128 or $P(B)$ for $P(B \mid A)$ into the second, and we obtain the *special multiplication rule*

$$P(A \cap B) = P(A) \cdot P(B) \qquad \qquad \star$$

This formula can be used, for example, to find the probability of getting two heads in a row with a balanced coin or the probability of drawing two aces in a row from an ordinary deck of 52 playing cards *provided the first card is replaced before the second is drawn.* For the two flips of the coin we get $\frac{1}{2} \cdot \frac{1}{2} = \frac{1}{4}$ and for the two aces we get $\frac{4}{52} \cdot \frac{4}{52} = \frac{1}{169}$, since there are four aces among the 52 cards. (Had the first card not been replaced before the second card was drawn, the probability of getting two aces in a row would have been $\frac{4}{52} \cdot \frac{3}{51} = \frac{1}{221}$; this distinction will be discussed further in Chapter 7, as it is important in *statistics*, where we speak of sampling "with replacement" or of sampling "without replacement.") The following are two further applications of the special multiplication rule: If the probability that a person will make a mistake in his income tax return is 0.12, then the probability that two totally unrelated persons (who do not use the same accountant) will both make a mistake is $(0.12)(0.12) = 0.0144$; if the probability that a person will choose blue as his favorite color is 0.24, then the probability that neither of two totally unrelated persons will choose blue is $(0.76)(0.76) = 0.5776$.

The special multiplication rule can easily be extended so that it applies to the occurrence of three or more independent events—*we simply multiply all of the respective probabilities.* For instance, the probability of getting *four heads* in a row with a balanced coin is $\frac{1}{2} \cdot \frac{1}{2} \cdot \frac{1}{2} \cdot \frac{1}{2} = \frac{1}{16}$, and the probability of first rolling two 1's and then some other number in three rolls of a balanced die is $\frac{1}{6} \cdot \frac{1}{6} \cdot \frac{5}{6} = \frac{5}{216}$. For dependent events the formulas become somewhat more complicated, as is illustrated in Exercise 14 on page 132.

EXERCISES

1. With reference to the example on page 126, find $P(E \mid T')$ and $P(T \mid E')$, and express *in words* what these probabilities represent.

2. If C is the event that a recent graduate of U.C.L.A. will take a job in California, B is the event that he will work for a bank, and E is the event that he will enjoy his work, state *in words* what probability is expressed by each of the following:
   (a) $P(E \mid B)$;
   (b) $P(B \mid C)$;
   (c) $P(E \mid C')$;
   (d) $P(B \cap E \mid C)$;
   (e) $P(E' \cap C \mid B')$;
   (f) $P(E \mid C \cap B)$.

3. If A and B are independent events and $P(A) = 0.35$ and $P(B) = 0.60$, find
   (a) $P(A \mid B)$;
   (b) $P(A \cap B)$;
   (c) $P(A \cup B)$;
   (d) $P(A' \cap B')$.

4. Given $P(A) = 0.4$, $P(B \mid A) = 0.3$, and $P(B' \mid A') = 0.2$, find
(a) $P(A')$; (c) $P(B)$; (e) $P(A \mid B)$.
(b) $P(B \mid A')$; (d) $P(A \cap B)$;

5. With reference to the illustration on page 127, verify that event Y is also independent of event X, namely, that $P(Y \mid X) = P(Y)$.

6. Given $P(A) = 0.30$, $P(B) = 0.50$, and $P(A \cap B) = 0.15$, verify that
(a) $P(A \mid B) = P(A)$; (c) $P(B \mid A) = P(B)$;
(b) $P(A \mid B') = P(A)$; (d) $P(B \mid A') = P(B)$.
(It is interesting to note that if any one of these four relations is satisfied, the other three must also be satisfied and the events A and B are independent.)

7. Each month, a brokerage house studies two groups of industries and rates the individual companies as being low risks or moderate-to-high risks. In a recent report it published its findings on 13 aerospace companies and 27 processors of foods with the overall results summarized as follows:

	Low Risk	Moderate-to-High Risk
Aerospace	4	9
Foods	16	11

If a person selects one of these companies at random to invest in its stock (i.e., each company has a probability of $1/40$ of being selected), U and R denote the events that the company he selects is a low risk or a moderate-to-high risk, while A and F denote the events that he selects an aerospace company or a processor of foods, determine each of the following probabilities:
(a) $P(U)$; (d) $P(A \cap R)$; (g) $P(U \mid F')$;
(b) $P(R)$; (e) $P(U \mid A)$; (h) $P(A \mid R)$;
(c) $P(A \cup R)$; (f) $P(R \mid F)$; (i) $P(F \mid U)$.

8. Supposing that in Exercise 7 the investor assigns each low-risk company a probability of 0.04 and each of the other companies a probability of 0.01, recalculate each of the nine probabilities of Exercise 7.

9. A company has 1,000 replacement parts for a given assembly. Twenty per cent of the parts are defective and the rest are good, 40 per cent were bought from external sources and the rest were made by the company itself, and of those bought from external sources 80 per cent are good. If a part is randomly selected from this stock, what is the probability that
(a) the part is company-made and good;
(b) the part is either defective or bought;
(c) the part is neither company-made nor good;
(d) the part is bought, given that it is defective?

10. Which of the following pairs of events are independent:
   (a) Getting sixes in two successive rolls of a die;
   (b) Being intoxicated while driving and having an accident;
   (c) Having a driver's license and owning a car;
   (d) Being a college professor and having green eyes;
   (e) Any two mutually exclusive events;
   (f) Being born in December and having flat feet?

11. Two business partners are, respectively, 45 and 53 years old. If the probability that a person aged 45 will live at least another 20 years is 0.66 and the probability that a person aged 53 will live at least another 20 years is 0.47, what is the probability that both partners will still be alive 20 years hence? What assumption do we have to make? Is it reasonable?

12. As we indicated on page 130, the probability that any number of independent events will occur is given by the product of their respective probabilities. Use this rule to find:
   (a) The probability of getting five *heads* in a row with a balanced coin;
   (b) The probability that a fairly good marksman will hit the target four times in a row, given that the probability of his hitting the target on any one try is 0.80;
   (c) The probability of drawing (with replacement) four *hearts* in a row from an ordinary deck of 52 playing cards.

13. One critical operation in assembling a delicate electronic device requires that a skilled operator fit one part to another precisely. If the operator succeeds in matching the parts on his first attempt, he moves on to the next assembly; otherwise, he repeats his (independent) attempts until he gets a match. What is the probability that an operator with a constant match probability of 2/3 will succeed in matching the parts in a given assembly (a) on the fourth attempt, and (b) within four attempts? (Note that in this example the set of all possible outcomes is *not finite*, and when this is the case we must modify the third postulate of probability so that it applies to the union of any number of mutually exclusive events; nevertheless, it is possible to solve this problem with the methods discussed so far in this chapter.)

14. The problem of determining the probability that any number of events will occur becomes more complicated when the events are *not independent*. For three events A, B, and C, for example, the probability that they will all occur is obtained by multiplying the probability of A by the probability of B *given A*, and then multiplying the result by the probability of C *given $A \cap B$*. For instance, the probability of drawing (without replacement) three aces in a row from an ordinary deck of 52 playing cards is

$$\frac{4}{52} \cdot \frac{3}{51} \cdot \frac{2}{50} = \frac{1}{5,525}$$

Clearly, there are only three aces among the 51 cards which remain after the first ace has been drawn, and only two aces among the 50 cards which remain after the first two aces have been drawn.

(a) Referring to the illustration on page 126, what is the probability that if three of the applicants are randomly selected for further interviews, they will all have had some previous experience?

(b) Symbolically or in words, give a rule for the probability that four events, A, B, C, and D will all occur.

(c) In a certain city, the probability of passing the test for a driver's license on the first try is 0.75; after that the probability of passing becomes 0.60, regardless of how often a person has failed. What is the probability of finally getting one's license on the fourth try?

(d) In the fall, the probability that a rainy day will be followed by a rainy day is 0.80 and the probability that a sunny day will be followed by a rainy day is 0.40. Assuming that each day is classified as being either rainy or sunny and that the weather on any given day depends only on the weather the day before, find the probability that a rainy day is followed by three more rainy days, then two sunny days, and finally another rainy day.

THE RULE OF BAYES

Although $P(A \mid B)$ and $P(B \mid A)$ may look alike, there may be a great difference between the probabilities which these symbols represent. In the example on page 126, $P(E \mid T)$ is the probability that the concern will hire someone with previous experience *given that he has had some formal training*, $P(T \mid E)$ is the probability that it will hire someone with formal training *given that he has had some experience*, and the values of these two probabilities are 2/5 and 1/3. Similarly, if C represents the event that a certain person committed a crime and G represents the event that he is judged guilty, then $P(G \mid C)$ is the probability that the person will be judged guilty *given that he actually committed the crime*, and $P(C \mid G)$ is the probability that the person actually did commit the crime *given that he has been judged guilty*—clearly, there is a big difference between the events to which these two conditional probabilities refer.

Since there are many problems which involve such pairs of conditional probabilities, let us try to find a formula which expresses $P(B \mid A)$ in terms of $P(A \mid B)$ for any two events A and B. Fortunately, we do not have to look very far; all we have to do is equate the two expressions for $P(A \cap B)$ on pages 128 and 129, and we get

$$P(A) \cdot P(B \mid A) = P(B) \cdot P(A \mid B)$$

and, hence,

$$P(B \mid A) = \frac{P(B) \cdot P(A \mid B)}{P(A)}$$

after dividing the expressions on both sides of the equation by $P(A)$. To illustrate the use of this formula, suppose that the records of a company selling encyclopedias door to door with a large sales force show that of the initial calls made in a certain area, 70 per cent are made by salesmen and the rest by saleswomen; furthermore, the records show that *on the first call* men close sales 3 per cent of the time, while the entire sales force as a whole closes sales *on the first call* 2.7 per cent of the time. *What is the probability that an encyclopedia sold by the company's sales force on the first call was sold by a man?* If we let A denote the event that an encyclopedia is sold on the first call and B denote the event that the sale is made by a man, the above information can be expressed by writing $P(B) = 0.70$, $P(A \mid B) = 0.03$, and $P(A) = 0.027$, so that substitution into the formula for $P(B \mid A)$ yields

$$P(B \mid A) = \frac{(0.70)(0.03)}{0.027} = \frac{0.021}{0.027} = \frac{7}{9}$$

This is the probability that the sale was made by a man given that it was made on the first call, and it follows by subtraction that the corresponding probability for a first-call sale having been made by a woman is $1 - \frac{7}{9} = \frac{2}{9}$.

The formula which we used in this last example is a very simple version of the *Rule of Bayes* (or *Bayes' Theorem*); although there is no question about its *validity*, questions have been raised about its *applicability*. This is due to the fact that it involves a "backward" or "inverse" sort of reasoning—namely, *reasoning from effect to cause*. In our example we asked for the probability that the sale was "caused" by a man, and we could use the same formula to calculate the probability that a given airplane accident was "caused" by structural failure (see Exercise 3 on page 136), or to determine the probability that a certain company's success in getting a government contract was "caused" by its major competitor's failure to bid (see Exercise 1 on page 136).

When there are more than two possible "causes," it is best to analyze the situation by means of a tree diagram like that of Figure 5.15, where the various possible "causes" of A are labeled $B_1, B_2, \ldots$, and B_k. With reference to this diagram we can say that $P(B_i \mid A)$ is the probability that event A is reached via the ith branch of the tree, for $i = 1, 2, \ldots$, or k, and that its value is given by the *ratio* of the probability associated with the ith branch, namely, $P(B_i) \cdot P(A \mid B_i)$, to the *sum* of the probabilities associated with *all* of the branches of the tree. Symbolically,

$$P(B_i \mid A) = \frac{P(B_i) \cdot P(A \mid B_i)}{P(B_1) \cdot P(A \mid B_1) + P(B_2) \cdot P(A \mid B_2) + \cdots + P(B_k) \cdot P(A \mid B_k)} \qquad \star$$

for $i = 1, 2, \ldots,$ or k.

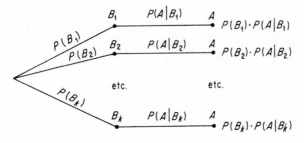

FIGURE 5.15   Tree diagram

To illustrate this more general form of the Rule of Bayes, suppose that a management consultant is asked for his opinion as to whether an executive's dissatisfied secretary quit her job mainly because she did not like the work, because she felt that she was underpaid, or because she did not like her boss. Unable to get any *direct* information about the secretary, he takes the following data from a large-scale corporate morale and motivation study: *Among all dissatisfied secretaries, 20 per cent are dissatisfied mainly because they dislike their work, 50 per cent because they feel they are underpaid, and 30 per cent because they dislike their boss. Furthermore, the corresponding probabilities that they will quit are, respectively, 0.60, 0.40, and 0.90.*

Picturing this situation as in Figure 5.16, we find that the probabil-

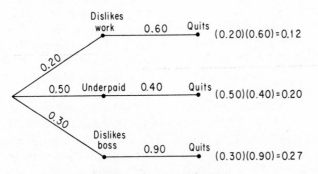

FIGURE 5.16   Tree diagram

ities associated with the three branches of the tree are, respectively, $(0.20)(0.60) = 0.12$, $(0.50)(0.40) = 0.20$, and $(0.30)(0.90) = 0.27$, and that they add up to 0.59. Thus, the probability that a dissatisfied secretary has quit *mainly because she didn't like her work* is 0.12/0.59 or approximately 0.20, the probability that she has quit *mainly because she felt that she was underpaid* is 0.20/0.59 or approximately 0.34, and the probability that she has quit *mainly because she did not like her boss* is 0.27/0.59 or approximately 0.46. It follows that *the management consultant's best bet is to say that the secretary quit mainly because she did not like her boss.*

To solve this problem by means of Bayes' formula (i.e., without reference to a tree diagram like that of Figure 5.16), we let A represent the event that a dissatisfied secretary quits her job, while B_1, B_2, and B_3 represent the respective events that she is dissatisfied mainly because she dislikes her work, because she feels that she is underpaid, and because she does not like her boss. Thus, being given the information that $P(B_1) = 0.20$, $P(B_2) = 0.50$, $P(B_3) = 0.30$, $P(A \mid B_1) = 0.60$, $P(A \mid B_2) = 0.40$, and $P(A \mid B_3) = 0.90$, we find that the formula yields

$$P(B_1 \mid A) = \frac{(0.20)(0.60)}{(0.20)(0.60) + (0.50)(0.40) + (0.30)(0.90)} = \frac{12}{59}$$

and corresponding expressions with the *same denominator* but the numerators $(0.50)(0.40)$ and $(0.30)(0.90)$ for $P(B_2 \mid A)$ and $P(B_3 \mid A)$. The results are, of course, identical with the ones obtained before.

EXERCISES

1. There is a fifty-fifty chance that Firm X will bid for the construction of a new city hall. Firm Y submits a bid and the probability that it will get the job is 2/3 provided Firm X does not bid; if Firm X submits a bid, however, the probability that Firm Y will get the job is only 1/5. If Firm Y gets the job, what is the probability that Firm X did not bid?

2. Bin P contains two defective and three good parts, and bin Q contains two defective and one good part. Someone selects a bin at random (with equal probabilities) and then randomly chooses one part from that bin. If the part chosen is a good part, what is the probability that it came from bin P?

3. The probability that an airplane accident due to structural failure is diagnosed correctly is 0.72, and the probability that an airplane accident which is *not* due to structural failure is diagnosed incorrectly as being due to structural failure is 0.12. If 40 per cent of all airplane accidents are due to structural failure, what

is the probability that an airplane accident which is diagnosed as being due to structural failure is actually due to this cause?

4. In-basket Q contains four "routine" and three "action" letters while in-basket R contains five "routine" and two "action" letters. One letter is randomly selected from Q and put into R, then a letter is randomly drawn from R. If the letter drawn from R is an "action" letter, what is the probability that the letter moved from Q to R was also an "action" letter?

5. A man has taken a vaccine from either (storage) unit P (which contains 30 current and 10 outdated vaccines), or from unit Q (which contains 20 current and 20 outdated vaccines), or from unit R (which contains 10 current and 30 outdated vaccines), but he is twice as likely to have taken it from unit P as from unit Q and twice as likely to have taken it from unit Q as from unit R. If the vaccine the man took was an outdated one, what is the probability that he took it from unit R?

6. A distributor of phonograph records employs three stock clerks, K, L, and M, who pull records from bins and stack them for subsequent verification and packaging. K makes a mistake in an order (gets the wrong records or the wrong quantity) one time in 100, L makes a mistake in an order 10 times in 100, and M makes a mistake in an order five times in 100. Of all the orders delivered for verification, K, L, and M fill, respectively, 50, 30, and 20 per cent. If a mistake is found in a particular order, what are the respective probabilities that the order was filled by K, L, and M?

7. The probabilities that a brewery will decide to sponsor the televising of football games, a soap opera, or a news program are, respectively, 0.50, 0.30, and 0.20. If they decide on the football games, the probability that they will get a high rating is 0.60; if they decide on a soap opera, the probability that they will get a high rating is 0.30; and if they decide on a news program, the probability that they will get a high rating is 0.20. If it turns out that they do get a high rating, what is the probability that they chose a soap opera?

A WORD OF CAUTION

Many fallacies involving probabilities are due to inappropriate assumptions concerning the equal likelihood of events. Consider, for example, the following situation:

Among three identical file trays one contains two current records, one contains one current and one dead record, and the other contains two dead records. After taking one of these trays at random, a clerk randomly takes one record from it. If this record is a current one, what is the probability that the other record on this tray is also a current one?

Without giving the matter too much thought, it may seem reasonable to say that this probability is $1/2$. After all, the current record must have come from the first or second tray. For the first tray the other record is a current one, for the second tray the other record is a dead one, and it would seem reasonable to say that these two possibilities are equally likely. Actually, this is not the case: The correct value of the probability is $2/3$, and the reader can verify this by drawing an appropriate tree diagram or by applying the Rule of Bayes. (When drawing a tree diagram showing the six possible outcomes corresponding to which of the six records is actually chosen, it will be convenient to label the two current records on the first tray C_1 and C_2 and the two dead records on the third tray D_1 and D_2.)

6

EXPECTATIONS,
GAMES,
AND DECISIONS

MATHEMATICAL EXPECTATION

When we say that in Florida a couple can expect to have 1.46 children, that a person living in the United States can expect to eat 161.5 pounds of meat and 20.1 apples a year, or that a resident of Geneva, Switzerland, can expect to go to the movies 25.2 times a year, it must be obvious that we are not using the word "expect" in its colloquial sense. Two of these events cannot possibly occur, and it would certainly be very surprising if a person actually did eat 161.5 pounds of meat and 20.1 apples during a calendar year. So far as the Florida couples are concerned, some of them will have no children, some will have one child, some will have two children, some will have three, . . . , and the 1.46 figure must be interpreted as an *average*, or as we shall call it here, a *mathematical expectation*.

Originally, the concept of a mathematical expectation arose in connection with games of chance, and in its simplest form it is given by the *product* of the amount a player stands to win and the probability that he will win. Thus, if we stand to win $5 if a balanced coin comes up *tails*, our mathematical expectation is $5(1/2) = \$2.50$. Similarly, if we consider buying one of 1,000 raffle tickets issued for a prize (e.g., a television set) worth $480, our mathematical expectation is 480(0.001) = 0.48 or 48 cents; thus, it would be foolish to pay more than 48 cents for the ticket unless the proceeds of the raffle went to a worthy cause (or the difference could be credited to whatever pleasure a person might derive from placing a bet). Note that in this example 999 of the tickets will not pay anything at all, one ticket will pay $480 (or the equivalent in merchandise), so that altogether the 1,000 tickets pay $480, or *on the average* 48 cents per ticket.

So far, we have considered only examples in which there was a single "payoff," namely, one prize or a single payment. To demonstrate how the concept of a mathematical expectation can be generalized, let us change the raffle for the television set so that there is also a second prize (say, a record player) worth $120 and a third prize (say, a radio) worth $40. Now we can argue that 997 of the tickets will not pay anything at all, one ticket will pay the equivalent of $480, another will pay the equivalent of $120, while a third will pay the equivalent of $40; altogether, the 1,000 raffle tickets will pay $480 + $120 + $40 = $640, or *on the average* 64 cents per ticket—this is the *mathematical expectation* for each ticket. Looking at the problem in a different way, we could argue that if the raffle were repeated many times, we would win nothing 99.7 per cent of the time and win each of the three prizes 0.1 per cent of the time. On the average we would thus win

$$0(0.997) + 480(0.001) + 120(0.001) + 40(0.001) = 64 \text{ cents}$$

which is the sum of the products obtained by multiplying each amount by the corresponding probability. Generalizing from this example, let us now make the following definition:

If the probabilities of obtaining the amounts $a_1, a_2, a_3, \ldots,$ or a_k, are, respectively, $p_1, p_2, p_3, \ldots,$ and p_k, then the mathematical expectation is

$$E = a_1p_1 + a_2p_2 + a_3p_3 + \cdots + a_kp_k$$

Each amount is multiplied by the corresponding probability, and the mathematical expectation, E, is given by the sum of all these products. So far as the a's are concerned, it is important to remember that they are *positive* when they represent profits, winnings, or gains (i.e., amounts which we receive), and that they are *negative* when they represent losses, penalties, or deficits (i.e., amounts which we have to pay). For instance, if we bet $5 on the flip of a coin (that is, we either win $5 or lose $5 depending on the outcome), the amounts a_1 and a_2 are $+5$ and -5, the probabilities are $p_1 = 0.50$ and $p_2 = 0.50$, and the mathematical expectation is

$$E = 5(0.50) + (-5)(0.50) = 0$$

This is what it should be in an *equitable game*, that is, in a game which does not favor either player.

To consider another example, suppose that Mr. Brown is interested in investing in a piece of property for which the probabilities are respectively 0.22, 0.36, 0.28, and 0.14 that he will sell it at a profit of $2,500, that he will sell it at a profit of $1,000, that he will break even, or that he will sell it at a loss of $1,500. If we substitute all these figures into the formula for E, we get

$$E = 2,500(0.22) + 1,000(0.36) + 0(0.28) + (-1,500)(0.14)$$
$$= \$700$$

and this is his *expected profit*.

A DECISION PROBLEM

When we are faced by uncertainties, mathematical expectations can often be used to great advantage in making decisions. Generally speaking, if we have to choose between several alternatives, it is considered "rational" to select the one with the "most promising" mathematical expectation: the one which *maximizes expected profits, minimizes expected costs, maximizes expected tax advantages, minimizes expected losses,* and so on.

Although this approach to decision making has great intuitive appeal and sounds very logical, it is not without complications—there are many problems in which it is hard, if not impossible, to assign values to all of the a's (amounts) and p's (probabilities) in the formula for E on page 140. To illustrate some of these difficulties, let us consider the following problem: Mr. Knight, the advertising manager of a toy company, must decide whether to terminate an advertising program intended to popularize a new toy or to authorize funds for its continuation. If the advertising program is continued and turns out to be successful (i.e., the new toy becomes popular), this will yield a net profit of $200,000 for Mr. Knight's company; if the advertising program is continued but proves to be unsuccessful, this will result in a net loss of $120,000. If the advertising program is terminated but another company successfully promotes a similar toy, this will entail a net loss of $80,000 to Mr. Knight's company (partly for being put at a competitive disadvantage). And finally, if the advertising program is terminated and nobody else finds it possible to promote a similar new toy, there will be a net profit of $20,000 (which will result from the overseas sale of patent rights to the toy). Schematically, all this information can be summarized as follows:

	Advertising Program Is Successful	Advertising Program Is Not Successful
Advertising Program Is Continued	$200,000	−$120,000
Advertising Program Is Terminated	−$80,000	$20,000

Obviously, it will be better to continue the advertising program only if the toy can be made popular, and Mr. Knight's decision will therefore have to depend on his own evaluation of the chances that this will be the case. Suppose, for instance, that (on the basis of many years' experience) he judges that the odds are 3 to 2 *against* the success of the program; in other words, he assigns the advertising program's success the probability $\frac{2}{2+3}$ = 0.40, and its failure the probability $\frac{3}{3+2}$ = 0.60. He can then argue that *if the advertising program is continued*, the company's *expected gain* is

$$200,000(0.40) + (-120,000)(0.60) = \$8,000$$

and *if the advertising program is terminated*, the company's *expected gain* is

$$(-80,000)(0.40) + 20,000(0.60) = -\$20,000$$

Since an *expected gain* of $8,000 is obviously preferable to an *expected loss* of $20,000, it stands to reason that Mr. Knight should not hesitate in deciding to continue the advertising program. *Or should he?* What if he was wrong in assessing the odds as 3 to 2 against the advertising program's success? What if the odds should have been 4 to 1 against the advertising program's success, or perhaps only 2 to 1? The point we are trying to make is that *one should use mathematical expectations as criteria for making decisions only if one is good at assessing odds, namely, if one's probabilities (or probability estimates) are "correct" (or at least close).* As the reader will be asked to verify in Exercise 11 on page 148, Mr. Knight should decide to terminate the advertising program when the odds against its success are 4 to 1, and the whole situation is a toss up when the odds against the advertising program's success are 2 to 1.

Continuing the analysis, let us suppose that the 3 to 2 odds against the advertising program's success are correct, but let us add another alternative: If the advertising program is continued for another month at a cost of $40,000, it will then be known *for sure* whether the advertising program will be successful and the toy will become popular. Mr. Knight is thus faced with the problem of having to decide whether it is worthwhile to spend this substantial sum of money before reaching a final decision. One way of handling this kind of problem is to determine what is called the *expected value of perfect information.* If he knew for sure whether or not the advertising program would be successful, Mr. Knight could act accordingly and assure his company either a gain of $200,000 (if the advertising program will be successful) or a gain of $20,000 (if the advertising program will not be successful). Using the same probabilities as before, 0.40 and 0.60, he finds that the *expected gain* of the company would be

$$200,000(0.40) + 20,000(0.60) = \$92,000$$

and this is what is called the *expected value of perfect information.* In general, *the expected value of perfect information is the amount one can expect to gain (profit or win) in any given situation provided one always makes the right decision.* So far as Mr. Knight is concerned, we have shown that he can increase his company's *expected gain* from $8,000 (corresponding to his decision to continue the advertising program) to $92,000 (corresponding to his delaying his decision for a month at a cost of $40,000). Since the increase in the expected gain exceeds $40,000, the cost of getting the perfect information, it stands to reason that he should decide to continue the advertising program for another month. In fact, it would have been worthwhile to continue the advertising program for another month (and then know for sure whether it will be successful) so long as the extra cost is less than $92,000 − $8,000 = $84,000.

The way in which we have studied this problem is referred to as a *Bayesian analysis.* In this kind of analysis, probabilities are assigned to the conditions about which uncertainties exist (the so-called "states of Nature," which in our example are the success or lack of success of the advertising program); *then, the alternative which is ultimately decided upon is the one which has the greatest expected profit or gain.* As we saw in our example, a Bayesian analysis can also include the possibility of delaying any final action until further information is obtained. This is of special importance in statistics, where we generally deal with sample data obtained from surveys or experiments, and where we may have to decide how large a sample to take, whether a given sample is adequate for reaching a decision, or whether further observations should be made.

Having shown in this example how mathematical expectations might be used as a basis for rational decisions, let us examine briefly what Mr. Knight might have done if he had no idea whatsoever about the advertising program's chances for success. To suggest one possibility, suppose that Mr. Knight is a *confirmed optimist:* Looking at the situation through rose-colored glasses, he sees that if the advertising program is continued the company might gain as much as $200,000, whereas the decision to terminate the program would lead at best to a gain of $20,000. Always expecting the best (in the sense of wishful thinking), Mr. Knight would thus decide to continue the advertising program, and we might say that by doing so he is *maximizing the company's maximum gain.* (In other words, he is choosing the alternative for which the company's greatest possible gain is a maximum.)

Now suppose that Mr. Knight is a *confirmed pessimist:* Looking at the situation through dark-colored glasses, he notes that if the advertising program is continued the company might lose as much as $120,000, whereas the decision to terminate the program could lead at worst to a loss of $80,000. Always expecting the worst (in the sense of resignation or fear), Mr. Knight would thus decide to terminate the advertising program, and we might say that by doing so he is *minimizing the company's maximum losses.* (In other words, he is choosing the alternative for which the company's greatest losses are a minimum, and we refer to this as the *minimax criterion.*)

There are various other ways in which decisions can be made in the absence of any knowledge about the probabilities concerning the various "states of Nature." Suppose, for example, that Mr. Knight is the kind of person who is always afraid to *lose out on a good deal.* This might lead him to the argument that if he decided to continue the advertising program and it turned out to be unsuccessful, his company would have been better off by $20,000 − (−$120,000) = $140,000 if he had terminated the program in the first place. Also, if he decided to terminate the program and it turned out that it could have been successful, his company would have been better off by $200,000 − (−$80,000) = $280,000 if he had continued the advertising program. These differences are generally referred to as *opportunity losses* (or *regrets*), and the whole situation can be pictured as follows:

	Advertising Program Is Successful	Advertising Program Is Not Successful
Advertising Program Is Continued	0	$140,000
Advertising Program Is Terminated	$280,000	0

To explain the two zeros, note that, when the advertising program is continued and it is successful in popularizing the new toy, *there is no loss of opportunity;* the same is true also in the case where the advertising program is terminated and it turns out that the toy could not have been successfully popularized anyhow. Now, if Mr. Knight were the kind of person who always wants to hold his opportunity losses to a minimum, he would probably apply the *minimax criterion* (i.e., choose the alternative for which the greatest opportunity loss is a minimum) and decide that the advertising program should be continued.

There are also situations where the various criteria which we have discussed are outweighed by other considerations. Suppose, for instance, that Mr. Knight has learned that he will be fired unless the advertising program turns out to be a success. In that case he would be foolish (though, perhaps, unselfish) not to continue the program, which at least would give him a chance. On the other hand, if he discovers that he will be fired if the company loses more than $100,000 on the advertising program, he could *play it safe* by terminating the program. Other examples of situations in which extraneous factors play important roles are given in Exercises 17 and 18 on page 151, and it is hoped that this will make it clear that *there is no universal rule or criterion which will always lead to the best possible decisions.*

Earlier in this section, we pointed out that it is sometimes difficult to base decisions on mathematical expectations, since this requires knowledge of all the a's (amounts) and p's (probabilities) in the formula for E on page 140. As we indicated on page 142, different probabilities assigned to the "states of Nature" can lead to different decisions (see also Exercise 15 on page 150), and in Exercise 12 on page 148 the reader will be asked to show that the same is true also for changes in the "payoffs," namely, changes in the a's. We did not worry about this in the example in the text, assuming that the figures in the table on page 142 were correct. Generally speaking, though, the problem of assigning "cash values" to the consequences of one's decisions can pose serious difficulties, and this is true, especially, when the consequences involve such intangibles as the overall effects of a bankruptcy, the pleasure a person may get from playing cards even if he does not win, the possible side effects of a new drug, the emotional effects of a broken home, the satisfaction a salesman may get from making a sale, and so on.

EXERCISES

1. As part of a promotional scheme, a soap manufacturer offers a first prize of $50,000 and a second prize of $10,000 to someone willing to try a new product (distributed without charge) and send in his name on the label. The winners will be drawn at random in front of a large television audience.

    (a) What would be each entrant's mathematical expectation, if 1,500,000 persons were to send in their names?

    (b) Would this make it worthwhile to spend the 8 cents postage it costs to send in an entry?

2. A jeweler wants to "unload" 5 watches that cost him $80 each and 45 watches that cost him $12 each. If he wraps these watches in identically shaped un-marked boxes and lets each customer take his pick, find (a) each customer's mathematical expectation and (b) the jeweler's expected profit per customer, if he charges $25 for the privilege of taking a pick.

3. A contractor is bidding on a road construction job which promises a profit of $42,000 with a probability of 0.60 or a loss of $16,000 (due to faulty estimates, strikes, late delivery of materials, etc.) with a probability of 0.40. What is the contractor's mathematical expectation?

4. An urn contains eight red beads and four white beads, and a player is to draw two beads at random without replacement.

    (a) What are the probabilities that he will draw 0, 1, or 2 red beads?

    (b) If he wins $10 for each red bead and $5 for each white bead, what is his expectation?

    (c) If the player receives $3 for each red bead he draws, how much should he be "penalized" for each white bead he draws to make the game equitable?

5. A grab-bag contains 5 packages worth $1 apiece, 5 packages worth $3 apiece, and 10 packages worth $5 apiece. Is it rational to pay $4 for the privilege of selecting one of these packages at random?

6. The probability that Mr. Jones will sell his house at a loss of $1,000 is 3/18, the probability that he will break even is 5/18, the probability that he will sell it at a profit of $1,000 is 7/18, and the probability that he will sell it at a profit of $2,000 is 3/18. What is his expected profit?

7. If in shooting dice a person bets $1 on the "Field," he wins $1 on the roll of 3, 4, 9, 10, or 11, he wins double on 2, triple on 12, and otherwise he loses. What is the expectation of such a "Field" bet?

8. An urn contains two black beads and three red beads. If two players, A and B, take turns drawing one bead without replacement, with A going first, what is Player A's expectation if whoever draws a red bead first wins $10? (*Hint:* Draw a tree diagram showing the various possible outcomes.)

9. (*Mathematical expectations and subjective probabilities*) The following example illustrates how mathematical expectations can be used to determine subjective probabilities: Suppose we let a friend choose his own birthday present by either accepting an outright gift of $10 or by accepting a gamble on the outcome of an election, where he is to receive $40 if his candidate wins and nothing if he loses. Now then, if he feels that his candidate's chances of winning are given by the probability p, the mathematical expectation of the gamble is $E = 40p + 0(1 - p) = 40p$, and if he prefers this to the outright gift of $10, we can argue that $40p > 10$ and, hence, that $p > \dfrac{10}{40}$ $\left(\text{i.e., that } p \text{ is greater than } \dfrac{1}{4}\right)$. Similarly, if he prefers the outright gift of $10, we can argue that $40p < 10$ and, hence, that $p < \dfrac{10}{40}$ $\left(\text{i.e., that } p \text{ is less than } \dfrac{1}{4}\right)$. Finally, if he cannot make up his mind, we can argue that $40p = 10$, and hence, he really feels that his candidate's chances of winning are given by the probability $p = \dfrac{1}{4}$.

   (a) A recent college graduate is faced by a decision which cannot wait, namely, that of accepting or rejecting a job paying $7,800 a year. What can we say about the probability which he assigns to landing his only other prospect, a job paying $11,700 a year, if he decides to take the $7,800 job?

   (b) An insurance company agrees to pay the promoter of a rodeo $4,000 in case the event has to be canceled because of rain. If the company's actuary feels that a fair net premium for this insurance would be $640, what probability does he assign to the prospect that the rodeo will have to be canceled because of rain?

   (c) A playwright is offered the option of either taking an immediate cash payment of $5,000 for his script, or gambling on the success of the play— in which case he will receive $20,000 if it is a success and only a token fee of $1,000 if it fails. What can we say about the probability which he assigns to the success of the play if he is willing to take the risk?

10. (*Mathematical expectations and the measurement of utility*) Mathematical expectations can also be used to measure the utility which a person assigns to anything of value (even intangibles). Suppose, for instance, that a music lover claims that he would "give his right arm" for a ticket to a concert which has been sold out for weeks. To see how far he might go, suppose furthermore that we propose the following deal: *For $10 we will let him draw one of 10 sealed envelopes, 9 of which contain a dollar bill while the other one contains a ticket to the concert.* If the utility, or "cash value," he assigns to the ticket equals U, the mathematical expectation of the gamble is

$$U \cdot (0.1) + (\$1) \cdot (0.9)$$

and if he considers this gamble worth *at least* $10, we can argue that

$$U \cdot (0.1) + (\$1) \cdot (0.9) > \$10$$

and, hence, that $U \cdot (0.1) > \$9.10$ and $U > \$91$. Thus, we have found that he feels this concert ticket is worth *at least* $91, and if we varied the odds and the amounts (the number of envelopes and the amounts which they contain), we could narrow it down more than that.

(a) Mr. Green has the choice of staying home and reading a good book or going to a party. If he goes to the party he might have a terrible time (to which he assigns a utility of 0), or he might have a wonderful time (to which he assigns a utility of 40). If he feels that the odds against his having a good time are 8 to 2 and he decides not to go, what can we say about the utility which he assigns to staying home and reading a good book?

(b) Mr. Jones would love to beat Mr. Brown in an upcoming tennis tournament, but his chances are nil unless he takes $400 worth of extra lessons, which (according to the tennis pro at his club) will give him a fifty-fifty chance. If Mr. Jones assigns the utility U to his beating Mr. Brown and the utility $-\frac{1}{5} U$ to his losing to Mr. Brown, find U if Mr. Jones decides that it is just about worthwhile to spend the $400 on extra lessons.

(c) It is a well-known fact that the utility which a person assigns to money is not necessarily its *monetary* value. To a college student who needs $20 to take his girlfriend out for the evening, having $20 would be worth *more than 20 times as much* as having only $1; on the other hand, to someone who has been very successful in business, the second million may very well *not be worth as much* as the first. If the utility which a certain person assigns to money equals its monetary value up to $10, what can we say about the utility he assigns to $100, if he *turns down* the privilege of paying $6 for drawing one of 20 sealed envelopes of which 10 contain $1, nine contain $2, and the other one contains $100?

11. Referring to the example on page 141, find the toy company's expected gains corresponding to the advertising program being continued and being terminated, if

(a) the odds against its success are 4 to 1;

(b) the odds against its success are 2 to 1.

12. Referring to the example on page 141, suppose that if the advertising program is abandoned and no competitor finds it possible to promote a similar toy, a net profit of $70,000 (instead of $20,000) will be realized from the overseas sale of patent rights to the toy. Recalculate the company's *expected profits* corresponding to the program being continued or being terminated, and determine whether this will affect Mr. Knight's *original* decision. (His odds against the success of the program were 3 to 2.)

13. The Board of Regents of a university is faced with the problem of having to decide whether to authorize funds for the construction of a new football

stadium. They are told that if the new stadium is built and the university has a good football team, there will be a profit of $410,000; if the new stadium is built and the university has a poor football team, there will be a deficit of $100,000; if the old stadium is used and the university has a good football team, there will be a profit of $200,000; and if the old stadium is used and the university has a poor football team, there will be a profit of $20,000 (mostly from games played away from home).

(a) Present all this information in a table like the one on page 142.

(b) Believing their athletic director when he tells them that the odds are 2 to 1 that they will *not* have a good football team, what should the regents decide so as to maximize the expected profit?

(c) Believing the sport's editor of the local newspaper who tells them that the odds are 3 to 2 that they will *not* have a good football team, what should the regents decide so as to maximize the expected profit?

(d) If Regent Moore is a confirmed pessimist, which way would he be inclined to vote? Explain your answer.

(e) If Regent Wilson is a confirmed optimist, which way would he be inclined to vote? Explain your answer.

(f) Construct a table like the one on page 145 showing the opportunity losses associated with the various possibilities. What action should the regents take if they wanted to hold the greatest possible opportunity losses to a minimum? Using the odds of part (b), what should the regents do so as to minimize their *expected opportunity losses?*

(g) Using the odds of part (b), what is the expected value of perfect information? Would it be worthwhile to pay an "infallible" forecaster $10,000 to tell them for sure whether the university will have a good football team?

14. An executive is planning to attend a sales convention in San Diego, California, and he must send in his hotel reservation immediately. The convention is so large that it is held in part at Hotel A and in part at Hotel B, but the executive does not know whether the particular session he wants to attend will be scheduled for Hotel A or Hotel B. He is planning to stay only one day, which would cost him $21 at Hotel A and $18 at Hotel B, but it will cost him an extra $5 for cab fare if he stays at the wrong hotel.

(a) Present all this information in a table like that on page 142.

(b) Where should he make his reservation if he wants to *minimize his expected expenses* and feels that the odds are 5 to 1 that the session he wants to attend will be held at Hotel A?

(c) Repeat part (b) when the odds are 2 to 1 instead of 5 to 1.

(d) Repeat part (b) when the odds are 4 to 1 instead of 5 to 1.

(e) Where would he make his reservation if he were a confirmed optimist?

(f) Where would he make his reservation if he were a confirmed pessimist?

(g) Referring to the odds of part (b), what is the *expected value of perfect information?* Would it be worthwhile to spend $1.50 on a long-distance call to find out where the session will be held?

(h) Where should he make his reservation if he wants to minimize the greatest possible loss of opportunity?

15. A wildcatter drilling for oil is faced with the problem of whether to stop or to continue drilling. It is not possible to place monetary values on either the loss from stopping the drilling when oil in fact lies below, or the gain from continuing to drill and striking oil. However, the driller measures the relative values involved in the situation in terms of crude "satisfaction" units as follows:

	No Oil	Oil
Stop Drilling	100	−500
Continue Drilling	−300	1,000

Thus, his satisfaction in stopping now if no oil will be found is 100 units, in continuing to drill when no oil will be found is −300 units, and so on.

(a) What action should the driller take if he wishes to maximize his expected satisfaction and he feels that the probability of striking oil is 2/3?

(b) What action should the driller take if he wishes to maximize his expected satisfaction and he feels that the probability of striking oil is 1/10?

(c) What action would the driller take if he were a confirmed optimist?

(d) What action would the driller take if he were a confirmed pessimist?

(e) What is the *expected value of perfect information* when the probability of striking oil is 2/3?

16. At a time when the Republican party has not decided where to hold its national convention but has narrowed its choice to either San Francisco or Miami, a women's specialty store in San Francisco is offered either 1,000 or 2,000 novelty bottles of perfume in the form of the party elephant at a cost of $2 each. If the store buys 1,000 bottles and the convention goes to Miami, it expects to sell all of them at $5 each; if it buys 2,000 bottles, however, it expects to be able to sell only 1,000 bottles at $5 each and to be forced to sell the remainder at 50 cents each. If the convention comes to San Francisco the store expects to sell 2,000 bottles at $5 each, but if it buys only 1,000, it will have to rebuy and pay $4 each for the additional 1,000 bottles.

(a) If the store wants to maximize its expected profit on this item, what should its initial order be if it appears that there is a fifty-fifty chance the convention will come to San Francisco? If the odds are 3 to 2 that it will go to Miami?

(b) If the store has no information whatever about the chances the two cities have of getting the convention, which initial order will maximize its minimum profit?

(c) Assuming the odds of part (a), what is the expected value of perfect information?

17. A contractor has to choose between two jobs. The first promises a profit of $80,000 with a probability of 3/4, or a loss of $20,000 (due to strikes and other delays) with a probability of 1/4; the second promises a profit of $120,000 with a probability of 1/2, or a loss of $30,000 with a probability of 1/2.

(a) Which job should the contractor choose so as to maximize his expected profit?

(b) What job might the contractor choose if his business is in bad shape and he will go broke unless he can make a profit of at least $100,000 on his next job?

18. A manufacturer has produced for sale a large lot of items among which an unknown proportion p are defective. He can decide (1) not to put the lot on the market at all or (2) sell the lot with a double-your-money-back guarantee on all defective items.

(a) What should the manufacturer do if he has no information whatsoever about p and he wants to minimize his maximum losses?

(b) Discuss the reasonableness of using the minimax criterion in this kind of situation.

19. A retailer has shelf space for four highly perishable items which are destroyed at the end of the day if they are not sold. The unit cost of the item is $2, the selling price is $4, and the profit is thus $2 per item sold.

(a) If nothing is known about the possible demand for the item, how many should the retailer stock so as to minimize the maximum possible losses? Comment on the appropriateness of this criterion in the given problem.

(b) How many items should the retailer stock so as to maximize his expected profit, if it is known that the probabilities of the demand for 0, 1, 2, 3, or 4 items are, respectively, 0.10, 0.30, 0.40, 0.10, and 0.10?

GAMES OF STRATEGY

The decision problem of the preceding section may well have given the impression that Mr. Knight was playing a game with *Nature* as his opponent. In fact, on page 143 we referred to the two alternatives concerning the feasibility of popularizing the toy as "states of Nature." Each of the "players" in the game had the choice of two moves: Mr. Knight had the choice of continuing or terminating the advertising program, and Nature controlled whether or not the toy could be popularized. Depending on the choice of their moves, there were certain *payoffs*—the figures shown in the table on page 142.

This analogy is not at all farfetched; the problem we have been discussing is typical of the kind of situation treated in the *Theory of Games*, a relatively new branch of mathematics which has stimulated considerable interest in recent years. This theory is not limited to parlor games, as its name might suggest, but it applies to any kind of competitive situation which might arise in business, in social interactions (between nations, individuals, political parties, etc.), or in the conduct of a war.

To introduce some of the basic concepts of the Theory of Games, let us begin by explaining what we mean by a *zero-sum two-person game*. The "two-person" means that there are two players (or, more generally, two parties with conflicting interests), and the "zero-sum" means that whatever one player wins the other one loses. Thus, in a zero-sum game there is no "cut for the house" as in professional gambling, and no capital is created or destroyed during the course of play.

Games are also classified according to the number of *strategies* (moves, choices, or alternatives) each player has at his disposal. For instance, if each player has to choose one of two alternatives, we say that they are playing a 2×2 game (where 2×2 reads "two by two"); if one player has three possible moves while the other has four, we say that they are playing a 3×4 game, and so on. In this book we shall consider only *finite games*, that is, games in which each player has a finite, or fixed, number of possible moves.

It is customary in Game Theory to refer to the two players as Player A and Player B, with the possible strategies (moves, choices, or alternatives) of Player A labeled I, II, III, IV, . . . , and those of Player B labeled 1, 2, 3, 4, ... The amounts of money which change hands when the players choose their respective strategies are usually shown as in the following kind of table, called a *payoff matrix*:

		Player A	
		I	II
Player B	1	5	−6
	2	6	8

Here *positive amounts* represent payments which Player A makes to Player B, and *negative amounts* (for instance, the −6 in the above table) represent payments which Player B makes to Player A. Thus, if Player A chooses Strategy I in this example and Player B chooses Strategy 1, then Player A has to pay 5 (say, dollars) to Player B; if Player A chooses Strategy II and Player B chooses Strategy 1, then Player B has to pay 6 (dollars) to Player A.

In actual practice, games such as baseball, chess, bridge, or Monopoly are described by listing the respective rules according to which these games are played. We describe the pieces (or other kinds of equipment) being used, the way in which they are manipulated, and sometimes we say whether there are any penalties or rewards depending on what happens during the course of play. All this is important, of course, if we actually want to play one of these games, but its analysis in the Theory of Games requires only that we list the strategies available to each player and the corresponding payoff amounts. Although it does not really matter, we shall assume here that all payoffs are in dollars; in actual practice, they could be expressed in terms of any goods or services, units of utility (desirability, or satisfaction), or even in terms of life or death (as in Russian roulette). It will also be assumed that *each player has to choose his strategy without knowledge of what his opponent has done or is planning to do, and that once a player has made his choice, it cannot be changed.*

The objectives of the Theory of Games are to determine *optimum strategies* (i.e., strategies which are most profitable to the respective players), and the corresponding payoff, which is called the *value* of the game. To illustrate, let us consider the game which is characterized by the following *payoff matrix:*

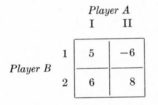

As can be seen by inspection, it would be foolish for Player B to choose Strategy 1, since Strategy 2 will yield more than Strategy 1 *regardless of the choice made by Player A.* (Strategy 2 yields 6 rather than 5 when Player A chooses Strategy I, and it yields 8 rather than −6 when Player A chooses Strategy II.) In a situation like this we say that Strategy 1 *is dominated by* Strategy 2 (or that Strategy 2 *dominates* Strategy 1), and it stands to reason that any strategy which is dominated by another should be eliminated (ignored or crossed out). If we do this in our example, we find that Player B's *optimum strategy* is Strategy 2, the only one left, and that Player A's *optimum strategy* is Strategy I, which makes him lose $6 rather than $8; the value of the game is the corresponding payoff, namely, $6.

To consider another example, suppose that the payoffs of a 3 × 2 zero-sum two-person game are as shown in the following table:

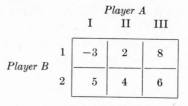

Player A

	I	II	III
1	−3	2	8
2	5	4	6

Player B

Thus, Player A has three different moves, Player B has two, and all but one of the payoffs go from Player A to Player B; the −3 indicates that if Player A chooses Strategy I and Player B chooses Strategy 1, then Player B must pay $3 to Player A. In this game neither strategy of Player B dominates the other (5 is greater than −3, 4 is greater than 2, yet 6 is less than 8), but it can be seen that the third strategy of Player A is dominated by each of the other two; that is, Player A's third strategy is worse than either of the other two *regardless of what Player B decides to do.* (Clearly, 8 is greater than −3 as well as 2, *and* 6 is greater than 5 as well as 4.) Thus, we can cross out the third column of the table, getting

−3	2
5	4

and we now find that Strategy 2 of Player B dominates his Strategy 1. (Evidently, 5 is greater than −3, and 4 is greater than 2.) It follows that Player B's *optimum choice* is Strategy 2, and since Player A would obviously prefer to lose $4 rather than $5, his *optimum choice* is Strategy II. The value of this game is $4, the payoff corresponding to Strategies 2 and II.

The process of discarding dominated strategies can be of great help in finding the solution of a game (i.e., in finding optimum strategies and the value of a game), but *what do we do when no dominances exist?* To illustrate one possibility, consider the following 3 × 3 zero-sum two-person game,

Player A

		I	II	III
	1	−2	5	−3
Player B	2	1	3	5
	3	−3	−7	11

where, as before, each player must select one of his strategies without knowledge of the other's choice. As can easily be verified, there are no dominances among the strategies of either player, but if we look at the problem from Player A's point of view, we might argue as follows: If he chooses Strategy I, the worst that can happen is that he loses $1; if he chooses Strategy II, the worst that can happen is that he loses $5; and if he chooses Strategy III, the worst that can happen is that he loses $11. Looking at the problem from this rather *pessimistic* point of view, it would seem advantageous to Player A if he *minimized his maximum losses* by choosing Strategy I. Using the terminology introduced on page 144, we are thus suggesting that Player A apply the *minimax criterion* to the losses he might incur.

If we apply the same kind of argument to select a strategy for Player B, we find that if he chooses Strategy 1, the most he can lose is $3; if he chooses Strategy 2, the worst that can happen is that he wins $1; and if he chooses Strategy 3, the most he can lose is $7. Thus, Player B would *minimize his maximum losses* (or *maximize his minimum gain*, which is the same) by choosing Strategy 2.

The selection of Strategies I and 2, appropriately called *minimax strategies*, is really quite reasonable. By choosing Strategy I, Player A makes sure that his opponent can win at most $1, and by choosing Strategy 2, Player B makes sure that he actually does win this amount. Thus, the value of the game is $1, which means that it favors Player B, but we could easily make it "equitable" by charging Player B $1 for the privilege of playing the game, while letting Player A play for free.

A very important aspect of minimax strategies is that they are completely "spy proof" in the sense that neither player can profit from any knowledge about the other's choice of strategies. Even if Player B announced publicly that he is going to choose Strategy 2, it would still be best for Player A to choose Strategy I, and the same is true for Player B if Player A announced that he is going to choose Strategy I.

Unfortunately, the method by which we solved this last example does not work for every finite zero-sum two-person game, but at least there exists a criterion by which we can decide for any given game whether minimax strategies are really spy proof. What we have to look for are *saddle points*, namely, pairs of strategies for which the corresponding entry in the payoff matrix is *the smallest value of its row and the greatest value of its column.* In the preceding example, the entry which corresponds to Strategies I and 2 *is* the smallest value of its row (1 is less than 3 or 5) and the greatest value of its column (1 is greater than -2 or -3), and, hence, it *is* a saddle point. Note that there can be more than one saddle point in a given game (see Exercise 5 on page 160), but in that case it does not matter which one we use in selecting the optimum strategies of the two players. Games which have a saddle point are said to be *strictly determined*.

If there is no saddle point, the method of the preceding example will not work, and we shall have to look for other ways of determining optimum strategies for the two players. To illustrate, let us consider the 2 × 2 zero-sum two-person game which is characterized by the following payoff matrix:

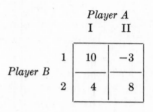

Since the smallest values of the two rows are −3 and 4, which are *not* the greatest values of their respective columns, *there is no saddle point*, but if we applied the same sort of reasoning as in the preceding example, we might argue that so far as Player A is concerned, a maximum loss of $8 is preferable to a maximum loss of $10, and, hence, that Strategy II is preferable to Strategy I. Similarly, Player B might choose Strategy 2, arguing that a minimum gain of $4 is preferable to a possible loss of $3. If Players A and B actually used these *minimax strategies*, the payoff would be $8, which should come as a very pleasant surprise to Player B—it is more than the minimum gain of $4 which he tried to assure for himself by choosing Strategy 2. So far as Player A is concerned, things turned out as expected, but had he known that Player B always chooses minimax strategies (and, hence, that he would choose Strategy 2), he could have chosen Strategy I and thus held his losses down to $4. This would have worked nicely, unless B had been smart enough to reason that this is precisely what Player A intends to do; he could then have played Strategy 1 and won $10. *This argument can be continued ad infinitum.* If Player A thought that Player B would try to outsmart him by choosing Strategy 1, he could in turn try to outsmart Player B by choosing Strategy II and winning $3; if Player B thought that this is precisely what Player A would do, he would only have to switch to Strategy 2 to assure himself a payoff of $8; and so on, and so on.

An important aspect of this example is that the minimax strategies are *not spy proof*, and that one player can outsmart the other if he knows how his opponent will react in a given situation. To avoid this possibility, it appears that each player should somehow *mix up his strategies intentionally*, and the best way of doing this is by introducing an element of chance into his final selection. Suppose, for instance, that Player B used a gambling device (dice, cards, numbered slips of paper, or random numbers) which leads to the choice of Strategy 1 with the probability x, and to the choice of Strategy 2 with the probability $1 - x$. He could then argue as follows: If *Player A chooses Strategy I*, he (Player B) can *expect* to win

$$E = 10x + 4(1 - x)$$

and *if Player A chooses Strategy II*, he (Player B) can *expect* to win

$$E = -3x + 8(1 - x)$$

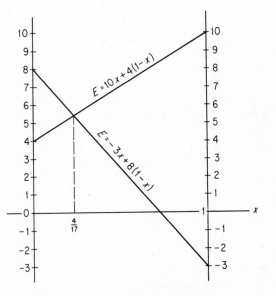

FIGURE 6.1   Expected winnings of Player B

Graphically, this situation is described in Figure 6.1, where we have plotted the two lines whose equations are

$$E = 10x + 4(1 - x) \quad \text{and} \quad E = -3x + 8(1 - x)$$

for values of x from 0 to 1. (We plotted these two lines by connecting the respective values of E which correspond to $x = 0$ and $x = 1$.)

Now suppose that we apply the *minimax criterion* to the expected winnings of Player B. (Actually, we shall *maximize minimum gains* rather than *minimize maximum losses*, but this amounts to the same thing, and it doesn't really matter whether we refer to the criterion as *minimax* or *maximin*.) By studying Figure 6.1, we find that *the least* Player B can expect to win (i.e., the smaller of the two values of E for any given value of x) is *greatest* where the two lines intersect, and to find the corresponding value of x, we have only to put $10x + 4(1 - x)$ equal to $-3x + 8(1 - x)$ and solve for x. Thus, we get

$$10x + 4(1 - x) = -3x + 8(1 - x)$$
$$10x + 4 - 4x = -3x + 8 - 8x$$
$$17x = 4$$

and, finally,

$$x = \frac{4}{17}$$

This means that if Player B labels 4 slips of paper "Strategy 1," 13 slips of paper "Strategy 2," shuffles them thoroughly, and then acts according to which kind he randomly selects, he will be applying the *minimax principle* to his *expected winnings*, and he will be assuring for himself *expected winnings* of

$$10 \left(\frac{4}{17}\right) + 4 \left(\frac{13}{17}\right) = 5\frac{7}{17}$$

or \$5.41 to the nearest cent.

If a player's ultimate choice is left to chance, his overall strategy is referred to as *randomized* or *mixed*. Note that the *optimum mixed strategy* (with $x = \frac{4}{17}$) assures Player B expected winnings of \$5.41, whereas the direct choice of one of his *pure strategies* (Strategy 1 or Strategy 2) guarantees him at best minimum winnings of \$4. So far as Player A is concerned, the analysis is very much the same; as the reader will be asked to show in Exercise 6 on page 160, Player A can minimize his maximum expected losses by choosing between Strategies I and II with respective probabilities of 11/17 and 6/17, and he will thus be holding his expected losses down to $5\frac{7}{17}$ or \$5.41. The amount \$5.41 to which Player B can *raise his expected winnings* and to which Player A can *hold down his expected losses* is again called the *value* of the game.

The examples of this section were all given without any "physical" interpretation, and this was done mainly because many games that are of any practical importance involve so many possible moves (strategies) that their analysis requires a tremendous amount of work. In any case, we were interested mainly in introducing some of the mathematical concepts that are basic to the Theory of Games. Had we studied the "real" decision problem of the preceding section as if it were a 2 × 2 zero-sum two-person game, we would have arrived at the conclusion that Mr. Knight should randomize his decision and choose between continuing and terminating the advertising program with respective probabilities of 5/21 and 16/21 (see Exercise 7 on page 160). His company's *expected losses* would then be $13,333, and this is much better than the potential losses of $120,000 and $80,000 to which he is exposed by either of his pure strategies. Of course, this assumes that Mr. Knight has no idea about the probability that the advertising program will be a success. It must be remembered, also, that this game-theoretical analysis of the problem assumes that Nature (which controls whether or not the toy can be successfully promoted) is a *malevolent opponent*, who is trying to make things as difficult as possible for Mr. Knight and his company. Whether or not this kind of assumption is reasonable can only be judged separately for each individual problem. In this case, let us say that Mr. Knight (being the advertising manager of the company) *should* have some idea about the feasibility of the project and, hence, that the problem could be solved by the methods of the preceding section.

EXERCISES

1. Determine for each of the following games whether one strategy of Player B dominates the other:

A

(a) B

10	−17	0	15	4	−40	11
5	−30	−10	3	0	−52	7

A

(b) B

10	−17	0	15	4	−40	11
5	−17	−10	15	4	−52	7

A

(c) B

10	−17	0	15	4	−40	11
5	7	−10	3	0	−52	7

2. Verify the fact that each of the three games of Exercise 1 has the *same* solution, namely, the same optimum strategies for each player and the same value.

3. Solve each of the following games; that is, find the best strategy for each player and the value of the game:

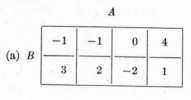

A

(a) B

−1	−1	0	4
3	2	−2	1

A

(b) B

−1	−2	8
7	5	−1
6	0	12

4. Solve each of the following games; that is, find optimum strategies for the two players and the value of the game:

A

(a) B

7	0	1
5	3	2
−2	7	0

A

(b) B

2	0	7	1
2	2	1	2
5	4	4	3

5. If a 3×3 zero-sum two-person game has a saddle point corresponding to Strategies I and 3 and another corresponding to Strategies III and 1, explain why there must also be a saddle point corresponding to Strategies I and 1 and another corresponding to Strategies III and 3, and that all four of these saddle points must have the same payoff.

6. Referring to the example on page 156, show that Player A's optimum strategy is to choose between Strategies I and II with respective probabilities of 11/17 and 6/17.

7. Referring to the illustration of the preceding section and the table on page 142, show that Mr. Knight's optimum strategy is to randomize his decision and to choose between continuing or terminating the advertising program with respective probabilities of 5/21 and 16/21.

8. Red forces Blue into playing the following game:

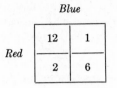

Blue

Red

12	1
2	6

Find Red's and Blue's optimum strategies and the value of the game.

9. If a large game (with many strategies for each player) does not have a saddle point, it can sometimes be reduced in size by first eliminating all dominated strategies. Solve the following game (i.e., find optimum strategies for the two players and the value of the game) by first reducing it to a 2 × 2 game:

A

B

1	−2	5
7	2	−3
4	−1	6

When stating the results, indicate dominated strategies by assigning them the probability 0.

10. Show that in the following game Player B's optimum strategy is his second strategy:

A

B

−2	200	200	200	200
0	0	0	0	0

It has been suggested that in a game like this it would be wholly irrational for Player B to choose his second strategy. Give examples where (a) this suggestion would be reasonable, and (b) this suggestion would not be reasonable.

11. A certain type of computer is made by only two companies who share the market for the machine equally. Both would prefer not to introduce a new model at this time, but both suspect that the other is readying a new model and that if it is introduced some sales will be lost to the competitor. If neither brings out a new model or if both bring out new models, the status quo will be maintained and both will continue to get their same relative share of the market. If, however, one brings out a new model and the other does not, there will be a loss of 10 per cent in share of the market to the competitor with the new machine. What is the best strategy for the two companies to use with respect to the introduction of a new model?

12. Country B has two island bases with installations worth $10 million and $30 million, respectively. Of these bases it can defend only one against an attack by Country A. Country A, on the other hand, can attack only one of these islands and take it successfully only if it is left undefended. Considering the "payoff" to Country B to be the total value of the installations held by Country B after the attack, find the optimum strategies for both countries and the value of this "game." Why wouldn't it be best for Country B simply to defend the island with the more valuable installation?

13. Two persons agree to play the following game: The first writes either 1 or 4 on a slip of paper and at the same time the second writes either 0 or 3 on another slip of paper. If the sum of the two numbers is odd, the first wins this amount in dollars; otherwise, the second wins $2.

    (a) What are the best strategies for the two players, and what is the value of the game?

    (b) Show that if 25 cents is subtracted from each payoff in this game, the game will be fair (equitable) but the best strategies of the two players remain unchanged.

    (c) Referring again to the game as originally stated, if $1 were added to each payoff to the first player when the sum of the numbers is odd, and otherwise the second player wins only $1, what would be the best strategies of the two players and the value of the game?

14. Referring to Exercise 14 on page 149, how should the executive randomize his choice between the two hotels so as to minimize his maximum expected expenses?

15. With reference to Exercise 15 on page 150, how should the wildcatter randomize his decision so as to maximize his minimum expected satisfaction?

16. Because of various difficulties, the supplier of glue used in the manufacture of a laminated fiberboard product can guarantee the manufacturer only that it will deliver on schedule the required quantity of either Glue Q or Glue R (but not some of both). Because of time requirements, however, the manufacturer must set up his production process prior to knowledge of which glue will be available with no later change possible if he is to meet contractual obligations. Both glues can be used with any one of six production methods open to the company, but for technical reasons the profit per piece differs substantially from one method to another for the same glue. The estimated unit profits (in cents) for Methods 1–6 using Glue Q are 96, 146, 135, 160, 125, and 154, respectively, while the corresponding figures for Glue R are 255, 116, 175, 195, 235, and 202 cents. Which production method should the manufacturer use if he wants to maximize his minimum unit profit? Is this a reasonable criterion to use in this kind of situation?

17. Suppose that in Exercise 16 the manufacturer has the choice of only three production methods for which the unit profits (in cents) using Glue Q are, respectively, 120, 100, and 190. If the corresponding figures for Glue R are 205,

180, and 165 cents, what production method should the manufacturer use so as to maximize his minimum expected unit profit?

18. Suppose that in Exercise 16 there are three glues, Q, R, and S, and that the manufacturer has the choice of only two production methods for which the unit profits (in cents) using Glue Q are, respectively, 70 and 150. If the corresponding figures for Glue R are 100 and 80 cents, while those for Glue S are 140 and 60 cents, how should the manufacturer decide which method to use so as to maximize his minimum expected unit profit?

19. There are two gas stations in a certain block, and the owner of the first station knows that if neither station lowers its prices he can expect a net profit of $100 on any given day. If he lowers his prices while the other station does not, he can expect a net profit of $140; if he does not lower his prices but the other station does, he can expect a net profit of $70; and if both stations participate in this "price war," he can expect a net profit of $80. The owners of the two gas stations decide independently what prices to charge on any given day, and it is assumed that they cannot change their prices after they discover those charged by the other.

(a) Should the owner of the first gas station charge his regular prices or should he lower them, if he wants to maximize his minimum net profit?

(b) Assuming that the above profit figures apply also to the second gas station, how might the owners of the gas stations collude so that each could expect a net profit of $105? (Note that this "game" is *not* zero-sum, so that the possibility of collusion opens entirely new possibilities.)

A WORD OF CAUTION

One of the greatest difficulties in applying the methods of this chapter (and more general methods) to realistic problems in statistics, business management, and economics, in general, is that *we seldom know the exact values of all risks that are involved.* That is, we seldom know the exact values of the payoffs corresponding to the various eventualities, and we seldom have sufficient information about the values of all relevant probabilities. For instance, if a manufacturer has to decide whether to market a new drug right away, how can he put a cash value on the damage that might be caused by not waiting for a more thorough evaluation of the side effects of the drug, or the lives that might be lost by not marketing the drug? Similarly, if a management consultant has to decide whether to recommend the marketing of a new detergent, how can he possibly take into account all the effects which this advice (good or bad) might have on himself, on the company which produces the detergent, and ultimately on the consumer of the product?

The fact that we seldom have adequate information about all pertinent probabilities provides considerable obstacles in finding suitable decision criteria. For instance, in the example which dealt with Mr. Knight's decision as to whether to continue or terminate the advertising program, is it reasonable to base the decision on optimism, pessimism, or the principle of minimizing maximum expected losses? This last criterion we borrowed from the Theory of Games, where it is reasonable because the theory applies to conflicts between players who, so to speak, are out to cut each other's throat. But is it really reasonable in this case to look upon Nature as a malevolent opponent?

The difficulties we have just pointed out present serious obstacles to the application of the methods of this chapter, but they are nevertheless important because they tend to clarify the logic which underlies statistical thinking and decision making in general. The problems of statistical decision making which we shall introduce in later chapters will, indeed, be easier to grasp if we formulate them (at least tacitly) as games in which all sorts of things can happen, in which each interested party including Nature has various moves, and in which there are all sorts of consequences on which we may or may not be able to put appropriate values.

7

PROBABILITY DISTRIBUTIONS

PROBABILITY FUNCTIONS

The two tables shown below give some of the probabilities associated with two simple experiments; one consists of the roll of a die and the other consists of three flips of a balanced coin. Their purpose is to illustrate what we mean by a *probability function,* namely, *a correspondence which assigns probabilities to numerical descriptions of the outcomes of an experiment.* The first of these tables was easily obtained on the basis of the assumption that each face of the die has a probability of 1/6; the second was obtained by considering as equally likely the eight possible outcomes TTT, HTT, THT, TTH, HHT, HTH, THH, and HHH, where H stands for "head" and T for "tail."

Number of Points Rolled with a Die	Probability	Number of Heads Obtained in Three Flips of a Coin	Probability
1	1/6	0	1/8
2	1/6	1	3/8
3	1/6	2	3/8
4	1/6	3	1/8
5	1/6		
6	1/6		

Whenever possible, we try to express probability functions by means of formulas which enable us to calculate the probabilities associated with the various numerical descriptions of the outcomes. With the usual functional notation we can thus write

$$f(x) = \frac{1}{6} \quad \text{for } x = 1, 2, 3, 4, 5, 6$$

for the first of the above examples; $f(1)$ represents the probability of rolling a 1, $f(2)$ represents the probability of rolling a 2, and so on.

Using the *factorial notation* (see page 105) we can similarly write for the second example

$$f(x) = \frac{3/4}{x!(3 - x)!} \quad \text{for } x = 0, 1, 2, 3$$

As can easily be verified, this will, indeed, give the probabilities listed in the right-hand column of the second table.

It is customary to refer to the number of points rolled with the die in the first example, the number of heads obtained in the second example, and numerical descriptions of experiments in general, as *random variables*. The value a random variable assumes in a given experiment is controlled by chance; that is, the probability that a random variable takes on a particular value is given by an appropriate probability function. To be rigorous, let us say that *a random variable is a function defined over the sample space of an experiment,* and the probability that any one of its values will occur is given by a suitable probability function. Note also that since a random variable has to assume one of its values, the sum of the probabilities in a probability function must always equal 1. Of course, the values of a probability function, being probabilities, must be positive or zero.

THE BINOMIAL DISTRIBUTION

There are many applied problems in which we are interested in the probability that an event will take place x times in n "trials," or in other words, x times out of n, while the probability that it will take place in any one trial is some constant number p and the trials are independent. We may thus be interested in the probability of getting eight 5's in 50 rolls of a die, the probability that eight out of 25 newly franchised restaurants will go bankrupt within two years, the probability that a sample of 50 out-of-town visitors to Disneyland will include six from Illinois, and so forth. Referring to the occurrence of any of these events as a *success* (which is a holdover from the days when probability theory was applied only to games of chance), we are interested in each case in the probability of getting "*x* successes in *n* trials." To handle problems of this kind we use the following formula, which will be derived in Technical Note 4 on page 211:

The probability of getting x successes in n independent trials is given by

$$f(x) = \binom{n}{x} p^x (1 - p)^{n-x} \quad \text{for } x = 0, 1, 2, \ldots, n \qquad \bigstar$$

where p is the constant probability of a success for each trial and $\binom{n}{x}$ *is a binomial coefficient as defined on page 110.*

It is customary to say that the number of successes in *n* trials is a random variable having the *binomial probability distribution* or, simply, the *binomial distribution*. The terms "probability distribution" and "probability function" are often used interchangeably, although some persons make the distinction that the term "probability distribution" refers to *all* the probabilities associated with a random variable, and not only those given directly by its probability function. (The following illustrates this rather technical distinction: The probabilities of getting, say, "one *or* two successes" or "at least three successes" are part of the *probability distribution* of the number of successes; although these probabilities are easily obtained by adding some of the value of the corresponding *probability function*, strictly speaking, they are not part of it.)

To illustrate the use of the above formula, let us first calculate the probability of getting 6 heads and 9 tails in 15 flips of a balanced coin. Substituting $x = 6$, $n = 15$, $p = \frac{1}{2}$, and $\binom{15}{6} = 5{,}005$ (see Table X), we get

$$f(6) = 5{,}005 \left(\frac{1}{2}\right)^6 \left(1 - \frac{1}{2}\right)^{15-6} = \frac{5{,}005}{32{,}768}$$

or approximately 0.15. Similarly, to find the probability of getting four busy signals in 12 telephone calls placed to numbers randomly selected from a directory, let us suppose that the probability of getting a busy signal under these conditions is 0.20. Then, substituting $x = 4$, $n = 12$, $p = 0.20$, and $\binom{12}{4} = 495$ (see Table X), we get

$$f(4) = 495(0.20)^4(1 - 0.20)^{12-4}$$
$$= 495(0.0016)(0.1677216)$$

or approximately 0.13.

In actual practice, problems involving binomial distributions are often solved by referring to special tables (see the Bibliography) or by using one of the two approximations which will be discussed later in this chapter. The direct application of the formula for the binomial distribution would be much too involved, for instance, to calculate the probability that 312 of 5,000 cars stopped at a road block will have defective brakes, the probability that 420 of 800 persons entering a department store will make at least one purchase, or the probability that 220 of 4,000 persons drafted by the army cannot pass a literacy test.

THE HYPERGEOMETRIC DISTRIBUTION

If we sample *without replacement* from a finite population, the probability of obtaining a certain kind of element will change after each observation, and successive trials are *not independent*. Suppose, for example, that a company ships alarm clocks from its factory in cartons of 12 clocks each. When they arrive at the retailer's warehouse, an inspector randomly selects two alarm clocks from each carton, and *the whole carton is accepted if and only if the two clocks are in good working condition*; otherwise, each alarm clock is inspected individually. (By "random" we mean that each clock has the same chance of being selected.) It is evident that this kind of sampling inspection involves certain risks, for it is possible (though unlikely) that a carton will be accepted without further inspection even though 10 of the 12 clocks are defective. Of course, this is an extreme case; more realistically, perhaps, it may be of interest to know the probability that a carton will be accepted without further inspection even though, say, 3 of the 12 clocks are defective. This means that we would be interested in the probability of getting two successes (nondefective clocks) in two trials, and we might be tempted to argue that since 9 of the 12 clocks are nondefective, the probability of a success is $\frac{9}{12} = \frac{3}{4}$, and hence, the probability of "two successes in two trials" is

$$\binom{2}{2}\left(\frac{3}{4}\right)^2\left(1 - \frac{3}{4}\right)^{2-2} = 1 \cdot \frac{9}{16} \cdot 1 = 0.5625$$

in accordance with the formula for the binomial distribution.

This result would be correct if each clock was *replaced* before the next one is chosen from the same carton, but if sampling is *without replacement*, the probability that the second clock is nondefective *given that the first clock was nondefective* is 8/11 (instead of 9/12). For sampling without replacement, successive trials are *not independent*, and (using the multiplication rule on page 129) we find that the correct value of the probability of getting two nondefective clocks is $\frac{9}{12} \cdot \frac{8}{11} = \frac{6}{11}$ or approximately 0.5455. The difference between 0.5625 and 0.5455 may not be very large, and it may not be of critical significance, but the binomial probability function is not the correct one in this instance.

The correct probability function for this kind of problem is that of the *hypergeometric distribution*. It applies whenever n elements are randomly selected *without replacement* from a set containing a elements of one kind (successes) and b elements of another kind (failures), and we are interested in the probability of getting x successes and $n - x$ failures. The formula for this probability function is

$$f(x) = \frac{\binom{a}{x} \binom{b}{n-x}}{\binom{a+b}{n}} \quad \text{for } x = 0, 1, 2, \ldots, \text{ or } n \qquad \bigstar$$

(*subject to the restriction that x cannot exceed a and $n - x$ cannot exceed b*). If we apply this formula to our illustration, where we had $x = 2$, $n = 2$, $a = 9$, and $b = 3$, we get

$$f(2) = \frac{\binom{9}{2} \binom{3}{0}}{\binom{12}{2}} = \frac{36 \cdot 1}{66} = \frac{6}{11}$$

which is identical with the result obtained before.

To prove the formula for the hypergeometric distribution, we have only to observe that the x successes can be selected from among a possibilities in $\binom{a}{x}$ ways, the $n - x$ failures can be selected from among b possibilities in $\binom{b}{n-x}$ ways, so that the x successes *and* $n - x$ failures can be selected in $\binom{a}{x} \cdot \binom{b}{n-x}$ ways in accordance with the rule on page 102. The total number of ways in which we can select n elements from a set of $a + b$

elements is $\binom{a+b}{n}$, and according to the special rule for equiprobable events on page 120, the desired probability is given by the ratio of $\binom{a}{x} \cdot \binom{b}{n-x}$ to $\binom{a+b}{n}$.

To consider another example in which we use the formula for the hypergeometric distribution, suppose that as part of an air pollution survey an inspector decides to examine the smokestacks of 9 of the 18 factories in a given city. He suspects that 6 of the 18 factories have smokestacks which emit excessive amounts of pollutants, and he wants to know the probability that *if his suspicion is correct,* his sample will catch *at least* 4 of the 6. The probability which he wants to know is given by $f(4) + f(5) + f(6)$, where each term in this sum is to be calculated by means of the formula for the hypergeometric distribution with $a = 6$, $b = 12$, and $n = 9$. Substituting these values together with $x = 4$, $x = 5$, and $x = 6$, we get

$$f(4) = \frac{\binom{6}{4}\binom{12}{5}}{\binom{18}{9}} = \frac{15 \cdot 792}{48,620} = 0.244$$

$$f(5) = \frac{\binom{6}{5}\binom{12}{4}}{\binom{18}{9}} = \frac{6 \cdot 495}{48,620} = 0.061$$

$$f(6) = \frac{\binom{6}{6}\binom{12}{3}}{\binom{18}{9}} = \frac{1 \cdot 220}{48,620} = 0.005$$

and it follows that the probability that the sample inspection will catch at least 4 of the 6 "bad" factories is $0.244 + 0.061 + 0.005 = 0.310$. Perhaps, this should make the inspector examine more than 9 of the factories, and it will be left to the reader (in Exercise 15 on page 175) to find the probability that at least four of the "bad" factories will be caught if he examines the smokestacks of $n = 12$.

In the beginning of this section we introduced the hypergeometric distribution in connection with a problem in which we *erroneously* used the binomial distribution. Actually, when n is small compared to $a + b$, the binomial distribution often provides a very good *approximation* to the hypergeometric distribution. It is generally agreed that this approximation can be used so long as n constitutes less than 5 per cent of $a + b$; this is good because the binomial distribution has been tabulated much more extensively than the hypergeometric distribution, and it is generally easier to use.

THE POISSON DISTRIBUTION

If n is large and p is small, binomial probabilities are often approximated by means of the formula

$$f(x) = \frac{(np)^x \cdot e^{-np}}{x!} \qquad \text{for } x = 0, 1, 2, 3, \ldots \qquad \bigstar$$

which is that of the *Poisson distribution*. Here e is the number $2.71828 \ldots$ used in connection with natural logarithms, and the values of e^{-np} may be obtained from Table IX at the end of the book. Note also that for this distribution the set of all values taken on by the random variable is *infinite;* practically speaking, this is of no significance since the probabilities become negligible (very close to zero) after the first few values of x.

To illustrate the use of the above formula, suppose that a fire insurance company insures 10,000 houses which are widely dispersed throughout the United States. Assuming that the probability of any one house being destroyed by fire during any one year is $p = \frac{1}{5,000}$, let us find the probabilities that $0, 1, 2, 3, \ldots$, of the 10,000 houses will be destroyed by fire during a given year. Ruling out the use of the formula for the binomial distribution for practical reasons, we substitute $n = 10,000$, $p = \frac{1}{5,000}$, and hence, $np = 2$ into the formula for the Poisson distribution, getting $f(0) = \frac{2^0 e^{-2}}{0!} = 0.135$, $f(1) = \frac{2^1 e^{-2}}{1!} = 0.270$, $f(2) = \frac{2^2 e^{-2}}{2!} = 0.270$, and $f(3) = \frac{2^3 e^{-2}}{3!} = 0.180$ for the first four values of x. Continuing this way, we would obtain the values shown in the following table, where we stopped with $x = 8$, since the probabilities for $x = 9, 10, 11, \ldots$, are all 0.000 rounded to three decimals (i.e., they are all less than 0.0005):

Number of Houses Destroyed by Fire	Probability
0	0.135
1	0.270
2	0.270
3	0.180
4	0.090
5	0.036
6	0.012
7	0.003
8	0.001

A histogram of this distribution is shown in Figure 7.1.

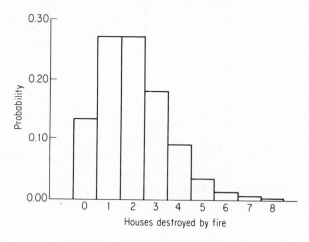

FIGURE 7.1 A Poisson distribution

The Poisson distribution also has many important applications which have no direct connection with the binomial distribution. In that case, np is replaced by the parameter λ (*lambda*), and we calculate the probability for getting x successes by means of the formula

$$f(x) = \frac{\lambda^x \cdot e^{-\lambda}}{x!} \qquad \text{for } x = 0, 1, 2, 3, \ldots \qquad \bigstar$$

As the reader will be asked to verify in Exercise 9 on page 192, λ can be interpreted as the *expected*, or *average*, number of successes. Thus, if a finance company receives on the average three bad checks per day, the probability that it will receive exactly *one* bad check on any given day is

$$f(1) = \frac{3^1 \cdot e^{-3}}{1!} = \frac{3 \cdot 0.050}{1} = 0.150$$

where the value of e^{-3} was obtained from Table IX. Similarly, if the number of accidents per week at a busy intersection is a random variable having the Poisson distribution with $\lambda = 4.8$, the probability that there will be exactly three accidents in any given week is

$$f(3) = \frac{(4.8)^3 \, e^{-4.8}}{3!} = \frac{(110.592)(0.008)}{6}$$

or approximately 0.15.

EXERCISES

1. Use the formula for the binomial distribution to find
   (a) the probability of rolling exactly two 4's in six rolls of a balanced die;
   (b) the probability of rolling at most two 4's in six rolls of a balanced die;
   (c) the probability of getting exactly three heads in nine flips of a balanced coin;
   (d) the probability of getting at most three heads in nine flips of a balanced coin.

2. A multiple-choice test consists of 10 questions and three answers to each question (of which only one is correct). If each question is answered by rolling a balanced die until 1, 2, or 3 appears and scoring the question accordingly, find
   (a) the probability of getting exactly four correct answers;
   (b) the probability of answering each question incorrectly;
   (c) the probability of passing the test by this method, if it takes at least seven correct answers to pass the test.

3. If 5 per cent of all new golf balls used at a driving range are badly mauled the first time they are hit, what are the probabilities that of four new golf balls hit by different golfers (a) exactly one is badly mauled and (b) at most one is badly mauled?

4. A professor of marketing gives each of the 12 students in his seminar a list of 100 case studies, of which half are terribly dull. If each of the students randomly and independently selects one case to read, what are the probabilities that
   (a) either no one or everyone selects a terribly dull case;
   (b) exactly 10 students select a terribly dull case;
   (c) at least 10 students select a terribly dull case?

5. If 60 per cent of all children in a certain age group suffer side-effects (e.g., fever and rash) from an injection of a measles vaccine without an accompanying injection of gamma globulin in the other arm, find the probabilities that among five children thus vaccinated

(a) at least one child suffers these side-effects;

(b) exactly three suffer these side-effects;

(c) at least three suffer these side-effects.

6. If 20 per cent of all women who admit a vacuum cleaner salesman into their homes will end up buying one, what is the probability that among six women who admit such a salesman into their home at most one will end up buying a vacuum cleaner?

7. In the game of "chuck-a-luck" three dice are thrown and a player bets on the occurrence of a number which he can choose. If he *wins* $1 if his number appears on only one die, $2 if his number appears on two dice, $3 if his number appears on all three dice, and he *loses* $1 if his number appears on none of the dice, find the probabilities of his winning $1, $2, $3, and the probability of his losing $1. Also determine how much a player can *expect* to win (in the sense of a mathematical expectation).

8. Find the constant probability that a census enumerator will find any one family on his list at home in the afternoon, if the probability that all four of the families he plans to visit on an afternoon are out equals 16/81.

9. If the probability that a freshman entering a very large state university will not finish the first year is 0.35, what is the probability that at least three of five freshmen (whose names are randomly selected from the registrar's files) will finish the first year?

10. On page 169 we showed that if 3 of the 12 clocks in a carton are defective, the probability that two randomly selected clocks are nondefective is 6/11. What would this probability have been if

(a) only two of the clocks had been defective;

(b) four of the clocks had been defective?

11. Referring to the illustration on page 168, where 3 of the 12 clocks in the carton are defective, what are the probabilities that among 4 clocks randomly selected from this carton

(a) none are defective;

(b) exactly two are defective;

(c) at least two are defective?

12. A secretary is supposed to send three of nine letters by special delivery. If she gets them all mixed up and randomly puts special delivery stamps on three of them, what are the following probabilities:

(a) She puts all the special delivery stamps on wrong letters?

(b) She puts all the special delivery stamps on the right letters?

(c) She puts two of them on letters which were supposed to go by special delivery and the other on a letter which was not supposed to go by special delivery?

13. Among the 90 employees of a factory 60 are union members while the others are not. If four of the employees are chosen by lot to serve on a committee to resolve a labor-management dispute, what is the probability that two of them will be union members while the other two are not? By how much would we have been off in this example if we had erroneously used the binomial probability function with $p = \dfrac{60}{90} = \dfrac{2}{3}$ to calculate the probability of two successes in four trials?

14. In a file of 400 invoices exactly 6 contain errors. If an auditor randomly selects 3 invoices from this file, what is the probability that none of them contains an error? By how much would we have been off in this example if we had erroneously used the binomial probability function with $p = \dfrac{6}{400} = 0.015$ to calculate the probability of zero successes in 3 trials?

15. With reference to the illustration on page 170, find the probability that at least four of the "bad" factories will be caught if the inspector examines the smokestacks of 12 of the 18 factories (of which 6 are "bad").

16. A buyer randomly selects two replacement batteries for a flashlight, taking them from a drugstore bin containing 80 batteries of which 60 were made in Hong Kong and the rest in Taiwan. Find the probability distribution of the number of batteries he gets that are made in Hong Kong. Also draw a histogram of this distribution.

17. Use the Poisson distribution to find an approximation for the probability of getting 4 successes in 100 independent trials when $p = 0.05$.

18. A large shipment of textbooks contains 3 per cent with imperfect bindings. Use the Poisson distribution to find an approximation for the probability that, among 200 books randomly selected from this shipment, exactly 5 have imperfect bindings.

19. To meet specifications, each can of mixed nuts must contain at least one cashew nut. If the cans of mixed nuts coming from a certain processing plant are such that actually 1.2 per cent are without cashew nuts, what is the probability that among 300 cans delivered from this processing plant exactly 2 are without cashew nuts?

20. Verify the probabilities given in the table on page 172 for $x = 4$, $x = 5$, $x = 6$, $x = 7$, and $x = 8$. Also verify that for $x = 9$ the probability is less than 0.0005.

21. Given that the switchboard of a department store has on the average four incoming calls per minute, use the Poisson distribution with $\lambda = 4$ to find the probabilities that
   (a) there will be no incoming calls during any given minute;
   (b) there will be exactly two incoming calls during any given minute;
   (c) there will be exactly five incoming calls during any given minute.

22. Suppose that in the inspection of sheet metal produced in continuous rolls the number of imperfections spotted by an inspector during a period of 10 minutes is a random variable having the Poisson distribution with $\lambda = 1.4$. Find the probabilities that during a 10-minute period an inspector will find (a) no imperfections, (b) one imperfection, and (c) three imperfections.

23. The records of a take-out and delivery restaurant in San Francisco specializing in Mexican food show that the number of orders which cannot be delivered on any given day (because they are refused, the doorbell is unanswered, the address is incorrect, etc.) is a random variable having the Poisson distribution with $\lambda = 4$. Find the probabilities that

(a) on any given day there are exactly two orders that cannot be delivered;

(b) on any given day there are at most two orders that cannot be delivered.

24. If medical records show that there are on the average 0.4 cases of malaria per month in a certain county, what are the respective probabilities that there will be 0, 1, 2, 3, or 4 cases during a given month? If these probabilities are rounded to three decimals, explain why there is no need, for example, to calculate the probability for five cases. Also draw a histogram of this distribution.

PRIOR PROBABILITIES AND POSTERIOR PROBABILITIES

There are many situations in which we must modify probability judgments in the light of new evidence, that is, after obtaining additional or different kinds of information. Suppose, for instance, that Mr. Adam and Mr. Brown are planning to form a corporation to build and operate a number of gas stations, and that Mr. Adam (subjectively) evaluates the probability that any new gas station will show a profit during the first year as 1/2, while Mr. Brown feels that this probability is 2/3. Suppose, furthermore, that we may want to invest some money in this corporation, and hence, we would like to know who is right or, at least, who is more likely to be right. If an expert whom we consult feels that in view of Mr. Adam's experience and past performance he is *three times as likely* to be right as Mr. Brown, he is, in fact, assigning the event B_1 that Mr. Adam is right the probability $P(B_1) = \dfrac{3}{4}$ and the event B_2 that Mr. Brown is right the probability $P(B_2) = \dfrac{1}{4}$. Since the entire operation is still in the planning stage, we have no *direct* information on the performance of the gas stations, and we refer to $P(B_1) = \dfrac{3}{4}$ and $P(B_2) = \dfrac{1}{4}$ as the *prior probabilities* of the events B_1 (that Mr. Adam is right, i.e., that the probability for a first-year profit is 1/2) and B_2 (that Mr. Brown is right, i.e., that the probability for a first-year profit is 2/3).

Now suppose that the corporation opens six gas stations and that five of them show a profit during the first year while the sixth shows a loss. *How will this affect our opinion about the earlier appraisals, namely, the probabilities of 1/2 and 2/3, given by Mr. Adam and Mr. Brown?* If we let A denote the event that five of six new gas stations (put into operation by this corporation) show a profit during the first year, we are thus asking for the *conditional* probabilities $P(B_1 \mid A)$ and $P(B_2 \mid A)$. These are not difficult to find if the conditions underlying the binomial distribution are met, for in that case substitution into the formula on page 167 yields the following probabilities required for the use of Bayes' rule on page 135:

$$P(A \mid B_1) = \binom{6}{5} \left(\frac{1}{2}\right)^5 \left(1 - \frac{1}{2}\right)^{6-5} = \frac{3}{32}$$

and

$$P(A \mid B_2) = \binom{6}{5} \left(\frac{2}{3}\right)^5 \left(1 - \frac{2}{3}\right)^{6-5} = \frac{64}{243}$$

Then, substitution into the formula

$$P(B_1 \mid A) = \frac{P(B_1) \cdot P(A \mid B_1)}{P(B_1) \cdot P(A \mid B_1) + P(B_2) \cdot P(A \mid B_2)}$$

gives

$$P(B_1 \mid A) = \frac{\frac{3}{4} \cdot \frac{3}{32}}{\frac{3}{4} \cdot \frac{3}{32} + \frac{1}{4} \cdot \frac{64}{243}} = \frac{2{,}187}{4{,}235} = 0.52$$

and it follows that $P(B_2 \mid A) = 1 - 0.52 = 0.48$. The probabilities $P(B_1 \mid A)$ and $P(B_2 \mid A)$ are referred to as the *posterior probabilities* of the events B_1 and B_2, and it is of interest to note how we have combined the direct evidence (five of the six gas stations showed a profit during the first year) with the original subjective evaluation of the expert, and how the weight of the direct evidence has given increased merit to Mr. Brown's claim. *Whereas his claim was originally assigned a probability of 1/4, it is now assigned a probability which is almost 1/2.*

To consider another example in which we modify probabilities in the light of new evidence by using the rule of Bayes, suppose that three engineers (who are otherwise considered about equally reliable in their claims) make the conflicting statements that a new computer which they designed can be expected to break down on the average 1.4, 2.0, and 3.1 times per week while in normal use. An actual experiment shows that during a week's operation the computer broke down four times. If we let B_1 denote the event that the first engineer is right, B_2 the event that the second engineer is right, B_3 the event that the third engineer is right, and A the event that the computer actually broke down four times during a week of normal use, the probabilities we want to determine are the three *prior probabilities* $P(B_1)$, $P(B_2)$, $P(B_3)$ and the three *posterior probabilities* $P(B_1 \mid A)$, $P(B_2 \mid A)$, and $P(B_3 \mid A)$. First, the three prior probabilities are simply $P(B_1) = P(B_2) = P(B_3) = \dfrac{1}{3}$, for we stated that the engineers are "otherwise considered about equally reliable in their claims." Then, if we assume that the number of times the computer breaks down (in a week of normal use) is a random variable having a Poisson distribution with $\lambda = 1.4$ if the first engineer is right, $\lambda = 2.0$ if the second engineer is right, and $\lambda = 3.1$ if the third engineer is right, substitution into the formula on page 172 yields

$$P(A \mid B_1) = \frac{(1.4)^4 \cdot e^{-1.4}}{4!} = 0.040$$

$$P(A \mid B_2) = \frac{(2.0)^4 \cdot e^{-2.0}}{4!} = 0.090$$

and

$$P(A \mid B_3) = \frac{(3.1)^4 \cdot e^{-3.1}}{4!} = 0.173$$

and Bayes' rule on page 135 with $k = 3$ gives

$$P(B_1 \mid A) = \frac{\frac{1}{3}(0.040)}{\frac{1}{3}(0.040) + \frac{1}{3}(0.090) + \frac{1}{3}(0.173)} = 0.13$$

In Exercise 1 on page 182 the reader will be asked to show that corresponding calculations lead to $P(B_2 \mid A) = 0.30$ and $P(B_3 \mid A) = 0.57$, and it should really not come as a surprise that after the computer was observed to break down four times in a week the claim of the third engineer should be regarded with more credibility than those of the other two. The reader may have observed that in order to use Bayes' rule in the first example we had to assume that either Mr. Adam or Mr. Brown was right (nothing else was possible), and to use Bayes' rule in the second example we, similarly, had to assume that one of the three engineers was right. This may not seem very reasonable, for why couldn't the probability that one of the gas stations will make a profit in its first year be, say, 2/5 instead of 1/2 or 2/3, and why couldn't the computer break down on the average, say, 3.6 times instead of 1.4, 2.0, or 3.1 times per week? All we can say about this here is that, *without making the assumption that no other alternatives exist, this kind of analysis cannot be made.* Incidentally, the kind of reasoning which we used in our two illustrations is referred to as Bayesian, and it is currently gaining in favor among statisticians.

A BAYESIAN DECISION PROBLEM

Let us consider a very large company which routinely pays thousands of invoices submitted by its suppliers. They are collected in batches of 1,000 before payment is made, and since there is always the possibility of an error, the company must decide whether or not to check (verify) each invoice in a batch. Clearly, the proportion of invoices containing errors will vary from batch to batch, and for the sake of simplicity let us assume that for any given batch it can take on only the values $p = 0.001$, $p = 0.01$, or $p = 0.02$. Also, the company's records show that *on the average* an invoice containing an error overcharges the company by $5, and it is known that it costs $35 to eliminate all errors from a batch of 1,000 invoices by means of a computer.

All this information is summarized in the following table, where the entries in the second column show that it costs the company $35 to check and eliminate all errors from a batch of 1,000 invoices *regardless of how many of them are in error.* The entries in the first column are the *expected costs* (overcharges) which result from not checking a batch before payment is made. If $p = 0.001$, the expected number of erroneous invoices per batch is $1,000(0.001) = 1$, and hence, the *expected cost* (overcharge) is $1 \cdot 5 = \$5$.

Cost or Expected Cost per Batch

		Without Checking	*With Checking*
	0.001	$5	$35
Proportion of Invoices Containing Errors	0.010	$50	$35
	0.020	$100	$35

Similarly, for $p = 0.01$ the expected number of erroneous invoices per batch is $1,000(0.01) = 10$ and for $p = 0.02$ it is $1,000(0.02) = 20$, so that the corresponding costs are $10 \cdot 5 = \$50$ and $20 \cdot 5 = \$100$.

If the proportion of erroneous invoices is p, the *expected number* of erroneous invoices per batch is $1,000p$, the *expected cost* per batch is $1,000p \cdot 5 = 5,000p$ dollars, and this *equals* the cost of checking all of the invoices before payment is made when $5,000p = 35$, namely, when $p = \dfrac{35}{5000} = 0.007$. In other words, the "break-even" value of p is 0.007 and one can expect it to cost less *not to check* when $p < 0.007$ (in our example when $p = 0.001$); similarly, one can expect it to cost less *to check* when $p > 0.007$ (in our example when $p = 0.01$ or 0.02).

Now, if the value of p were known for each batch before it is decided whether or not it should be checked, the solution of the problem would be easy—when $p = 0.001$ the company pays the invoices without checking, and when $p = 0.01$ or $p = 0.02$ the company has each invoice checked before payment is made, for in either case this *minimizes the expected cost*. In practice, of course, the value of p is not known in advance, and this is where the importance of probability in decision making becomes evident. If someone felt "pretty sure" that $p = 0.001$ for a given batch (i.e., if he assigned $p = 0.001$ a very high subjective probability for this batch), he would *not* check the batch; conversely, if someone felt "pretty sure" that $p = 0.02$ for a given batch, he would have it checked. Since different "feelings" about the correct value of p can thus affect the decision whether or not to check a batch, let us look more closely into the problem of assigning probabilities to the different values which p can take on.

One way of handling this is to assign *prior probabilities* to the various values of p (0.001, 0.01, and 0.02, in our example) on the basis of the company's records extending back over a long period of time. If these records show that when batches of invoices were actually checked, p equaled 0.001, 0.01, and 0.02, respectively, 70 per cent of the time, 20 per cent of the time, and 10 per cent of the time, then the expected cost associated with the decision *not to check* is

$$\$5(0.70) + \$50(0.20) + \$100(0.10) = \$23.50$$

where the expected costs of \$5, \$50, and \$100 were read off the table on page 179. Since this value is less than \$35 (the cost of checking a batch), the company can minimize its expected cost *relative to the given prior probabilities* by deciding not to check. Another way of looking at this problem is to argue that *on the average* the proportion of erroneous invoices is

$$(0.70)(0.001) + (0.20)(0.01) + (0.10)(0.02) = 0.0047$$

and since this is less than the break-even value of 0.007, the expected cost is minimized by deciding not to check.

If no further information having a bearing on this problem can be obtained, the above would seem to provide a reasonable solution. Of course, it has the disadvantage that one's "feelings" about the possible values of p, the prior probabilities, are based on the company's records (*historical* data) and hence tell us nothing *directly* about a particular batch with whose disposition we are concerned. One way to proceed in this situation is to take a *sample* of the 1,000 invoices contained in the batch, observe the number of invoices containing errors, and modify the prior probabilities by means of the method indicated in the preceding section. Thus, let us suppose that 25 invoices are randomly selected (at negligible cost) from the batch in question, and that only one of them contains an error. If we let A denote the event of getting "one success in 25 trials" (i.e., one erroneous invoice in the sample of 25), B_1 denote the event that the actual proportion of erroneous invoices in the lot is 0.001, and B_2 and B_3 denote the events that the corresponding proportions are 0.01 and 0.02, we first want to determine the probabilities $P(A \mid B_1)$, $P(A \mid B_2)$, $P(A \mid B_3)$, and then by means of Bayes' rule the probabilities $P(B_1 \mid A)$, $P(B_2 \mid A)$, $P(B_3 \mid A)$. If sampling is *without replacement*, as is usually the case, we should really calculate the first three probabilities by means of the formula for the hypergeometric distribution. Since the sample of 25 invoices constitutes but a small portion of the entire batch, however, we can use the formula of the binomial distribution as an approximation, getting

$$P(A \mid B_1) = \binom{25}{1} (0.001)^1 (0.999)^{24} = 0.0244$$

and, similarly, $P(A \mid B_2) = 0.1965$ and $P(A \mid B_3) = 0.3080$. (To check these figures the reader would have to use logarithms.)

Combining these probabilities with the prior probabilities $P(B_1) = 0.70$, $P(B_2) = 0.20$, and $P(B_3) = 0.10$, and substituting into the formula for Bayes' rule with $k = 3$ (as on page 135), we get

$$P(B_1 \mid A) = \frac{(0.70)(0.0244)}{(0.70)(0.0244) + (0.20)(0.1965) + (0.10)(0.3080)}$$

$$= \frac{0.01708}{0.08718}$$

$$= 0.196$$

$$P(B_2 \mid A) = \frac{(0.20)(0.1965)}{0.08718}$$

$$= 0.451$$

and, hence, $P(B_3 \mid A) = 1 - 0.196 - 0.451 = 0.353$. Finally, using these *posterior probabilities* to estimate the expected cost of the decision *not* to check, we proceed as on page 180, getting

$$\$5(0.196) + \$50(0.451) + \$100(0.353) = \$58.83$$

and since this exceeds $35, the cost of checking the whole batch, the expected cost will be minimized by deciding to check the whole batch.

It is interesting to note that the decision was actually reversed by considering the sample data as well as the prior information, but this should not come as a surprise—whereas the prior probabilities favored $p = 0.001$, the sample proportion $\frac{1}{25} = 0.04$ exceeded even $p = 0.02$ and, hence, shifted the weight of the combined evidence beyond the break-even value of 0.007.

EXERCISES

1. Verify the probabilities $P(B_2 \mid A) = 0.30$ and $P(B_3 \mid A) = 0.57$ obtained on page 178.

2. In planning the operations of a new restaurant, one expert claims that only one out of four waitresses can be expected to stay with the establishment for more than a year, while a second expert claims that it would be correct to say one out of five. In the past, the two experts have been about equally reliable, so that in the absence of direct information we would assign their judgments equal weight. What posterior probabilities would we assign to their claims if it were found that among six waitresses actually hired for the restaurant only one stayed for more than a year?

3. The landscaping plans for a new motel call for a row of palm trees along the driveway. The landscape designer tells the owner that if he plants *Washingtonia filifera*, 20 per cent of the trees will fail to survive the first heavy frost, the manager of the nursery which supplies the trees tells the owner that 10 per cent of the trees will fail to survive the first heavy frost, and the owner's wife tells him that 50 per cent of the trees will fail to survive the first heavy frost.

(a) If the owner feels that in this matter the landscape designer is 10 times as reliable as his wife while the manager of the nursery is 9 times as reliable as his wife, what prior probabilities should he assign to these percentages?

(b) If 10 of these palm trees are planted and 4 fail to survive the first heavy frost, what posterior probabilities should the manager assign to the three percentages?

4. Discussing the sale of a large estate, one broker expresses the feeling that a newspaper ad should produce three serious inquiries about the estate, a second broker feels that it should produce five serious inquiries, and a third broker feels that it should produce eight.

(a) If in the past the second broker has been about twice as reliable as the first and the first has been about three times as reliable as the third, what prior probabilities should we assign to their claims?

(b) How would these probabilities be affected if the ad actually produced six inquiries and it can be assumed that the number of inquiries is a random variable having the Poisson distribution with $\lambda = 3$, $\lambda = 5$, and $\lambda = 8$ according to the three claims?

5. A coin dealer receives a shipment of five ancient gold coins from abroad, and, on the basis of past experience, he feels that the probabilities that 0, 1, 2, 3, 4, or all 5 of them are counterfeits are, respectively, 0.80, 0.05, 0.02, 0.01, 0.02, and 0.10. Since modern methods of counterfeiting have become greatly refined, the cost of authentication has risen sharply, and the dealer decides to select one of the five coins at random and send it away for authentication. If it turns out that the coin is a forgery, what posterior probabilities should he assign to the possibilities that 0, 1, 2, 3, or all 4 of the remaining coins are counterfeits?

6. Light bulbs are usually packaged in pairs, and the manager of a supermarket knows from experience that the probabilities that such a package will contain 0, 1, or 2 defective bulbs are, respectively, 0.96, 0.03, 0.01. If he randomly selects a package of bulbs from his shelves and then randomly selects one of the two bulbs and finds that it is defective, what is the probability that the other bulb is also defective?

7. Suppose that in the example on page 179 none of the 25 invoices in the sample had been erroneous. How would this have affected the posterior probabilities and the ultimate decision?

8. A large electrical supply house is closing out a number of items, among them a lot of four components of a certain kind at a price of $40 for the entire lot, "as is—all sales final." A man can resell all good components at $20 each, but each defective component represents a complete loss of $10. Based on his familiarity both with components of this kind and with the seller, the man assigns prior probabilities of 0.1, 0.5, 0.2, 0.1, and 0.1, respectively, to the events that there are 0, 1, 2, 3, and 4 defective components in the lot.

(a) If no inspection is possible, should the man buy the lot?

(b) If the seller will let the man unpack and inspect one of the four components and he finds that it is defective, should he buy the lot?

9. With reference to Exercise 8, suppose that the man can resell the good components for only $15 while everything else remains unchanged.

(a) If no inspection is possible, should the man buy the lot?

(b) If the seller will let the man unpack and inspect two components and he finds that one is good and the other is defective, should he buy the lot?

(c) With reference to part (b), should he buy the lot if both parts inspected are good?

10. A company has received a lot of 1,000 ball bearings which are used in mechanical devices which the company manufactures. For the sake of simplicity, we assume that the proportion of defective bearings in the lot is either $p = 0.01$, $p = 0.03$, or $p = 0.05$. The lot may be put directly into production without inspection, but each defective bearing entering production costs the company $5 (largely for labor to replace it with a good one at final inspection). Alternately, the company can submit the lot to a rapid man-machine screening procedure at a cost of 7 cents per item. In this procedure, the lot is 100 per cent inspected and all defective items found are replaced with good ones. It is considered that the efficiency of this screening is such that all screened lots entering production will be 99 per cent free of defective items. Based on previous experience the company assigns probabilities of 0.6, 0.3, and 0.1, respectively, to the events that the lot contains 1 per cent, 3 per cent, and 5 per cent defective bearings.

(a) In the absence of any further information, should the company put the lot directly in production or have it screened?

(b) Suppose that it is possible for the company to make a preliminary inspection of the lot prior to deciding how it should be handled. Specifically, suppose that the company inspects at a negligible cost a sample of 10 bearings from the lot and finds one defective bearing in the sample. Should the company put the lot directly in production, or have it screened?

11. In buying a lot of three electronic components, a man must decide whether to insure his purchase or not. The components cost $30 each and he can resell the good ones for $60 each, but defectives represent a complete loss of $30 each to him. If he wishes, he can buy a "one-defective deductible" insurance policy at a cost of $25. Under the policy, he collects nothing if there is no defective or one defective in the lot. However, if there are two defectives in the lot, he is given his money back on one of them and it is replaced free with a good one; if all the components are defective, he is given his money back on two of them and these two are replaced free with good ones. Based on his experience in similar situations in the past, the man assigns probabilities of 0.4, 0.3, 0.2, and 0.1 to the events that the lot contains 0, 1, 2, or 3 defectives, respectively.

(a) Based on the man's probabilities, what is the insurance company's expected profit if he buys the policy?

(b) If no inspection is possible, should the buyer take out the insurance policy?

(c) Suppose that, before deciding whether or not to buy the insurance policy, the buyer is allowed to inspect one component, which he randomly selects and finds to be good. Should he still buy the insurance policy?

THE MEAN OF A PROBABILITY DISTRIBUTION

Let us now apply the concept of a mathematical expectation to the two illustrations on page 165. The first dealt with the number of points rolled with a balanced die, and if we multiply 1, 2, 3, 4, 5, and 6 by their respective probabilities, which all equaled 1/6, we find that the expected number of points is

$$1\left(\frac{1}{6}\right) + 2\left(\frac{1}{6}\right) + 3\left(\frac{1}{6}\right) + 4\left(\frac{1}{6}\right) + 5\left(\frac{1}{6}\right) + 6\left(\frac{1}{6}\right) = 3\frac{1}{2}$$

Of course, we cannot get $3\frac{1}{2}$ in any one roll of a die, but this is the *average* value we should get if we rolled the die a great many times. Considering now the second probability distribution, which dealt with the number of heads obtained in three flips of a balanced coin, we find that the *expected* number of heads is $1\frac{1}{2}$. This value, which must again be interpreted as an average, is the sum of the products obtained by multiplying 0, 1, 2, and 3 by their respective probabilities of 1/8, 3/8, 3/8, and 1/8.

The expected values which we have calculated in these two examples are referred to as the *means* of the respective probability distributions, using this term in very much the same sense as in Chapter 3. In general, if we are given a probability distribution, that is, the probabilities $f(x)$ which are associated with the various values of x (the numerical descriptions of an experiment), *we define the mean of this probability distribution as*

$$\mu = \Sigma\, x \cdot f(x) \qquad\qquad \bigstar$$

with the summation extending over all values of x. This average, which is denoted by the Greek letter μ (*mu*) to distinguish it from the mean of an actual set of data, is, in fact, the value which we can *expect* for the particular numerical description of the experiment.

To study another example, let us consider the probability function whose histogram is shown in Figure 7.2; it gives the probabilities (rounded to four decimals) that among six shoppers interviewed in a department store, 0, 1, 2, 3, 4, 5, or 6 will charge their purchases, when the probability that any one of them will charge his or her purchases is 0.80. Multiplying 0, 1, 2, . . ., and 6 by the corresponding probabilities and adding the products thus obtained, we find that the mean of this distribution is $\mu = 4.7998$ or approximately 4.8.

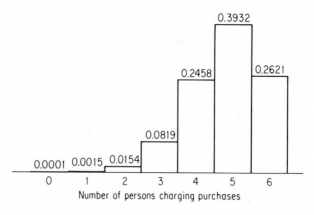

FIGURE 7.2 Probability distribution

When a random variable can take on a very large set of values, the calculation of μ can be quite tedious. For instance, if we wanted to know how many among 400 persons who see a certain television commercial can be expected to purchase the product advertised, when the probability that any one of them will purchase the product is 0.06, we would first have to calculate the 401 probabilities corresponding to 0, 1, 2, . . . , and 400 purchasing the product. However, if we think for a moment, we might argue that in the long run 6 per cent of all persons who see the commercial will purchase the product, 6 per cent of 400 is 24, and hence, we can *expect* 24 of the 400 persons to buy the product. Similarly, it stands to reason that if a balanced coin is flipped 1,000 times, we can expect to get heads 50 per cent of the time—that is, we can *expect* 500 heads in 1,000 flips of a balanced coin. These *expectations* are, indeed, correct, and we were able to find them so easily because there exists the special formula

$$\mu = n \cdot p \qquad\qquad ★$$

for the mean of a *binomial distribution,* that is, the distribution of the number of successes in n trials, when the probability of success is p for each trial and the trials are all independent. In words, the mean of a binomial distribution is simply the product of the number of trials and the probability of success for an individual trial.

Using this special formula, we can now verify the value obtained for the mean of the distribution of Figure 7.2 and the value obtained for the mean of the second distribution on page 165. For the probability function of Figure 7.2 we have $n = 6, p = 0.80, \mu = 6(0.80) = 4.80$, and the difference between this result and the one obtained before is the small rounding error of 0.0002; for the distribution of the number of heads obtained in three flips of a balanced coin we have $n = 3$, $p = \frac{1}{2}$, $\mu = 3 \left(\frac{1}{2}\right) = 1\frac{1}{2}$, and this agrees with the value obtained before.

It is important to remember, of course, that the formula $\mu = n \cdot p$ applies only to binomial distributions. Fortunately, there are other special formulas for other special distributions; for the *hypergeometric distribution*, for example, the formula is

$$\mu = \frac{n \cdot a}{a + b} \qquad \bigstar$$

so that in the example on page 170 the inspector can *expect* to catch

$$\mu = \frac{9 \cdot 6}{6 + 12} = 3$$

of the six "bad factories" among the nine which he inspects. Also, for the Poisson distribution, $\mu = \lambda$, as we indicated on page 172. (Proofs of all these formulas may be found in any textbook on mathematical statistics.)

THE VARIANCE OF A PROBABILITY DISTRIBUTION

In Chapter 3 we saw that there are many problems in which we must describe the *variability* of a distribution (i.e., its spread or dispersion) as well as its mean or some other measure of central location. *This is true also of probability distributions.* The most widely used measures of variability are the *variance* and its square root, the *standard deviation*, which measure the spread of a set of data (looked upon as a population) by averaging the squared deviations from the mean. When dealing with probability distributions, we measure variability in almost the same way—instead of averaging the squared deviations from the mean, we find their *expected value.* Thus, if x is some numerical description of an experiment and the mean of the corresponding probability distribution is μ, then the deviation from the mean is $x - \mu$, and we define the *variance of the probability distribution* as the value which we *expect* for the squared deviation from the mean, namely, as

$$\sigma^2 = \Sigma (x - \mu)^2 \cdot f(x) \qquad \bigstar$$

where the summation extends over all values of x. As in Chapter 3, the square root of the variance, σ, is called the *standard deviation* of the probability distribution.

To illustrate the calculation of the variance of a probability distribution, let us refer again to the second of the two probability distributions on page 165, the one dealing with the number of heads obtained in three flips of a balanced coin. Since the mean of this distribution was shown to be $\mu = 3 \left(\dfrac{1}{2} \right) = 1.50$, we can arrange the calculations as in the following table:

Number of Heads, x	Probability	Deviation from Mean	Square of Deviation from Mean	$(x - \mu)^2 \cdot f(x)$
0	1/8	−1.50	2.25	0.28125
1	3/8	−0.50	0.25	0.09375
2	3/8	0.50	0.25	0.09375
3	1/8	1.50	2.25	0.28125

$$\sigma^2 = 0.75000$$

The values in the last column are obtained by multiplying each squared deviation from the mean by the corresponding probability, and their sum is the variance of the distribution. Thus, the standard deviation of this probability distribution is $\sigma = \sqrt{0.75}$, which equals approximately 0.87. Following these same steps, the reader will be asked to show in Exercise 8 below that the variance of the first probability distribution on page 165 is $\sigma^2 = \dfrac{35}{12}$, and that the one for the probability distribution of Figure 7.1 equals approximately 2.

As in the case of the mean, the calculation of the variance or the standard deviation of a probability distribution can often be simplified when dealing with special kinds of probability distributions. For instance, for the *binomial distribution* we have the formula

$$\sigma^2 = n \cdot p \cdot (1 - p) \qquad \bigstar$$

which we shall not prove, but which can easily be verified for our various examples. For the distribution of the number of heads obtained in three flips of a balanced coin we have $n = 3$, $p = 0.50$,

$$\sigma^2 = 3(0.50)(0.50) = 0.75$$

and this agrees with the value we obtained before. Similarly, for the probability distribution of Figure 7.2 we have $n = 6$, $p = 0.80$, and

$$\sigma^2 = 6(0.80)(0.20) = 0.96$$

namely, the value the reader is asked to verify by the long method in Exercise 4 (page 191).

Intuitively speaking, the variance and the standard deviation of a probability distribution measure its spread or its dispersion: When σ is small the probability is high that we will get a value close to the mean, and when σ is large we are more likely to get a value far away from the mean. This important idea is expressed rigorously in a theorem called Chebyshev's Theorem, which we mentioned briefly in Chapter 3. Formally, this theorem, which is due to the Russian mathematician, P. L. Chebyshev (1821–1895), can be stated as follows:

Given a probability distribution with the mean μ and the standard deviation σ, the probability of obtaining a value within k standard deviations of the mean is at least $1 - 1/k^2$.

Thus, the probability of getting a value within *two* standard deviations of the mean (a value on the interval from $\mu - 2\sigma$ to $\mu + 2\sigma$) is at least 3/4, the probability of getting a value within *five* standard deviations of the mean is at least 24/25, and the probability of getting a value within *ten* standard deviations of the mean is at least 99/100. (The quantity k in the theorem can be any positive number, although the theorem becomes trivial when k is 1 or less.)

If in some actual problem we have $\mu = 25$ and $\sigma = 1$, we can say that the probability of getting a value between 20 and 30 is at least 0.96 (corresponding to $k = 5$); had σ been larger, say, $\sigma = 2.5$, we would have been able to assert only that the probability of getting a value between 20 and 30 is at least 0.75 (corresponding to $k = 2$).

Changing around the argument, we can also say that *the probability of getting a value which differs from the mean by more than k standard deviations is less than $1/k^2$.* We thus have a probability less than 0.25 of getting a value which differs from the mean by more than *two* standard deviations, and a probability less than 0.04 of getting a value which differs from the mean by more than *five* standard deviations.

To give a concrete application, suppose we obtained 146 heads and 254 tails in 400 flips of a coin and we are wondering whether this is sufficient evidence to raise the question of whether the coin is really balanced. If the coin is balanced, we are dealing with a binomial distribution having $n =$

400 and $p = \frac{1}{2}$, and substitution into the special formulas on pages 186 and 188 yields

$$\mu = n \cdot p = 400 \left(\frac{1}{2}\right) = 200$$

and

$$\sigma = \sqrt{np(1 - p)} = \sqrt{400 \left(\frac{1}{2}\right) \left(\frac{1}{2}\right)} = 10$$

Using $k = 5$, Chebyshev's Theorem asserts that the probability of getting anywhere from 150 to 250 heads is at least 0.96 *or* that the probability of being off by more than 50 heads from the expected 200 is less than 0.04. Since this probability is quite small, it would seem reasonable to go ahead and question the "honesty" of the coin.

It is of interest to note that Chebyshev's Theorem applies also to frequency distributions, that is, distributions of actual data. In that case we can assert that *the proportion of the total number of cases (values or items) which fall within k standard deviations of the mean is at least $1 - 1/k^2$*. Thus, at least 75 per cent of all the items must fall within *two* standard deviations of the mean, and at least 96 per cent must fall within *five* standard deviations of the mean.

EXERCISES

1. The following table gives the probabilities that a woman who enters a shoe store will buy 0, 1, 2, 3, or 4 pairs of shoes:

Pairs of shoes, x:	0	1	2	3	4
Probability, f(x):	0.41	0.37	0.16	0.05	0.01

Calculate the mean and the variance of this distribution.

2. The following table gives the probabilities that a real estate salesman will sell a piece of property to the first, second, third, . . . , or seventh customer to whom it is offered:

Number of customer, x:	1	2	3	4	5	6	7
Probability, f(x):	0.41	0.25	0.16	0.09	0.05	0.03	0.01

Calculate the mean and the standard deviation of this distribution.

3. As can easily be verified by means of the formula for the binomial distribution (or by listing all 32 possibilities), the probabilities of getting 0, 1, 2, 3, 4, or 5 heads in five flips of a balanced coin are, respectively, 1/32, 5/32, 10/32, 10/32, 5/32, and 1/32. Find the mean and the standard deviation of this binomial distribution using

   (a) the given probabilities and the definition formulas for μ and σ;

   (b) the special formulas for the mean and the standard deviation of a binomial distribution.

4. Find the mean and the variance of the distribution pictured in Figure 7.2 using (a) the probabilities shown in the diagram and the definition formulas for μ and σ^2 and (b) the special formulas for the mean and the variance of a binomial distribution (with $n = 6$ and $p = 0.80$).

5. Find the mean and the standard deviation of the (binomial) distribution of

   (a) the number of heads obtained in 676 tosses of a balanced coin;

   (b) the number of 1's obtained in 606 rolls of a balanced die;

   (c) the number of defectives in a sample of 600 parts made with a certain machine, when the probability that any one of the parts is defective is 0.04;

   (d) the number of cars with faulty brakes among 900 cars stopped at a road block, when the probability that any one of the cars has faulty brakes is 0.10;

   (e) the number of persons (among 800 invited) who will attend a political fund-raising dinner, when the probability that any one of them will attend is 0.15.

6. On page 170 we calculated the probabilities that an inspector will catch 4, 5, or 6 factories having smokestacks which emit excessive amounts of pollutants, if he checks on 9 of 18 factories, among which 6 have smokestacks that emit excessive amounts of pollutants.

   (a) Show that the corresponding probabilities for his catching 0, 1, 2, or 3 are 0.005, 0.061, 0.244, and 0.380.

   (b) Find the mean and the standard deviation of this hypergeometric distribution, and verify the result obtained for the mean by using the special formula on page 187.

7. On page 169 we showed that if a carton of 12 clocks contains 3 which are defective, the probability that a sample of 2 will contain *zero* defectives is 6/11.

   (a) Find the corresponding probabilities that the sample will contain one or two of the defective clocks.

   (b) Use the given probability and the ones obtained in part (a) to determine the mean of this hypergeometric distribution, and verify the result by using the special formula on page 187.

8. Show that

   (a) the variance of the first probability distribution on page 165 is 35/12;

(b) the variance of the Poisson distribution given on page 171 (and in Figure 7.1) is approximately 2. Note that this verifies the fact that the variance of the Poisson distribution is given by $\sigma^2 = \lambda$, where in this special case $\lambda = n \cdot p = 10,000 \cdot \dfrac{1}{5,000} = 2$.

9. On page 172 we showed that if a finance company receives on the average three bad checks per day, the probability that it will receive exactly one bad check on any given day is 0.150.

(a) Given that the corresponding probabilities for 0, 2, 3, . . . , and 10 bad checks are, respectively, 0.050, 0.224, 0.224, 0.168, 0.101, 0.050, 0.022, 0.008, 0.003, and 0.001, calculate the mean of this distribution and thus verify the formula $\mu = \lambda$.

(b) Use the given probabilities to calculate the variance of this Poisson distribution and thus verify the formula $\sigma^2 = \lambda$.

10. A student answers the 100 questions of a true-false test by flipping a balanced coin (*heads* is "true" and *tails* is "false").

(a) What does Chebyshev's Theorem tell us about the probability that he will get anywhere from 35 to 65 correct answers?

(b) What does Chebyshev's Theorem tell us about the probability that he will get fewer than 25 or more than 75 correct answers?

11. With reference to part (c) of Exercise 5, with what probability can we assert (according to Chebyshev's Theorem) that the sample will contain anywhere from 12 to 36 defectives?

12. With reference to part (d) of Exercise 5, with what probability can we assert (according to Chebyshev's Theorem) that anywhere from 72 to 108 of the 900 cars will have faulty brakes?

13. Show that we can assert each of the following with a probability of at least $\dfrac{35}{36} = 0.972$:

(a) In 400 flips of a balanced coin there will be anywhere from 140 to 260 heads, and hence, the *proportion* of heads will differ from 0.50 by 0.15 or less.

(b) In 10,000 flips of a balanced coin there will be anywhere from 4,700 to 5,300 heads, and hence, the *proportion* of heads will differ from 0.50 by 0.03 or less.

(c) In 1 million flips of a balanced coin there will be anywhere from 497,000 to 503,000 heads, and hence, the *proportion* of heads will differ from 0.50 by 0.003 or less.

This exercise illustrates the so-called *law of large numbers* according to which, as the number of trials becomes larger and larger, the proportion of successes approaches the probability of a success on an individual trial. Note that this law applies to the proportion of successes and *not* to the number of successes.

THE NORMAL DISTRIBUTION

When we first discussed histograms in Chapter 2, we pointed out that the frequencies, percentages, or proportions (and we may now add probabilities) that are associated with the various classes are represented by the *areas* of the rectangles. For example, the areas of the rectangles of Figure 7.3 represent the probabilities of getting 0, 1, 2, . . . , and 10 heads in 10

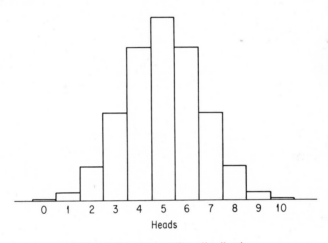

FIGURE 7.3 Probability distribution

tosses of a balanced coin or, better, they are equal or proportional to these probabilities. If we now look carefully at Figure 7.4, which is an enlargement of a portion of Figure 7.3, it is apparent that the area of rectangle *ABCD* is nearly equal to the shaded area under the continuous curve which we have drawn to approximate the histogram. Since the area of rectangle *ABCD* is equal to (or proportional to) the probability of getting three heads in 10 tosses of a balanced coin, we can say that this probability is also given by the shaded area under the continuous curve. More generally, *if a histogram is approximated by means of a smooth curve, the frequency, percentage, or probability associated with any given class (or interval) is represented by the corresponding area under the curve.*

If we approximate the distribution of 1967 family incomes in the United States with a smooth curve, as we did in Figure 7.5, we can determine what proportion of the incomes falls into any given interval by looking at the corresponding area under the curve. By comparing the shaded area of

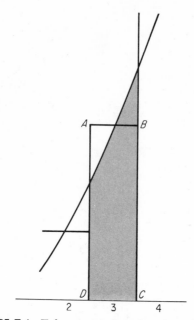

FIGURE 7.4 Enlargement of portion of Figure 7.3

Figure 7.5 with the total area under the curve (representing 100 per cent), we can judge that roughly 12 per cent of the families had incomes of $15,000 or more. It can similarly be seen from Figure 7.5 that about 41 per cent of the families had incomes under $7,000. We obtained these percentages by (mentally) dividing the corresponding areas under the curve by the total area under the curve.

Had we drawn Figure 7.5 so that the total area under the curve actually equaled 1, the proportion of the families belonging to any income group

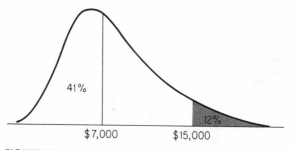

FIGURE 7.5 Curve approximating income distribution

would have been given directly by the corresponding area under the curve. Indeed, we shall refer to a curve as a *distribution curve* or as a *continuous distribution* if the area under the curve between any two values *a* and *b* (see Figure 7.6) *equals* the proportion of the cases falling between *a* and *b*.

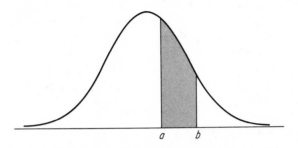

FIGURE 7.6   Continuous distribution

It follows that for a continuous distribution the total area under the curve must always be equal to 1. Repeating this definition in terms of probabilities, *we refer to a curve as a continuous distribution curve if the area under the curve between any two values a and b equals the probability of getting a value between a and b.*

For instance, if the curve of Figure 7.7 approximates the probability distribution of the number of housewives who prefer Package A to Package

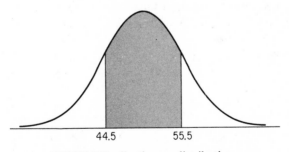

FIGURE 7.7   Continuous distribution

B in a random sample of 100 housewives (drawn from a very large population in which 50 per cent of the housewives prefer Package A and 50 per cent prefer Package B), the probability of getting anywhere from 45 to 55 housewives preferring Package A is given by the shaded area of Figure 7.7. Note that we shaded the area between 44.5 and 55.5 and *not* the area between 45 and 55. The number of housewives favoring Package A can only be a whole number, and in order to approximate the probability distribution with a continuous curve we must make the *continuity correction* of letting each integer k be represented by the interval from $k - 1/2$ to $k + 1/2$. Thus, 45 is represented by the interval from 44.5 to 45.5, ..., 55 is represented by the interval from 54.5 to 55.5, and "45 to 55 housewives" is represented by the interval from 44.5 to 55.5.

Since continuous distribution curves can always be looked upon as close approximations to histograms, we can define the mean and the standard deviation of continuous distributions (informally) in the following way: If a continuous distribution is approximated with a sequence of histograms having narrower and narrower classes, the means of the distributions represented by these histograms will approach a value which defines the mean of the continuous distribution. Similarly, the standard deviations of these distributions will approach a value which defines the standard deviation of the continuous distribution. Intuitively speaking, the mean and the standard deviation of a continuous distribution measure the identical features as the mean and the standard deviation of an ordinary frequency distribution (or a probability distribution), namely, its center and its spread. More rigorous definitions of these concepts cannot be given without the use of calculus.

Among the many continuous distributions used in statistics, the *normal curve*, or the *normal distribution*, is by far the most important. Its study dates back to eighteenth century investigations into the nature of experimental errors. It was observed that discrepancies between repeated measurements of the same physical quantity displayed a surprising degree of regularity; their patterns (distribution), it was found, could be closely approximated by a certain kind of continuous distribution curve, referred to as the "normal curve of errors" and attributed to the laws of chance. The mathematical properties of this kind of continuous distribution curve and its theoretical basis were first investigated by Pierre Laplace (1749–1827), Abraham de Moivre (1667–1745), and Carl Gauss (1777–1855).

The normal curve is a bell-shaped curve that extends indefinitely in both directions. Although this may not be apparent from a small drawing like the one of Figure 7.8, the curve comes closer and closer to the horizontal axis without ever reaching it, no matter how far we might go in either direction away from the mean. Fortunately, it is seldom necessary to extend the tails of the normal curve very far because the area under that part of the curve lying more than four or five standard deviations away from the mean is for most practical purposes negligible.

An important property of a normal curve is that it is completely determined by its mean and its standard deviation. In other words, the mathematical equation for the normal curve is such that we can determine the

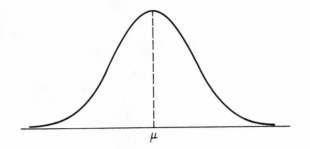

FIGURE 7.8   Normal distribution

area under the curve between any two points on the horizontal scale if we are given its mean and its standard deviation. In practice, we obtain areas under a normal curve by means of special tables, such as Table I on page 473. To use this table, we shall first have to explain what is meant by the normal curve in its *standard form*. Since the equation of the normal curve depends on μ and σ, we get different curves and, hence, different areas, for different values of μ and σ. For instance, Figure 7.9 shows the superimposed

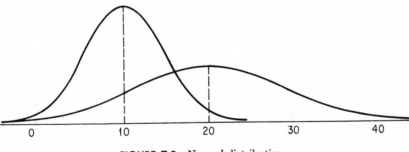

FIGURE 7.9   Normal distribution

graphs of two normal curves, one having $\mu = 10$ and $\sigma = 5$, and the other having $\mu = 20$ and $\sigma = 10$. As can be seen from this diagram, the area under the curve, say, between 10 and 12, is *not* the same for the two distributions.

As it would be impossible to construct separate tables of normal curve areas for each conceivable pair of values of μ and σ, we tabulate these areas only for the so-called *standard normal distribution* which has $\mu = 0$ and $\sigma = 1$. Then, we obtain areas under *any* normal distribution by performing the change of scale shown in Figure 7.10. All that is really necessary is to

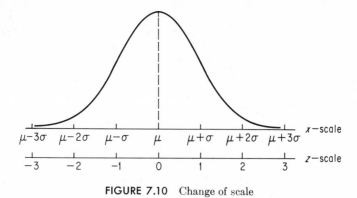

FIGURE 7.10 Change of scale

convert the units of measurement into *standard units* (see page 46) by means of the formula

$$z = \frac{x - \mu}{\sigma}$$

To find areas under normal curves whose mean and standard deviation are *not* 0 and 1, we have only to convert the x's (the values to the left of which, to the right of which, or between which we want to determine areas under the curve) into z's and then use Table I on page 473. *The entries in this table are the areas under the standard normal distribution between the mean ($z = 0$) and $z = 0.00, 0.01, 0.02, \ldots, 3.08, and 3.09$.* In other words, the entries in Table I are areas under the standard normal distribution curve like the one shaded in Figure 7.11. Note that Table I has no entries corresponding to negative values of z, but these are not needed by virtue of the *symmetry* of a normal curve about its mean. We can thus find the area under the standard normal distribution, say, between $z = -1.25$ and $z = 0$, by looking up the area between $z = 0$ and $z = 1.25$. As can be checked in Table I, this area is 0.3944.

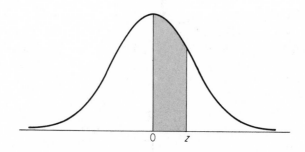

FIGURE 7.11 Normal distribution

Questions concerning areas under normal distributions arise in various ways, and the ability to find any desired area quickly can be a big help. For instance, although the table gives only the areas between the mean and selected values of z, we often have to find areas to the right of a given z, to the right or left of $-z$, between $-z$ and z, and so forth. Finding any one of these areas is easy, provided we remember exactly what part of the curve is referred to by a tabular value and the following property of the normal curve: Since the curve is symmetrical about its mean, the area to the right of the mean as well as the area to the left of the mean is 0.5000. With this knowledge we find, for example, that the probability of getting a z less than 1.64 (the area to the left of $z = 1.64$) is $0.5000 + 0.4495 = 0.9495$, and that the probability of getting a z greater than -0.47 (the area to the right of $z = -0.47$) is $0.5000 + 0.1808 = 0.6808$ (see also Figure

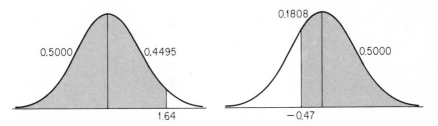

FIGURE 7.12 Normal distributions

7.12). Similarly, we find that the probability of getting a z greater than 0.76 is $0.5000 - 0.2764 = 0.2236$, and that the probability of getting a z less than -1.35 is $0.5000 - 0.4115 = 0.0885$ (see Figure 7.13). The prob-

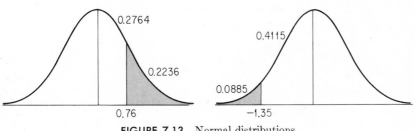

FIGURE 7.13 Normal distributions

ability of getting a z between 0.95 and 1.36 is $0.4131 - 0.3289 = 0.0842$, and the probability of getting a z between -0.45 and 0.65 is $0.1736 + 0.2422 = 0.4158$ (see Figure 7.14).

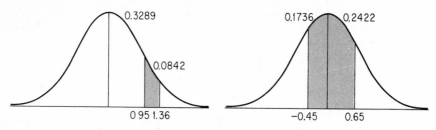

FIGURE 7.14 Normal distributions

There are also problems in which we are given areas under the normal curve and asked to find corresponding values of z. For instance, if we want to find a z which is such that the area to its right is 0.1000, it is apparent from Figure 7.15 that this z will have to correspond to an entry of 0.4000

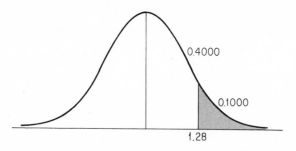

FIGURE 7.15 Normal distribution

in Table I. Referring to this table, we find that the closest value is $z = 1.28$.

We are now in the position to verify the remark made on page 46, namely, that for reasonably symmetrical bell-shaped distributions we can expect about 68 per cent of the data to fall within one standard deviation of the mean, about 95 per cent of the data to fall within two standard deviations of the mean, and over 99 per cent of the data to fall within three standard deviations of the mean. These figures apply to normal distributions, and it will be left to the reader to verify that 0.6826 of the area under the standard normal curve falls between $z = -1$ and $z = 1$, that 0.9544 of the area falls between $z = -2$ and $z = 2$, and that 0.9974 of the area falls between $z = -3$ and $z = 3$ (see Exercise 3 on page 203). This last result also provides a basis for our earlier remark that (although the tails extend indefinitely) the area under a normal curve lying more than four or five standard deviations from the mean is negligible.

To give an example in which we must first convert to standard units, let us suppose that a random variable has a normal distribution with $\mu = 24$

and $\sigma = 12$, and that we want to find the probability that it will assume a value between 17.4 and 58.8 (see Figure 7.16). Converting to standard units

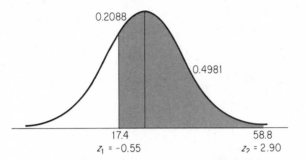

0.2088

0.4981

17.4
$z_1 = -0.55$

58.8
$z_2 = 2.90$

FIGURE 7.16 Normal distribution

we obtain

$$z_1 = \frac{17.4 - 24}{12} = -0.55 \quad \text{and} \quad z_2 = \frac{58.8 - 24}{12} = 2.90$$

and since the areas corresponding to these z's are 0.2088 and 0.4981, respectively, the probability is $0.2088 + 0.4981 = 0.7069$.

There are various ways in which we can test whether an observed distribution fits the overall pattern of a normal distribution. The test which we shall discuss here is not the best—it is largely subjective, but it has the decided advantage that it is extremely easy to perform. To illustrate this technique, let us refer to the seat-occupancy-time distribution, which can be presented in the following form as a "less than" percentage distribution:

Seat Occupancy Time (minutes)	Percentage
Less than 9.5	0
Less than 19.5	3
Less than 29.5	17
Less than 39.5	46
Less than 49.5	68
Less than 59.5	82
Less than 69.5	92
Less than 79.5	96
Less than 89.5	98
Less than 99.5	100

Note that in this table we are using the class boundaries instead of the class limits and that the percentages equal the cumulative frequencies of the table on page 13 because the total frequency happened to be 100. (Otherwise, we have to convert the cumulative frequencies into percentages by dividing each cumulative frequency by the total frequency and then multiplying by 100.)

Before we actually plot this cumulative distribution on the special paper illustrated in Figure 7.17, let us first investigate the scales of this kind of graph paper. As can be seen from Figure 7.17, the cumulative percentage scale is already marked off in the rather unusual pattern which makes the

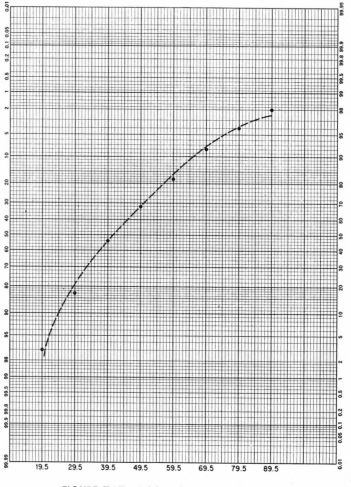

FIGURE 7.17 Arithmetic probability paper

paper suitable for our special purpose. The other scale consists of equal subdivisions that are not labeled; in our example they are used to indicate the class boundaries 19.5, 29.5, 39.5, . . . , and 89.5. This kind of graph paper (which is commercially available) is called *arithmetic probability paper*.

If we plot on this kind of paper the cumulative "less than" percentages which correspond to the class boundaries of a distribution and the points thus obtained lie very close to a straight line, we can consider this as evidence that the distribution follows the general pattern of a normal distribution. As can be seen from Figure 7.17, this is *not* the case in our example.

EXERCISES

1. Find the normal curve area which lies
   (a) to the right of $z = 1.76$;   (e) between $z = 1.18$ and $z = 1.39$;
   (b) to the left of $z = 1.05$;   (f) between $z = -1.84$ and $z = -0.44$;
   (c) to the right of $z = -0.13$;   (g) between $z = -2.33$ and $z = 0.97$;
   (d) to the left of $z = -1.14$;   (h) between $z = -0.98$ and $z = -0.63$.

2. Find the normal curve area which lies
   (a) between $z = 0$ and $z = 1.65$;   (e) to the right of $z = 0.44$;
   (b) between $z = -1.66$ and $z = 0$; (f) to the right of $z = -1.77$;
   (c) to the left of $z = 1.46$;   (g) between $z = 1$ and $z = 2$;
   (d) to the left of $z = -0.59$;   (h) between $z = -0.24$ and $z = 0.24$.

3. Find the normal curve area between $-z$ and z if
   (a) $z = 1$;   (c) $z = 3$;   (e) $z = 2.33$;
   (b) $z = 2$;   (d) $z = 1.96$;   (f) $z = 2.58$.

4. Find z if
   (a) the normal curve area between 0 and z is 0.4505;
   (b) the normal curve area to the right of z is 0.0392;
   (c) the normal curve area to the right of z is 0.9292;
   (d) the normal curve area to the left of z is 0.6480;
   (e) the normal curve area to the left of z is 0.0307;
   (f) the normal curve area between $-z$ and z is 0.5934.

5. In later chapters we shall let z_α denote the value of z for which the area under the normal curve *to its right* is equal to $\alpha(alpha)$. Find (a) $z_{.10}$, (b) $z_{.05}$, (c) $z_{.025}$, (d) $z_{.02}$, (e) $z_{.01}$, and (f) $z_{.005}$.

6. A random variable has a normal distribution with the mean $\mu = 71.8$ and the standard deviation $\sigma = 5.6$. What is the probability that this random variable will take on a value (a) less than 78.8, (b) greater than 60.6, (c) between 74.6 and 80.2, and (d) between 63.4 and 80.2?

7. A random variable has a normal distribution with the mean $\mu = 57.4$ and the standard deviation $\sigma = 8.4$. What is the probability that this random variable will take on a value (a) less than 70.0, (b) less than 51.1, (c) between 59.5 and 76.3, and (d) between 44.8 and 74.2?

8. A normal distribution has the mean $\mu = 78.0$. Find its standard deviation if 20 per cent of the area under the curve lies to the right of 86.4.

9. A random variable has a normal distribution with the standard deviation $\sigma = 21.5$. Find its mean if the probability that the random variable will take on a value less than 120.5 is 0.8849.

10. Convert the distribution of the weekly earnings of the 1,216 workers (given on page 8) into a cumulative "less than" percentage distribution, and use arithmetic probability paper to judge whether the shape of this distribution is roughly that of a normal curve.

11. Plot the cumulative "less than" percentage distribution of whichever data you grouped among those of Exercises 9, 11, or 13 on pages 16 and 17 on arithmetic probability paper, and judge whether the distribution has roughly the shape of a normal curve.

12. Plot the cumulative "less than" percentage distribution of whichever data you grouped among those of Exercises 12 and 14 on pages 16 and 17 on arithmetic probability paper, and judge whether the distribution has roughly the shape of a normal distribution.

APPLICATIONS OF THE NORMAL DISTRIBUTION

Let us now consider some applied problems in which we shall assume that the distribution of the data with which we are dealing can be approximated closely with a normal curve. Suppose, for instance, that the number of telephone calls made daily in a certain community between 3 P.M. and 5 P.M. is a random variable having a normal distribution with a mean of 584 and a standard deviation of 26. What we would like to know is the probability that there will be more than 600 calls between 3 P.M. and 5 P.M. on a given day. Note, first of all, that the number of telephone calls made in the community during the given period of time is a *discrete* random variable and that we shall, therefore, have to make the *continuity correction* referred to on page 196. Thus, 600 will have to be represented by the interval from 599.5 to 600.5, and the probability of more than 600 calls is represented by the area to the right of 600.5, the shaded area of Figure 7.18. Changing 600.5 into standard units, we obtain

$$z = \frac{600.5 - 584}{26} = 0.63$$

and the corresponding entry in Table I is 0.2357. Hence, the shaded area of Figure 7.18 is $0.5000 - 0.2357 = 0.2643$, and we find that the probability of more than 600 calls is approximately 0.26; the odds against it are slightly less than 3 to 1.

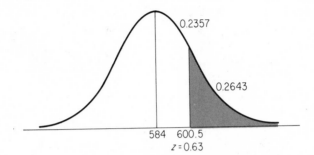

FIGURE 7.18 Normal distribution

To consider a somewhat different problem, suppose that a brass polish manufacturer wants to set his filling equipment so that in the long run only 3 cans in 1,000 will contain less than a desired minimum net fill of 31.4 ounces. It is known from experience that the filled weights are approximately normally distributed with a standard deviation of 0.2 ounces. At what level will the mean fill have to be set in order to meet this requirement?

In this problem we are given x, σ, and a percentage (a normal curve area), and we are asked to find μ. As can be seen from Figure 7.19, we will first

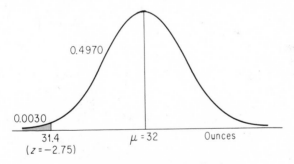

FIGURE 7.19 Normal distribution

have to find the z which is such that the normal curve area to its left equals 0.0030. Getting $z = -2.75$, we have

$$-2.75 = \frac{31.4 - \mu}{0.2}$$

and upon solving for μ we finally obtain $\mu = 32.0$. Thus, if the equipment is set to fill the cans on the average with 32.0 ounces (and if the distribution of the fills is approximately normal with a standard deviation of 0.2 ounces), in the long run only 3 cans in 1,000 will be below the desired minimum.

The normal distribution is sometimes introduced as a continuous distribution which provides a very close approximation to the binomial distribution when n, the number of trials, is very large and p, the probability of a success on an individual trial, is close to 0.50. Figure 7.20 contains the

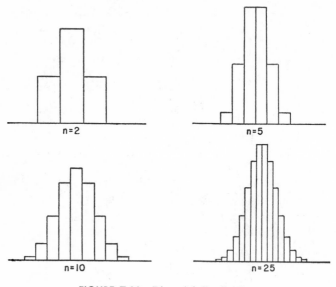

FIGURE 7.20   Binomial distributions

histograms of binomial distributions having $p = 0.50$ and $n = 2$, 5, 10, and 25, and it can be seen that with increasing n these distributions approach the symmetrical bell-shaped pattern of the normal distribution. In fact, a normal curve with the mean $\mu = np$ and the standard deviation $\sigma = \sqrt{np(1 - p)}$ can often be used to approximate a binomial distribution even when n is fairly small and p differs from 0.50, but is not too close to either 0 or 1. A good rule of thumb is to use this approximation only when np as well as $n(1 - p)$ are both greater than 5.

To illustrate this normal curve approximation to the binomial distribution, let us first consider the probability of getting 4 heads in 12 tosses of a balanced coin. Substituting $n = 12$, $x = 4$, $p = \dfrac{1}{2}$, and $\dbinom{12}{4} = 495$ into the formula on page 167, we get

$$495 \left(\frac{1}{2}\right)^4 \left(1 - \frac{1}{2}\right)^8 = \frac{495}{4,096}$$

or approximately 0.12. To find the normal curve approximation to this probability, we shall again have to use the *continuity correction* and represent 4 (heads) by the interval from 3.5 to 4.5 (see Figure 7.21). Since $\mu = 12\left(\frac{1}{2}\right) = 6$ and $\sigma = \sqrt{12 \left(\frac{1}{2}\right)\left(\frac{1}{2}\right)} = 1.732$, it follows that the values

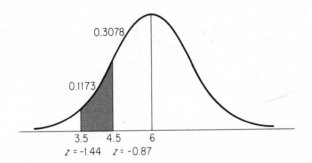

FIGURE 7.21   Normal curve approximation of binomial distribution

between which we want to determine the area under the normal curve are, in standard units,

$$\frac{3.5 - 6}{1.732} = -1.44 \quad \text{and} \quad \frac{4.5 - 6}{1.732} = -0.87$$

The corresponding entries in Table I are 0.4251 and 0.3078, and the required probability is $0.4251 - 0.3078 = 0.1173$. Note that the difference between this value and the one obtained with the formula for the binomial distribution is negligible.

The normal curve approximation to the binomial distribution is particularly useful in problems where we would otherwise have to use the formula for the binomial distribution repeatedly to obtain the values of many different terms. Suppose, for example, we want to know the probability of getting at least 12 replies to questionnaires mailed to 100 persons, when the probability that any one of them will reply is 0.18. In other words, we want to know the probability of getting at least 12 successes in 100 trials when the probability of a success on an individual trial is 0.18. If we tried to solve this problem by using the formula for the binomial distribution, we would have to find the sum of the probabilities corresponding to 12, 13, 14, . . . , and 100 successes (or those corresponding to $0, 1, 2, \ldots$, and 11). Evidently, this would involve a tremendous amount of work. On the other hand, using the normal curve approximation, we have only to find the shaded area of Figure 7.22, namely, the area to the right of 11.5. Note that

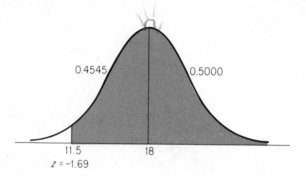

FIGURE 7.22 Normal curve approximation of binomial distribution

we are again using the continuity correction according to which 12 is represented by the interval from 11.5 to 12.5, 13 is represented by the interval from 12.5 to 13.5, and so on.

Since $\mu = 100(0.18) = 18$ and $\sigma = \sqrt{100(0.18)(0.82)} = 3.84$, we find that in standard units 11.5 becomes

$$\frac{11.5 - 18}{3.84} = -1.69$$

and that the desired probability is $0.4545 + 0.5000 = 0.9545$. This means that (in many mailings) we can expect to get at least 12 replies to 100 of these questionnaires *about 95 per cent of the time*, provided that 0.18 is the correct figure for the probability that any one person will reply. It is interesting to note that, rounded to two decimals, the *actual* value of this probability (obtained from an appropriate table) is 0.96.

EXERCISES

1. In a large suburban area, the monthly food expenditures of families with annual incomes between \$12,000 and \$15,000 are approximately normally distributed with a mean of \$176.55 and a standard deviation of \$27.30.
   (a) What proportion of these families have monthly food expenditures less than \$125?

(b) What proportion of these families have monthly food expenditures greater than $180?

(c) Above what value does one find the highest 10 per cent of these monthly food expenditures?

2. If the speeds recorded at a certain checkpoint have a mean of 47.5 miles per hour, a standard deviation of 4.6 miles per hour, and are roughly normally distributed, find

(a) what proportion of the speeds exceed 55 miles per hour;

(b) what proportion of the speeds are below 40 miles per hour;

(c) what proportion of the speeds are between 45 and 50 miles per hour.

(Since the speeds are presumably measured on a continuous scale, the continuity correction is not required.)

3. A soft-drink manufacturer wants to set his filling equipment so that in the long run only 3 bottles in 1,000 will contain more than a desired maximum fill of 32.6 ounces. Assuming that the fills are approximately normally distributed with a standard deviation of 0.2 ounces, find the mean fill needed to meet this requirement.

4. In a large class in business statistics, the final examination grades have a mean of 72.8 and a standard deviation of 15.0. Assuming that it is reasonable to treat the distribution of these grades (which, incidentally, are all *whole numbers*) as if it were a normal distribution, find

(a) what percentage of the grades should exceed 87;

(b) what percentage of the grades should be less than 32;

(c) the lowest A grade, if the highest 12 per cent of the grades are to be regarded as A's;

(d) the highest grade a student can get and yet fail the test, if the lowest 25 per cent of the grades are to be regarded as failing grades.

5. A manufacturer guarantees chromium strips in this way: "Widths approximately normally distributed; on the average 3 strips in 1,000 wider than 2 inches; product relative variation 10 per cent." Assuming the correctness of the claim and that relative variation is measured by the *coefficient of variation*, show that the mean of the widths is 1.57 inches.

6. A manufacturer must buy coil springs which will stand at least a 15 pound load. Supplier A guarantees that his springs stand an average load of 20.5 pounds with a standard deviation of 2.1 pounds, Supplier B guarantees that his springs stand an average load of 19.9 pounds with a standard deviation of 1.5 pounds, and Supplier C guarantees that his springs stand an average load of 18.2 pounds with a standard deviation of 1.1 pounds. All distributions are assumed to be approximately normal. On the basis of the guarantees, which springs seem preferable to the manufacturer, and why?

7. For the 2,400 employees of a large manufacturer, I.Q. is approximately normally distributed with the mean 112 and the standard deviation 12. It is known from experience that only persons with an I.Q. of at least 105 are intelligent enough

for a particular job and that persons with an I.Q. above 125 soon become bored and unhappy with it. On the basis of I.Q. alone, how many of the 2,400 employees would be suitable for the job? (Use the continuity correction.)

8. The average time required to perform Job A is 75 minutes with a standard deviation of 15 minutes, and the average time required to perform Job B is 100 minutes with a standard deviation of 10 minutes. Assuming normal distributions, what proportion of the time will Job A take longer than the average Job B, and what proportion of the time will Job B take less time than the average Job A?

9. Find the probability of getting 4 heads in 14 tosses of a balanced coin using
   (a) the formula for the binomial distribution;
   (b) the normal curve approximation.

10. Use the normal curve approximation to find the probability of getting fewer than 42 or more than 58 heads in 100 flips of a balanced coin.

11. A television network claims that its Monday night movie regularly has 36 per cent of the total viewing audience. If this claim is correct, what is the probability that, among 400 viewers reached by phone on Monday nights, more than 125 were watching the network's movie?

12. A manufacturer knows that on the average 2 per cent of the washing machines which he makes will require repairs within 60 days after they are sold. What is the probability that among 800 washing machines shipped by the manufacturer at least 20 will require repairs within 60 days after they are sold?

13. What is the probability of getting fewer than 80 responses to 1,000 invitations sent out to promote a real-estate development, if the probability that any one will respond to this solicitation is 0.10?

14. If 40 per cent of the customers of a service station use their credit cards, what is the probability that among 400 customers more than 250 pay cash?

A WORD OF CAUTION

When we use a specific probability function to describe a given situation, we must be sure that we are using the *right model*. That is, we must make sure that we do not use the hypergeometric distribution when we should be using the binomial distribution, that we are not using the Poisson distribution when we should be using some other distribution, and so on. More specifically, we must always make sure that the assumptions underlying the distribution which we choose are actually met. Thus, it would be a mistake to use the binomial distribution to determine, say, the probability that there will be five rainy days at a given resort during the first two weeks in August, or the probability that 8 of 100 persons (whose age ranges from 18 to 79) will be hospitalized at least once during the coming year. In the first case the successive trials are clearly *not independent*, and in the second case the probability of a success (being hospitalized) is *not the same* for each trial, that is, for each person.

Also, the fact that the normal distribution is the only continuous distribution which we have discussed may have given the erroneous impression that it is the only one that matters in the study of statistics. Although it is true that the normal distribution plays a very important role in many statistical problems, its indiscriminate use can lead to very misleading results. In later chapters we shall meet several other continuous distributions, among them the *t distribution*, the *chi-square distribution*, and the *F distribution*, which play important roles in problems of statistical inference.

TECHNICAL NOTE 4 (The Binomial Distribution)

To derive the formula for the binomial distribution, let us first observe that the probability of getting x successes in n trials and, hence, x successes and $n - x$ failures, *in some specific order* is $p^x(1 - p)^{n-x}$. There is one factor p for each success, one factor $1 - p$ for each failure, and the x factors p and $n - x$ factors $1 - p$ are all multiplied together by virtue of the assumption that the n trials are independent. Since this probability is evidently the same for each individual outcome representing x successes and $n - x$ failures (it does not depend on the order in which the successes and failures are obtained), the desired probability for x successes in n trials *in any order* is obtained by multiplying $p^x(1 - p)^{n-x}$ by the number of ways in which the x successes can be distributed among the n trials. In other words, $p^x(1 - p)^{n-x}$ is multiplied by the number of points (individual outcomes) of the sample space, each of which corresponds to the event of getting x successes and $n - x$ failures in some order. Hence, $p^x(1 - p)^{n-x}$ is multiplied by $\binom{n}{x}$, the number of ways in which x objects can be selected from a set of n objects, and this completes the proof.

8

SAMPLING AND SAMPLING DISTRIBUTIONS

RANDOM SAMPLING

Earlier in this book we distinguished between populations and samples, stating that a population consists of all conceivably or hypothetically possible instances (or observations) of a given phenomenon, while a sample is simply a part of a population. In preparation for the work which follows, let us now distinguish between two kinds of populations: "finite populations" and "infinite populations." A *finite population* is one which consists of a finite number, or fixed number, of elements (items, objects, measurements, or observations). Examples of finite populations are the net weight of the 24,000 cans of rust remover in a production lot, the scores made by the 650-man entering class of a technical school on the Engineer scale of the Strong Vocational Interest Blank, and the outside diameters of a lot of 10,000 precision ball bearings.

In contrast to finite populations, a population is said to be *infinite* if there is, at least hypothetically, no limit to the number of elements it can contain. The population which consists of the results obtained in all hypothetically possible rolls of a pair of dice is an infinite population, and so is the population which consists of all conceivably possible measurements which could be made of the length of a metal strip.

The purpose of most statistical investigations is to make valid generalizations on the basis of samples about both finite and infinite populations from which the samples came. The whole problem of when and under what conditions samples permit such generalizations has no easy solution. For instance, if we want to estimate the average amount of money people spend on their vacations, we would hardly take as our sample the amounts spent by the deluxe-class passengers on a 92-day ocean cruise, nor would we attempt to estimate wholesale prices of *all* farm products on the basis of the prices of apricots alone. In both of these cases we would reject estimates based on the suggested samples as ridiculous; but just which vacationers and which farm products we should include in our samples, and how many of them, is not intuitively clear.

In the theory which we shall study in most of the remainder of this book, it will always be assumed that we are dealing with *random samples*. This attention to random samples is due to the fact that they permit valid, or logical, generalizations and hence are widely used in actual practice. To begin with, let us ask the following three questions: (1) "How many distinct samples of size n can be taken from a finite population of size N?" (2) "How is a random sample to be defined?" (3) "How can a random sample be taken in actual practice?"

To answer the first question we have only to refer to the rule for combinations on page 106, according to which r objects can be selected from a set of n objects in $\binom{n}{r}$ ways. With a change of letters, we can thus say that the number of distinct samples of size n which can be drawn from a finite population of size N is $\binom{N}{n}$. For instance, $\binom{10}{2} = 45$ samples of size $n = 2$ can be drawn from a finite population of $N = 10$ elements, $\binom{100}{3} = 161{,}700$ samples of size $n = 3$ can be drawn from a finite population of size $N = 100$, and $\binom{100{,}000}{3} = 166{,}661{,}666{,}700{,}000$ samples of size $n = 3$ can be drawn from a finite population with $N = 100{,}000$ elements (see Exercises 1 and 2 on page 216).

In answer to the second question, a *simple random sample* (or more briefly, a *random sample*) from a finite population is *a sample which is chosen in such a way that each of the* $\binom{N}{n}$ *possible samples has the same probability of* $1 \Big/ \binom{N}{n}$ *of being selected*. For instance, if we have a finite population consisting of the $N = 5$ elements a, b, c, d, and e (which might be the heights, weights, or ages of five persons), there are $\binom{5}{3} = 10$ possible distinct samples of size $n = 3$, namely, those consisting of the elements abc, abd, abe, acd, ace, ade, bcd, bce, bde, and cde. Then, if we choose one of these samples using a technique whereby each of the possibilities listed has a probability of $1/10$, we would call this sample a random sample.

With regard to the third question of how to take a random sample in actual practice, we could, in simple cases like the one above, write each of the $\binom{N}{n}$ possible samples on a slip of paper, put these in a hat, shuffle them thoroughly, and then draw one without looking. Such a procedure is obviously impractical, if not impossible, in more realistically complex problems of sampling; we have mentioned it here only to make the point that the selection of a random sample should depend entirely on chance.

Fortunately, the same results can be attained without actually resorting to the tedious procedure of *listing all possible samples.* Merely *listing the N elements of a finite population*, we can achieve the same results (i.e., a random sample) by choosing the elements to be included in the sample one at a time, making sure that in each of the successive drawings each of the remaining elements of the population has the same chance of being selected. In this way each item of the population still has the probability n/N of being chosen (see Exercise 9 on page 217). For instance, to take a random sample of 20 past-due accounts from a file of 257 such accounts, we could write each account number on a slip of paper, put the slips in a box and mix them thoroughly, and then draw (without looking) 20 slips one after the other.

Even this relatively simple procedure is not necessary in practice, where the simplest way of taking a random sample is with the use of a table of *random digits* (or *random numbers*). Published tables of random numbers consist of pages on which the decimal digits 0, 1, 2, . . . , and 9 are set down in much the same fashion as they might appear if they had been generated by means of a chance or gambling device giving each digit the same probability of $1/10$ of appearing at any given place in the table. Some early tables of random numbers were copied from pages of census data or from tables of 20-place logarithms, but they were found to be deficient in various ways. Nowadays, such tables are made with the use of electonic computers, but it would be possible to generate a table with a perfectly constructed spinner like that shown in Figure 8.1.

Table VIII is an excerpt from a published table of random numbers, and we shall illustrate its use by considering the problem of taking a random sample of 10 printing firms from the 562 firms listed in the yellow pages of a large city telephone directory. Numbering the firms on the alphabetical

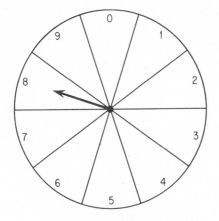

FIGURE 8.1 Spinner

list 001, 002, 003, . . . , 561, and 562, we arbitrarily pick a place from which to start in the table and then move in any direction, reading out three-digit numbers. For instance, if we arbitrarily choose the table on page 483 and read out the digits in the 25th, 26th, and 27th columns starting with the 31st row and going down, we find that the sample will consist of the 10 printing firms whose numbers are

290   280   193   278   024   469   371   338   009   166

In the selection of these numbers, those greater than 562 were ignored; also, had any number reoccurred, it would have been ignored. When lists are available and items are readily numbered, it is easy to obtain random samples from finite populations with the aid of tables of random numbers. Unfortunately, however, it is often impossible to proceed in the way we have just described. For example, if we wanted to estimate from a sample the average (mean) protein content of a carload of wheat, it would be impossible to number each of the millions of grains of wheat, choose correspondingly large random numbers, and then locate the grains corresponding to these numbers. In this and in many similar situations all one can do is proceed according to the dictionary definition of the word "random," namely, "haphazardly without definite aim or purpose." That is, we must not select or reject any element of a population because of its seeming typicalness or lack of it, nor must we favor or ignore any part of a population because of its accessibility or lack of it, and so on. Hopefully, we will thus get samples to which we can nevertheless apply statistical theory designed for the analysis of random samples.

To this point we have been discussing only random samples from finite populations. The concept of a random sample from an *infinite population* is somewhat more difficult to define, but a simple illustration will help to explain the basic characteristics of such a sample. Suppose we consider 10 tosses of a coin as a sample from the hypothetically infinite population consisting of all possible tosses of the coin. Then, if the probability of getting heads is the same for each toss and the 10 tosses are independent, we say that the sample is random. (Note that we would, in fact, be sampling from an infinite population if we sampled *with replacement* from a finite population, and that our sample would be *random* if in each draw all elements of the population had the same probability of being selected, and successive draws were independent.) Generally speaking, we assert that *the selection of each item in a random sample from an infinite population must be controlled by the same probabilities and that successive selections must be independent of one another.* Unless these conditions are satisfied (at least approximately), sets of observations should not be analyzed by means of techniques intended for random samples from infinite populations.

EXERCISES

1. How many different samples of size 2 can be selected from a finite population of (a) size 5, (b) size 12, and (c) size 25?

2. How many different samples of size 3 can be selected from a finite population of (a) size 5, (b) size 50, and (c) size 100?

3. On page 214 we listed the 10 possible samples of size 3 which can be drawn from the finite population which consists of the elements a, b, c, d, and e. If each of these samples is assigned the probability $1/10$, show that
   (a) the probability that any specific element will be contained in a sample is $3/5$;
   (b) the probability that any specific pair of elements will be contained in a sample is $3/10$.

4. List all possible samples of size 2 that can be drawn from the finite population whose elements are the numbers 1, 2, 3, 4, 5, 6, 7, and 8. Assigning each of these samples the same probability, show that the probability that any specific element of the population will be included in a sample is $1/4$.

5. Use random numbers to select a restaurant from among those listed in the yellow pages of your telephone directory (or that of a neighboring city).

6. Suppose you want three used-car dealers to bid on a used car you have for sale. Use random numbers to select these three used-car dealers from among those listed in the yellow pages of your telephone directory (or that of a neighboring city).

7. The laundry checks issued by a laundry are numbered serially from 1,000 to 1,500. Use random numbers to select a random sample of 25 of these laundry checks.

8. Explain why each of the following samples does not qualify as a random sample from the required population or might fail to give the desired information:

(a) To determine the average income of a city's residents, a research organization telephones persons randomly selected from the city's telephone directory.

(b) To determine the proportion of defectives, every tenth item produced by a machine is inspected.

(c) To study the religious affiliation of its students, a college sends a questionnaire to every fifth student on its alphabetically arranged list.

(d) To estimate the average annual income of its graduates 10 years after graduation, a college's alumni office sends questionnaires in 1971 to all members of the class of 1961, and the estimate is based on the questionnaires returned.

(e) To ascertain facts about tooth-brushing habits, the persons selected in a random sample of the residents of a community are asked how many times they brush their teeth each day.

(f) To study executive reaction to its copying machines, the Xerox Corporation hires a research organization to ask executives the question, "How do you like using Xerox copies?"

(g) A house-to-house survey is made to study consumer reaction to a new instant breakfast food, with no provisions for return visits in case no one is at home.

9. (a) Making use of the fact that among the $\binom{N}{n}$ samples of size n which can be drawn from a population of size N there are $\binom{N-1}{n-1}$ which contain a specific element, show that the probability that any specific element will be contained in the sample is n/N.

(b) Repeat (a) making use of the fact that the probability is $1/N$ that the specific element will be drawn first, the probability is $\dfrac{N-1}{N} \cdot \dfrac{1}{N-1}$ that the specific element will be drawn second, the probability is $\dfrac{N-1}{N} \cdot \dfrac{N-2}{N-1} \cdot \dfrac{1}{N-2}$ that the specific element will be drawn third, and so on.

SAMPLING DISTRIBUTIONS

Let us now introduce the concept of the *sampling distribution* of a statistic, which is probably the most basic concept of statistical inference.

Later on we shall also consider other statistics, but for now we shall focus our attention on the sample mean and its sampling distribution.

There are two ways in which we can approach the study of sampling distributions. One, based on appropriate mathematical theory, leads to what is called a *theoretical sampling distribution*; the other, based on repeated samples from the same population, leads to what is called an *experimental sampling distribution*. The latter will prove to be very useful in our study because it provides *experimental verification* of some necessary theorems, which cannot be derived formally at the level of this book.

Let us first give an example of a *theoretical sampling distribution of the mean*, specifically, that of the means of random samples of size $n = 2$ from the finite population of size $N = 5$, whose elements are the numbers 1, 3, 5, 7, and 9. The mean of this population is $\mu = \dfrac{1 + 3 + 5 + 7 + 9}{5} = 5$, its variance is

$$\sigma^2 = \frac{1}{5}[(1 - 5)^2 + (3 - 5)^2 + (5 - 5)^2 + (7 - 5)^2 + (9 - 5)^2]$$
$$= 8$$

and hence, its standard deviation is $\sigma = \sqrt{8}$. If we take random samples of size 2 from this population, the possible pairs of values we can get are 1 and 3, 1 and 5, 1 and 7, 1 and 9, 3 and 5, 3 and 7, 3 and 9, 5 and 7, 5 and 9, and 7 and 9, their respective means are 2, 3, 4, 5, 4, 5, 6, 6, 7, and 8, and (based on the assumption of random sampling which gives each sample the probability 1/10) we thus get

$\bar{x}$	Probability
2	1/10
3	1/10
4	2/10
5	2/10
6	2/10
7	1/10
8	1/10

This is the *theoretical sampling distribution of the mean* for random samples of size 2 from the given population, and its histogram is shown in Figure 8.2.

An examination of this sampling distribution reveals some pertinent information relative to the problem of estimating the mean of the population on the basis of a random sample of two items drawn from the population. For instance, it tells us that the probability is 6/10 that a sample mean

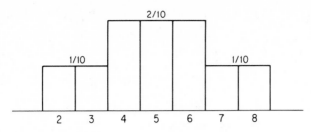

FIGURE 8.2 Theoretical sampling distribution of the means

will not differ from the population mean by more than 1, and that the probability is 8/10 that a sample mean will not differ from the population mean by more than 2; the first case corresponds to $\bar{x} = 4$, 5, or 6, and the second to $\bar{x} = 3, 4, 5, 6$, or 7.

Further useful information about this sampling distribution of the mean can be obtained by calculating its mean $\mu_{\bar{x}}$ and its standard deviation $\sigma_{\bar{x}}$, where we are using the subscript $\bar{x}$ to distinguish these parameters from those of the original population. Following the definitions of the mean and the variance of a probability distribution on pages 185 and 187, we obtain

$$\mu_{\bar{x}} = 2 \cdot \frac{1}{10} + 3 \cdot \frac{1}{10} + 4 \cdot \frac{2}{10} + 5 \cdot \frac{2}{10} + 6 \cdot \frac{2}{10} + 7 \cdot \frac{1}{10} + 8 \cdot \frac{1}{10}$$

$$= 5$$

and

$$\sigma_{\bar{x}}^2 = (2 - 5)^2 \frac{1}{10} + (3 - 5)^2 \frac{1}{10} + (4 - 5)^2 \frac{2}{10} + (5 - 5)^2 \frac{2}{10}$$

$$+ (6 - 5)^2 \frac{2}{10} + (7 - 5)^2 \frac{1}{10} + (8 - 5)^2 \frac{1}{10}$$

$$= 3$$

so that $\sigma_{\bar{x}} = \sqrt{3}$. Thus, we find that the mean of the sampling distribution of the $\bar{x}$ equals the mean of the population, and although the exact relationship is by no means obvious, the standard deviation of the sampling distribution of $\bar{x}$ is *smaller* than that of the population. These are fundamentally important relationships, and they will be stated formally later on. For now, let us merely take note of these relationships and take up the next problem—that of constructing an experimental sampling distribution of the mean—with the hope of thereby gaining some further insight into the "behavior" of sample means.

To give an example of an *experimental sampling distribution of the mean*, let us suppose that an operator of tow trucks for an automobile association wants to determine the average number of service calls he can expect on a Sunday afternoon. He needs this information to determine the number of tow trucks and drivers he has to make available. Let us suppose, furthermore, that in a sample of five Sunday afternoons he has received 13, 17, 22, 19, and 18 service calls. The mean of this sample is $\dfrac{13 + 17 + 22 + 19 + 18}{5} =$ 17.8, and in the absence of any other information he could use this figure as an estimate of μ, the true average number of service calls he can expect on a Sunday afternoon. However, at the same time he must acknowledge the fact that if his sample had consisted of five other Sunday afternoons, the mean would probably have been some number other than 17.8. Indeed, if he took several samples, each consisting of the number of service calls he received on five different Sunday afternoons, he might get such divergent values as 14.2, 18.4, 16.8, 12.8, 17.2, . . . , for the corresponding means.

In order to get some idea of how the means of such samples would vary purely as the result of chance, let us suppose that the number of service calls the tow-truck operator gets on a Sunday afternoon is a random variable having a Poisson distribution with the mean $\lambda = 16$ (see page 172), and let us perform an experiment which consists of *simulating* the drawing of 40 random samples, each consisting of five observations. How this can be done will be explained in Technical Note 5 and in Exercise 21 on page 230; for the time being, though, let us merely use the results which are shown in the following table:

Sample	Number of Service Calls	Sample	Number of Service Calls
1	13, 17, 22, 19, 18	21	15, 11, 14, 14, 18
2	15, 13, 16, 14, 21	22	22, 15, 13, 19, 11
3	16, 15, 8, 12, 23	23	20, 17, 11, 19, 15
4	16, 15, 11, 20, 13	24	15, 16, 16, 15, 17
5	17, 17, 16, 21, 14	25	15, 16, 17, 17, 16
6	18, 20, 14, 26, 18	26	13, 15, 15, 13, 18
7	13, 16, 17, 11, 6	27	12, 11, 19, 17, 16
8	17, 19, 15, 19, 16	28	15, 26, 19, 20, 15
9	11, 13, 18, 23, 18	29	15, 11, 21, 8, 17
10	12, 25, 21, 18, 8	30	12, 21, 10, 15, 16
11	21, 12, 14, 17, 16	31	10, 11, 9, 11, 11
12	19, 10, 15, 16, 18	32	12, 12, 24, 11, 5
13	20, 10, 15, 15, 19	33	16, 18, 14, 9, 11
14	16, 10, 26, 14, 20	34	17, 9, 18, 16, 9
15	18, 12, 13, 19, 9	35	11, 17, 19, 20, 17
16	15, 14, 21, 17, 11	36	10, 18, 19, 13, 20
17	12, 13, 13, 14, 12	37	14, 19, 16, 13, 21
18	18, 8, 21, 14, 15	38	18, 14, 23, 23, 14
19	20, 17, 16, 18, 19	39	17, 16, 11, 17, 11
20	19, 17, 16, 13, 15	40	16, 13, 10, 14, 20

Each of these samples contains five values, namely, the (simulated) number of service calls which the tow-truck operator received on five Sunday afternoons. Calculating the means of these 40 samples, we get

17.8	15.8	14.8	15.0	17.0	19.2	12.6	17.2	16.6	16.8
16.0	15.6	15.8	17.2	14.2	15.6	12.8	15.2	18.0	16.0
14.4	16.0	16.4	15.8	16.2	14.8	15.0	19.0	14.4	14.8
10.4	12.8	13.6	13.8	16.8	16.0	16.6	18.4	14.4	14.6

and their distribution is given by the histogram of Figure 8.3.

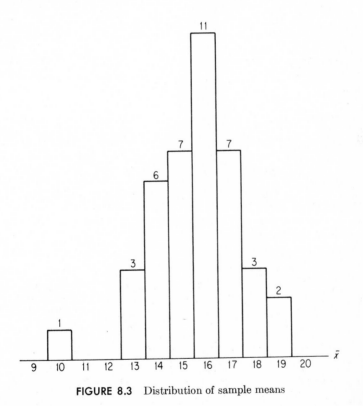

FIGURE 8.3 Distribution of sample means

As in the case of the theoretical sampling distribution constructed on page 218, inspection of this experimental sampling distribution of $\bar{x}$ tells us a great deal about the way in which sample means tend to scatter among themselves as the result of random sampling. For instance, the smallest mean is 10.4 and the largest is 19.2; furthermore, 25 out of 40 (or 62.5 per cent) of the means fall on the interval from 14.5 to 17.5, and 39 out of 40 (or 97.5 per cent) of the means fall on the interval from 12.5 to 19.5. Actually, though, the fundamental question involved here is *not* how much the 40 means vary among themselves, but how much they vary about the population mean μ, namely, the true average number of service calls the tow-truck operator receives on a Sunday afternoon. Directly, the histogram of Figure 8.3 does not tell us anything about this, but we *know* that the population mean is 16 since we simulated sampling from a population having a Poisson distribution with $\lambda = 16$. Thus, we can restate our description of the histogram in Figure 8.3 by saying that 62.5 per cent of the sample means were "off" (or in error) by at most 1.5, and that 97.5 per cent of them were "off" by at most 3.5. Furthermore, the poorest of them all was "off" by 5.6. This is the kind of information we really want—it tells us *how close we can expect a sample mean to be to the quantity it is supposed to estimate, namely, the mean of the given population.*

The two examples which we have given in this section are actually only *learning* devices designed to introduce the concept of a sampling distribution. In actual practice, one ordinarily takes only a single sample and, consequently, has only a single sample mean with which to estimate the mean of a population. There is no need to construct a sampling distribution in order to estimate μ on the basis of *one* sample or to evaluate the goodness of such an estimate. What should be clearly understood at this point is that (1) a single mean is only one of many (possibly, infinitely many) means which could be chosen, (2) the distribution of the totality of these means is called a sampling distribution of the mean, and (3) the many means which form this distribution scatter more or less widely (and, as we shall see, predictably) about the population mean μ.

In order to determine how good an estimate of μ a particular sample mean $\bar{x}$ might be, it is not necessary to have the entire sampling distribution before us. All we need are some essential facts about sampling distributions of means, specifically, those expressed by two theorems which will be discussed below. The following is the first of these theorems:

For random samples of size n taken from a population having the mean μ and the standard deviation σ, the theoretical sampling distribution of $\bar{x}$ has the mean μ and the standard deviation

$$\sigma_{\bar{x}} = \frac{\sigma}{\sqrt{n}} \cdot \sqrt{\frac{N - n}{N - 1}} \qquad \text{for finite populations of size } N$$

and

$$\sigma_{\bar{x}} = \frac{\sigma}{\sqrt{n}} \qquad \text{for infinite populations.}$$

The standard deviation of the theoretical sampling distribution of the mean, $\sigma_{\bar{x}}$, is generally called the *standard error of the mean,* and its role in statistics is fundamental since *it measures the extent to which sample means fluctuate, or vary, due to chance.* Clearly, some knowledge of this variability is essential in determining how "good" an estimate $\bar{x}$ is of the population mean μ. Intuition leads one (correctly) to feel that the smaller $\sigma_{\bar{x}}$ is (the less the $\bar{x}$'s are spread out) the better the estimate will be, and the larger $\sigma_{\bar{x}}$ is (the more the $\bar{x}$'s are spread out) the poorer the estimate will be. What determines the size of $\sigma_{\bar{x}}$, and hence the goodness of an estimate, can be seen from the above formulas. The formula for samples from finite populations shows among other things that (for fixed N) the standard error of the mean *increases* as the variability of the population increases, and that it *decreases* as the number of items in the sample increases. With respect to the latter, note that substitution into the formula yields $\sigma_{\bar{x}} = \sigma$ for $n = 1$, and $\sigma_{\bar{x}} = 0$ for $n = N$; in other words, $\sigma_{\bar{x}}$ takes on values between σ and 0 and is 0 only when the sample is, in fact, the entire population.

The second factor in the formula for $\sigma_{\bar{x}}$ for samples from finite populations is often omitted unless the sample constitutes a substantial portion of the population. For instance, if $n = 100$ and $N = 10,000$ (and the sample constitutes but 1 per cent of the population),

$$\sqrt{\frac{N-n}{N-1}} = \sqrt{\frac{10,000 - 100}{10,000 - 1}} = 0.995$$

which is so close to 1 that this factor can be ignored (for most practical purposes) in calculating the standard error of the mean. The factor $\sqrt{\dfrac{N-n}{N-1}}$ in the formula for $\sigma_{\bar{x}}$ for finite populations is called the *finite population correction factor,* for without it the two formulas for $\sigma_{\bar{x}}$ (for finite and infinite populations) are the same.

To get a feeling for the two formulas for $\sigma_{\bar{x}}$, let us return for a moment to the two sampling distributions discussed earlier in this section. In the first of these (on page 218), we constructed the theoretical sampling distribution of the mean for random samples of size $n = 2$ drawn (without replacement) from a finite population of size $N = 5$, for which $\mu = 5$ and $\sigma = \sqrt{8}$. Substituting these values into the first formula for $\sigma_{\bar{x}}$, the one for random samples from finite populations, we get

$$\sigma_{\bar{x}} = \frac{\sqrt{8}}{\sqrt{2}} \cdot \sqrt{\frac{5-2}{5-1}} = \sqrt{3}$$

and this agrees with the value which we obtained on page 219 for the standard deviation of the probability distribution of the means. In the second example (on page 220) we took 40 random samples of size 5 from a population having the Poisson distribution with the mean $\lambda = 16$ and, hence, the standard deviation $\sigma = \sqrt{\lambda} = \sqrt{16} = 4$ (see Exercise 8(b) on page 192). Thus, the standard error of the mean is

$$\sigma_{\bar{x}} = \frac{4}{\sqrt{5}} = 1.79$$

and it is impressive to see how close the standard deviation of the 40 sample means of our experiment comes to this theoretical value. Actually calculating the mean and the standard deviation of the distribution of the 40 means shown in Figure 8.3, we get, respectively, 15.575 and 1.78, which are *extremely close* to $\mu = 16$ and $\sigma_{\bar{x}} = 1.79$. In other words, *these values provide experimental verification of the theorem (on page 222) concerning the mean and the standard deviation of the sampling distribution of the mean.*

When we estimate the mean of a population, we often attach to it some probability about the possible size of our error. Using Chebyshev's Theorem, which we discussed on page 189, we can assert with a probability of at least $1 - 1/k^2$ that the mean of a random sample of size n will differ from the mean of the population from which it came by at most $k \cdot \sigma_{\bar{x}}$; in other words, *when we use the mean of a random sample to estimate the mean of a population, we can assert with a probability of at least $1 - 1/k^2$ that our error will be at most $k \cdot \sigma_{\bar{x}}$.* For instance, if we take a random sample of size $n = 64$ from an infinite population with $\sigma = 20$, we can assert with a probability of at least $1 - \frac{1}{2^2} = 0.75$ that the sample mean will be "off" (i.e., differ from the population mean) by at most $2 \cdot \frac{20}{\sqrt{64}} = 5$; similarly, if we estimate that the mean of another infinite population having $\sigma = 4$ is equal to the mean of a random sample of size 100, we can assert with a probability of at least $1 - \frac{1}{5^2} = 0.96$ that our estimate will be "off" by at most $5 \cdot \frac{4}{\sqrt{100}} = 2$.

THE CENTRAL LIMIT THEOREM

Chebyshev's Theorem applies to any set of data, and it is always possible to use it as in the examples of the preceding section. However, there exists another basic theorem of statistics, the *Central Limit Theorem*, which enables us in a great many instances to make much *stronger* probability statements than Chebyshev's Theorem does. In words, it can be formulated as follows:

> *If n (the sample size) is large, the sampling distribution of the mean can be approximated closely with a normal distribution.*

This theorem justifies the use of normal curve methods in a wide range of problems; it applies to infinite populations, and also to populations where n, though large, constitutes but a small portion of the population. It is difficult to say precisely how large n must be so that the Central Limit Theorem applies, but unless the population has a very unusual shape, $n = 30$ is usually regarded as sufficiently large. Note that the distribution of Figure 8.3 is fairly symmetrical and bell-shaped, even though the sample size was only $n = 5$.

The importance of the Central Limit Theorem can be illustrated by re-examining the two examples given at the end of the preceding section. Using Chebyshev's Theorem we were able to assert with a probability of at least 0.75 that the mean of a random sample of size 64 drawn from a population having $\sigma = 20$ will differ from the population mean μ by at most 5. Treating the sampling distribution of the mean as a normal distribution, on the other hand, we can say that the probability of getting a value within two standard deviations of the mean is $0.4772 + 0.4772 = 0.9544$ (see Figure 8.4). Using the Central Limit Theorem we can thus assert with a probability

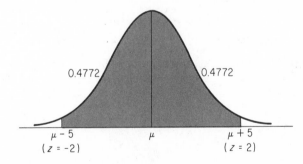

FIGURE 8.4 Sampling distribution of the mean

of 0.9544 that in this example a sample mean will be "off" by at most 5, while Chebyshev's Theorem enabled us to say only that this probability is at least 0.75. In the second example, Chebyshev's Theorem enabled us to assert with a probability of at least 0.96 that the mean of a random sample of size 100 drawn from a population having $\sigma = 4$ will differ from the population mean by at most 2. To use the Central Limit Theorem we would have to determine the normal curve area between $z = -5$ and $z = 5$ (i.e., the probability of getting a value within five standard deviations of the mean); this area cannot be looked up in Table I, but more extensive tables of normal curve areas show that it is actually greater than 0.999999. Again, this is a much *stronger* statement than the one based on Chebyshev's Theorem. (In applying the Central Limit Theorem in these two examples, we are treating the samples of size 64 and 100 as "large," but in view of what we said earlier, this is certainly justifiable unless the populations happen to be of a very unusual nature.)

If it is known that the distribution of the population, itself, can be approximated closely with a normal distribution, the Central Limit Theorem applies even when n is much smaller than 30. In the case of the sampling distribution which we constructed from the 40 sample means on page 221, the agreement between the experimental results and the results which we would expect according to the Central Limit Theorem is remarkably good.

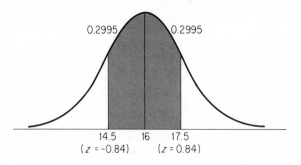

FIGURE 8.5 Sampling distribution of the mean

To show this, let us determine the percentage of means of random samples of size 5, drawn from a population with $\mu = 16$ and $\sigma = 4$, which can be expected to fall within 1.5 of the mean (i.e., on the interval from 14.5 to 17.5). Assuming that the Central Limit Theorem applies, we have to determine the normal curve area between

$$z = \frac{14.5 - 16}{1.79} = -0.84 \quad \text{and} \quad z = \frac{17.5 - 16}{1.79} = 0.84$$

since $\sigma_{\bar{x}} = 1.79$ was already determined on page 224. As can be seen from Figure 8.5 and Table I, the desired area is 0.5990; in other words, we can expect just about 60 per cent of the sample means to lie between 14.5 and

17.5. This is remarkably close to the 62.5 per cent which we actually obtained (see page 221). Similar calculations show that we can expect 95 per cent of the means to fall on the interval from 12.5 to 19.5, and this is very close to the 97.5 per cent which we actually obtained.

The main purpose of this chapter has been to introduce the concepts of random sampling and sampling distributions. We chose to introduce the latter by studying the sampling distribution of the mean, but we could just as well have studied the sampling distribution of any other statistic (e.g., the median, the range, or the standard deviation). In that case we would have studied the distribution of the medians of the 40 samples (the distribution of their ranges or the distribution of their standard deviations) and compared the results with the results we could expect according to appropriate theory.

EXERCISES

1. Random samples of size 2 are taken from the finite population which consists of the numbers 6, 7, 8, 9, 10, and 11.

   (a) Show that the mean of this population is $\mu = 8.5$ and that its standard deviation is $\sigma = \sqrt{\dfrac{35}{12}}$.

   (b) List the 15 possible random samples of size 2 that can be taken from this finite population and calculate their respective means.

   (c) Using the results of part (b) and assigning each possible sample a probability of 1/15, construct the sampling distribution of the mean for random samples of size 2 from the given finite population.

   (d) Calculate the mean and the variance of the probability distribution obtained in part (c) and verify the results with the use of the theorem on page 222.

2. Repeat parts (b), (c), and (d) of Exercise 1 for random samples of size 3 from the given population.

3. The finite population of Exercise 1 can be converted into an infinite population if we sample *with replacement*, that is, if we obtain a random sample of size 2 by first drawing one value and replacing it before drawing the second.

   (a) List the 36 possible samples of size 2 that can be drawn with replacement from the given population.

   (b) Determine the means of the 36 samples obtained in part (a) and, assigning each of the samples a probability of 1/36, construct the sampling distribution of the mean for random samples of size 2 from this infinite population.

   (c) Calculate the mean and the standard deviation of the probability distribution obtained in part (b) and compare them with the corresponding values expected according to the theorem on page 222.

4. Convert the 40 samples on page 220 into 10 samples of size 20 by combining Samples 1–4, 5–8, . . . , and 37–40. Averaging the means of the respective samples (given on page 221), find the means of the 10 new samples, calculate *their* mean and *their* standard deviation, and compare the results with the corresponding values expected according to the theorem on page 222.

5. Find the medians of the 40 samples on page 220 and calculate their mean and their standard deviation. Comparing the standard deviation of this experimental sampling distribution of the median with that of the corresponding experimental sampling distribution of the mean (which was 1.79), what do we learn about the relative "reliability" of the median and the mean in estimating the mean of the given population?

6. For large samples from populations having roughly the shape of a normal distribution, the *standard error of the median* is given by $\sqrt{\frac{\pi}{2}} \cdot \frac{\sigma}{\sqrt{n}}$. Compare the value obtained for the standard deviation of the experimental sampling distribution of the median in Exercise 5 with that given by this formula with $\sigma = 4$ and $n = 5$.

7. A very large population has roughly the shape of a normal distribution with mean μ and standard deviation σ. If μ is to be estimated by the *median* of a random sample of size n, how large must n be so that this estimate is equally as reliable as another estimate of μ based on the *mean* of a random sample of size 64?

8. When we sample from an infinite population, what happens to the standard error of the mean (and hence, to the size of the errors to which we are exposed when we use $\bar{x}$ to estimate μ) if
   (a) the sample size is increased from 64 to 576;
   (b) the sample size is increased from 60 to 240?

9. Show that if the mean of a random sample of size n is used to estimate the mean of an infinite population with the standard deviation σ, there is a *fifty-fifty chance* that the error is less than $0.67 \cdot \frac{\sigma}{\sqrt{n}}$. It has been the custom to refer to this quantity, or more precisely $0.6745 \cdot \frac{\sigma}{\sqrt{n}}$, as the *probable error of the mean;* nowadays, this term is used mainly in military applications.

10. Verify that if the Central Limit Theorem is applied to the example on page 220 (even though n is only 5), we can expect about 84 per cent of the sample means to fall on the interval from 13.5 to 18.5. What percentage of the 40 sample means actually fall on this interval?

11. The mean of a random sample of size $n = 36$ is used to estimate the mean of a population having a normal distribution with the standard deviation $\sigma = 15$.

What can we assert about the probability that the error will be less than 7.5

(a) using Chebyshev's Theorem;

(b) using the Central Limit Theorem?

12. The mean of a random sample of size $n = 144$ is used to estimate the mean of a population having the standard deviation $\sigma = 2.4$ inches. What can we assert about the probability that the error will be less than 0.25

(a) using Chebyshev's Theorem;

(b) using the Central Limit Theorem?

13. Agricultural statistics covering an extended period of time show that in a certain county the weights of all beef cattle brought to market can be closely approximated with a normal curve having a mean of 1,900 pounds and a standard deviation of 400 pounds. What is the probability that the mean weight of a random sample of 400 of these beef cattle is

(a) less than 1,870 pounds;

(b) greater than 1,950 pounds;

(c) between 1,860 and 1,910 pounds;

(d) either less than 1,840 pounds or more than 1,960 pounds?

14. If the weights of all men traveling by air on regularly scheduled commercial flights from a large metropolitan airport have a mean of 162 pounds and a standard deviation of 20 pounds, what is the probability that

(a) the combined gross weight of a random sample of 49 men on such a flight is more than 8,330 pounds;

(b) the combined gross weight of a random sample of 100 men on such a flight is at most 16,500 pounds?

(*The exercises which follow are based on the material discussed in Technical Note 5.*)

15. Using four random digits to represent the results obtained when tossing four balanced coins (0, 2, 4, 6, and 8 represent *heads* while 1, 3, 5, 7, and 9 represent *tails*), simulate an experiment consisting of 160 tosses of four coins. Compare the observed number of times that 0, 1, 2, 3, and 4 heads occurred with the corresponding expected frequencies which are 10, 40, 60, 40, and 10.

16. Repeat Exercise 15, letting 0, 1, 2, 3, and 4 heads be represented by the four-digit random numbers 0000–0624, 0625–3124, 3125–6874, 6875–9374, and 9375–9999, respectively.

17. Use one-digit random numbers (omitting 7, 8, 9, and 0) to simulate 240 rolls of a balanced die.

18. In a certain town, the probabilities of an adult passing his driving test on the first, second, third, fourth, or fifth try are, respectively, 0.44, 0.32, 0.16, 0.05, and 0.03. Using two-digit random numbers, simulate the experience of 50 adults in trying to get their driving license. If a person has to pay $2.50 the first time he takes the test and after that $1.00 for each additional try, what is the average amount paid by these 50 persons?

19. Referring to part (b) of Exercise 1, label the 15 possible Samples 1, 2, 3, . . . , 14, and 15, and use random numbers to simulate an experiment in which 100 random samples of size 2 are taken from the given population. Compare the distribution of the means of these samples with the corresponding theoretical sampling distribution obtained in part (c) of Exercise 1.

20. Referring to part (a) of Exercise 3, label the 36 possible Samples 1, 2, 3, . . . , 35, and 36, and then use random numbers to simulate an experiment in which 100 random samples of size 2 are taken from the given infinite population. Calculate the mean and the standard deviation of the means of the 100 samples and compare the results with the corresponding values expected according to the theorem on page 222.

21. To obtain the 40 samples given on page 220, we used three-digit random numbers and the following scheme:

Number of Service Calls	Probability	Random Numbers
5	0.001	000
6	0.003	001–003
7	0.006	004–009
8	0.012	010–021
9	0.021	022–042
10	0.034	043–076
11	0.050	077–126
12	0.066	127–192
13	0.082	193–274
14	0.093	275–367
15	0.099	368–466
16	0.099	467–565
17	0.093	566–658
18	0.083	659–741
19	0.070	742–811
20	0.056	812–867
21	0.043	868–910
22	0.031	911–941
23	0.022	942–963
24	0.014	964–977
25	0.009	978–986
26	0.006	987–992
27	0.003	993–995
28	0.002	096–997
29	0.001	998
30	0.001	999

Use this method to draw 50 random samples of size 4 from the given population (which has the mean $\mu = 16$ and the standard deviation $\sigma = 4$), calculate their mean and their standard deviation, and compare the results with those expected in accordance with the theorem on page 222.

A WORD OF CAUTION

A point worth repeating is that the experimental sampling distribution based on the 40 means and the theoretical sampling distribution for which we had to list all possible samples were meant to be *teaching aids,* designed to convey the concept of a sampling distribution. These examples do not reflect what we do in actual practice, where we must base an inference on *one* sample and not 40, and where there is seldom any need (or any advantage) to enumerate all possible samples. In Chapters 9 and 10 we shall delve more deeply into the problem of translating theory concerning sampling distributions into methods of evaluating the goodness of an estimate or the merits and disadvantages of a statistical decision procedure.

Another fact worth noting concerns the $\sqrt{n}$ appearing in the denominator of the formulas for the standard error of the mean. As we pointed out on page 223, it reflects the idea that if we get a larger sample and, hence, more information, the resulting generalizations should be subject to smaller errors, and in general, our methods should be more reliable and more precise. On the other hand, the formulas for $\sigma_{\bar{x}}$ also illustrate the fact that *gains in precision or reliability are not proportional to increases in the size of the sample.* That is, doubling the size of the sample does *not* double the reliability of $\bar{x}$ as an estimate of the mean of a population, and so on. As is apparent from the formula $\sigma_{\bar{x}} = \dfrac{\sigma}{\sqrt{n}}$ for samples from infinite populations, we must take four times as large a sample to cut the standard error in half, and nine times as large a sample to triple the reliability, namely, to divide the standard error by 3. This clearly illustrates the fact that *it seldom pays to take excessively large samples.* For instance, if we increase the sample size from 100 to 10,000 (probably at a considerable expense), the size of the errors to which we are exposed is reduced only by a factor of 10. Similarly, if we increase the sample size from 50 to, say, 20,000, the chance fluctuations to which we are exposed are reduced only by a factor of 20, and this is seldom worth the cost of taking 19,950 additional observations. This argument is not limited to the distribution of means; this "law of diminishing returns" concerning the information gained from samples applies also to most other statistics.

TECHNICAL NOTE 5 (Simulating Sampling Experiments)

Although we introduced random numbers originally to select random samples from finite populations, they are used for many other purposes. They serve to *simulate* almost any kind of gambling device; in fact, they can be used to simulate almost any situation involving an element of uncertainty or chance. For example, we can play the game of "Heads or Tails" without ever flipping a coin by letting the digits 0, 2, 4, 6, and 8 represent *heads* while the digits 1, 3, 5, 7, and 9 represent *tails*. Then, using, for instance, the 4th column of the table on page 483, we get 1, 5, 2, 0, 7, 5, 1, 0, 2, 5, . . . , and we interpret this as *tail, tail, head, head, tail, tail, tail, head, head, tail,* . . . .

We can similarly simulate the simultaneous flips of three coins by using, say, the first three columns of the table on page 484. Getting 550, 325, 467, 354, 352, 557, 747, 333, 550, 638, . . . , we interpret these results as getting, respectively, 1, 1, 2, 1, 1, 0, 1, 0, 1, 2, . . . , heads. If we did not want to use three columns of random numbers for this "experiment," we could make use of the results obtained on page 165, where we showed that the probabilities of getting 0, 1, 2, or 3 heads are, respectively, 1/8, 3/8, 3/8, and 1/8. Using the coding

Number of Heads	Random Digit
0	0
1	1, 2, 3
2	4, 5, 6
3	7

(where the digits 8 and 9 are ignored whenever they occur), we interpret the random numbers 1, 5, 2, 0, 7, 5, 1, 0, 2, 5, . . . , in the fourth column of page 483 as representing, respectively, 1, 2, 1, 0, 3, 2, 1, 0, 1, 2, . . . , heads in three flips of a balanced coin. Note that if we did not want to "waste" any digits, we could have performed this experiment also with three columns of random numbers and the coding shown in the following table:

Number of Heads	Random Digits
0	000–124
1	125–499
2	500–874
3	875–999

With this scheme, the random numbers 213, 109, 915, 657, and 359, for example, represent 1, 0, 3, 2, and 1 heads in five flips of three coins.

Proceeding as in this last example, we can simulate any kind of probability distribution, and this is usually much more satisfactory than tossing coins, drawing numbered slips out of a hat, rolling dice, or gambling with other kinds of physical models. Some applied problems in which random numbers are used to simulate actual operations of a business (or parts of a business) will be taken up in Chapter 16 under the heading of "Monte Carlo Methods."

9

DECISION MAKING:
ESTIMATION

PROBLEMS OF ESTIMATION

As we have suggested earlier, many problems of statistical inference consist of taking a sample from a population and using it to estimate some characteristic (parameter) of the population. Given as a single number, an estimate is a value intended to match, say, the average weekly sales of a gas station (the mean of a population), the proportion of a restaurant's customers who tip at least 15 per cent (a population proportion), or how much variation one can expect in the time it takes to drive from New York to Philadelphia (the standard deviation of a population). When we said "intended to match" and not "which matches," we meant just that—*it is possible for an estimate based on a sample to coincide with the population parameter it is supposed to estimate, but this is the exception rather than the rule.* This should be clear from our discussion of the sampling distribution of the mean in Chapter 8.

Referring to another example of estimating the mean of a population, suppose that an independent testing laboratory runs five tests on a newly developed fast-setting adhesive to estimate the average (mean) time it takes to set. If the set speeds which they obtain are 5.1, 4.8, 5.0, 5.0, and 5.1

seconds, they might simply use the sample mean $\bar{x} = 5.0$ seconds as an estimate of the "true" average time it takes the adhesive to set, and let it go at that. This kind of estimate is called a *point estimate*, as it consists of a single number, namely, a single point on the real number scale. Although this is the most common way in which estimates are expressed, it leaves room for many questions. One might wonder, for instance, how many tests were run (unless this is also specified), one might wonder by how much the set speeds varied from trial to trial, and one might be tempted to ask whether the point estimate may not be off by, say, 0.1 seconds or 0.5 seconds.

There are several ways in which some of this vagueness can be eliminated without requiring the "consumer" of the information to be a trained statistician. For instance, the laboratory might report the following:

> *The estimate of the mean set speed of the adhesive is 5.0 seconds, and we are 95 per cent sure (the probability is 0.95) that the error of this estimate is at most 0.15 seconds.*

How this "possible maximum error" was obtained will be discussed later, but let us note at this time that if the "possible maximum error" of 0.15 seconds is added to and subtracted from the mean of the sample, the laboratory's results could also have been reported as follows:

> *We are 95 per cent sure (the probability is 0.95) that the interval from 4.85 seconds to 5.15 seconds contains the "true" mean set speed of the adhesive.*

Estimates of this sort, in which we give an interval rather than a single number, are appropriately called *interval estimates*. The interval itself is called a *confidence interval*, and the probability with which we can assert that it does its job (i.e., contains the quantity it is supposed to estimate) is called the *degree of confidence*.

Throughout most of this chapter it will be assumed that our point estimates and our interval estimates are based only on direct observations or measurements. If this kind of information is to be supplemented with collateral information, say, a person's prior experience or subjective judgment, we will have to use some form of *Bayesian inference*, like that given in the section beginning on page 242.

THE ESTIMATION OF MEANS

To illustrate the various problems we face in the estimation of a population mean, let us consider the following example: Suppose that a company providing security guards wishes to estimate the average time it takes to patrol a certain size warehouse (check all doors and windows, call in from several stations, etc.), and that it has at its disposal the following times it took to do this job (in minutes):

24.0	48.3	39.8	36.4	47.9	29.0	52.5	38.4
51.4	29.9	34.5	35.5	43.3	46.5	28.7	41.6
39.7	42.6	47.2	49.8	34.5	38.0	28.9	39.2
42.1	38.5	41.3	21.0	38.5	33.8	19.9	32.6
32.2	37.6	44.1	26.7	37.8	57.1	32.4	44.3

The mean of this set of data (considered to be a random sample) is $\bar{x} = 38.2$ minutes, and in the absence of any other information it can serve as an estimate of μ, the *true* average time it takes to patrol such a warehouse.

This is all right, but to comply with the suggestion that point estimates should always be accompanied by information which makes it possible to judge their merits, let us return briefly to what we said in the preceding chapter about the sampling distribution of the mean. We know that the means of random samples will fluctuate from sample to sample, but we also know that the mean and the standard deviation of their distribution are μ and $\frac{\sigma}{\sqrt{n}}$, where μ and σ are the mean and the standard deviation of the (infinite) population from which the sample was obtained. Making use of the Central Limit Theorem, according to which the sampling distribution of the mean can be approximated closely with a normal curve (at least for large samples), we can now assert with the probability $1 - \alpha$, where α is the Greek letter *alpha*, that $\bar{x}$ will differ from μ either way by less than $z_{\alpha/2}$ standard deviations (see Exercise 5 on page 203 and also Figure 9.1). In other words, we can assert that $\bar{x}$ will differ from μ either way by less than $z_{\alpha/2} \cdot \frac{\sigma}{\sqrt{n}}$, and *since $\bar{x} - \mu$ is the error which we make when we use $\bar{x}$ as an*

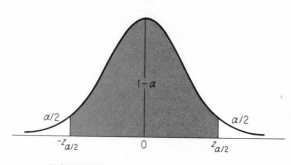

FIGURE 9.1 Area under normal curve

estimate of μ, we can assert with the probability 1 − α that the size of this error will be less than $z_{\alpha/2} \cdot \dfrac{\sigma}{\sqrt{n}}$. The two values which are most commonly used for α are 0.05 and 0.01, and as the reader was asked to show in parts (c) and (e) of Exercise 5 on page 203, $z_{.025} = 1.96$ and $z_{.005} = 2.58$.

The result we have obtained here involves one complication: To be able to say something about the possible size of the error we might make when using $\bar{x}$ as an estimate of μ, we must know σ, the standard deviation of the population. Since this is not the case in most practical situations, we have no choice but to replace σ with an estimate, usually the sample standard deviation s. In general, this is considered to be reasonable provided the sample size is sufficiently large, and by "sufficiently large" we mean 30 or more.

Returning now to our numerical example, where (as can easily be checked) the standard deviation of the 40 measurements is $s = 8.52$ minutes, we can now assert for $\alpha = 0.05$, namely, with the probability $1 − \alpha = 0.95$, that if we estimate the true average time it takes to patrol the given kind of warehouse as 38.2 minutes, the error will be less than

$$1.96 \cdot \frac{s}{\sqrt{n}} = 1.96 \cdot \frac{8.52}{\sqrt{40}} = 2.64 \text{ minutes}$$

Of course, the error of this estimate is either less than 2.64 minutes or it is not, *and we really don't know which*, but if we had to bet, 19 to 1 (95 to 5) would be fair odds that the error actually is less than 2.64 minutes.

When σ is not known and n is less than 30, the method explained above cannot be used, but there exists a modification of the formula for the maximum error which can be used for small samples provided the population from which we are sampling, or better, its distribution, has roughly the shape of a normal curve. This modification will be discussed later in this section on page 239.

An important feature of the formula for the maximum error is that it can also be used to determine the sample size that is needed to attain a desired degree of precision. Suppose we want to use the mean of a random sample to estimate the mean of a population, and we want to be able to assert with the probability $1 − \alpha$ that the error of this estimate will be less than some quantity E. We can thus write

$$E = z_{\alpha/2} \cdot \frac{\sigma}{\sqrt{n}}$$

and upon solving this equation for n we get

$$n = \left[\frac{z_{\alpha/2} \cdot \sigma}{E} \right]^2$$ ★

Note that this formula cannot be used unless we know (or can approximate) the standard deviation of the population whose mean we want to estimate.

To illustrate this technique, suppose we want to estimate the average clerical aptitude of a large group of persons (as measured by a certain standard test) and that we want this estimate to be off either way by at most 2.0 with a probability of 0.99. Suppose also that (on the basis of experience with similar data) it is reasonable to let σ equal 15.0. Substituting these values together with $z_{.005} = 2.58$ into the above formula for n, we obtain

$$n = \left[\frac{2.58(15.0)}{2.0} \right]^2 = 375$$

rounded up to the nearest whole number. Thus, a sample of size $n = 375$ will suffice for the stated purpose.

The error we make when using a sample mean to estimate the mean of a population is given by the difference $\bar{x} - \mu$, and the fact that the *magnitude* of this error is less than $z_{\alpha/2} \dfrac{\sigma}{\sqrt{n}}$ can be expressed by means of the inequality

$$-z_{\alpha/2} \frac{\sigma}{\sqrt{n}} < \bar{x} - \mu < z_{\alpha/2} \frac{\sigma}{\sqrt{n}}$$

(In case the reader is not familiar with inequality signs, let us point out that $a < b$ means "a is less than b," while $a > b$ means that "a is greater than b." Also, $a \leq b$ means "a is less than or equal to b," while $a \geq b$ means "a is greater than or equal to b.") Applying some simple algebra, we can rewrite the above inequality as

$$\bar{x} - z_{\alpha/2} \frac{\sigma}{\sqrt{n}} < \mu < \bar{x} + z_{\alpha/2} \frac{\sigma}{\sqrt{n}}$$ ★

and we can now assert with a probability of $1 - \alpha$ that the inequality is satisfied for any given sample, namely, that the interval from $\bar{x} - z_{\alpha/2} \dfrac{\sigma}{\sqrt{n}}$

to $\bar{x} + z_{\alpha/2} \dfrac{\sigma}{\sqrt{n}}$ actually contains the mean we are trying to estimate. An interval like this is called a *confidence interval* (as we pointed out on page 235), its endpoints are called *confidence limits,* and the probability $1 - \alpha$ with which we can assert that such an interval will "do its job" is called the *degree of confidence.* The values most commonly used for the degree of confidence are 0.95 and 0.99, so that the corresponding values of $z_{\alpha/2}$ are 1.96 and 2.58.

When σ is unknown and n is 30 or more, we proceed as before and estimate σ with the sample standard deviation s. The resulting $1 - \alpha$ *large-sample confidence interval* for μ becomes

$$\bar{x} - z_{\alpha/2} \frac{s}{\sqrt{n}} < \mu < \bar{x} + z_{\alpha/2} \frac{s}{\sqrt{n}} \qquad \bigstar$$

and if we apply this technique to our numerical example, where we had $n = 40$, $\bar{x} = 38.2$, and $s = 8.52$ (see page 236), we obtain the following 0.95 confidence interval for the *true* average time it takes to patrol the given kind of warehouse:

$$38.2 - 1.96 \cdot \frac{8.52}{\sqrt{40}} < \mu < 38.2 + 1.96 \cdot \frac{8.52}{\sqrt{40}}$$

$$35.56 < \mu < 40.84$$

Had we wanted the degree of confidence to be 0.99, we would have obtained the confidence interval

$$38.2 - 2.58 \cdot \frac{8.52}{\sqrt{40}} < \mu < 38.2 + 2.58 \cdot \frac{8.52}{\sqrt{40}}$$

$$34.72 < \mu < 41.28$$

and this illustrates the interesting fact that *the surer we want to be, the less we have to be sure of.* In other words, if we increase the degree of certainty (the degree of confidence), the confidence interval becomes *wider* and thus tells us *less* about the quantity we want to estimate.

So far we have assumed not only that the sample size was large enough to treat the sampling distribution of the mean as if it were a normal distribution, but that (when necessary) σ can be replaced with s in the formula for the standard error of the mean. To develop a corresponding theory that applies also to small samples, we shall now have to assume that the population from which we are sampling (or better, its distribution) has roughly the shape of a normal distribution. We can then base our methods on the statistic

$$t = \frac{\bar{x} - \mu}{s/\sqrt{n}} \qquad \qquad \bigstar$$

whose sampling distribution is called the *t distribution*. (More specifically, it is called the *Student-t distribution*, as it was first investigated by W. S. Gosset, who published his writings under the pen name of "Student.") The shape of this distribution is very much like that of the normal curve; it is symmetrical with zero mean, but there is a slightly higher probability for getting values falling into the two tails (see Figure 9.2). Actually, the

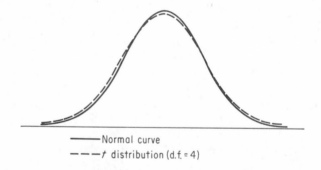

——— Normal curve
— — — *t* distribution (d.f. = 4)

FIGURE 9.2 Normal curve and *t* distribution

shape of the *t* distribution depends on the size of the sample or, better, on the quantity $n - 1$, which is called the *number of degrees of freedom.**

For the standard normal distribution we defined $z_{\alpha/2}$ as the value of z for which the area under the normal curve *to its right* equals $\alpha/2$, and for the *t* distribution we correspondingly let $t_{\alpha/2}$ be the value for which the area

---

* It is difficult to explain at this time why one should want to assign a special name to $n - 1$ which, after all, is only the sample size minus 1. However, we shall see in the next chapter that there are other applications of the *t* distribution, where the number of degrees of freedom is defined in a different way. The reason for the term "degrees of freedom" lies in the fact that if we know $n - 1$ of the deviations from the mean, then the nth is automatically determined (see argument on page 42). Since the sample standard deviation measures variation in terms of the squared deviations from the mean, we can thus say that this estimate of σ is based on $n - 1$ *independent quantities* or that we have $n - 1$ degrees of freedom.

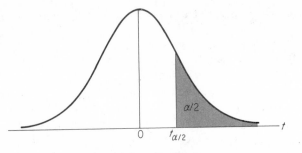

FIGURE 9.3 t distribution

under the curve to *its right* equals $\alpha/2$ (see Figure 9.3). Since the values of $t_{\alpha/2}$ depend on $n - 1$ (the number of degrees of freedom), their values must be looked up in special tables, such as Table II on page 474, which contains (among others) the values of $t_{.025}$ and $t_{.005}$ for 1 through 29 degrees of freedom. It is of interest to note that $t_{.025}$ approaches 1.96 and $t_{.005}$ approaches 2.58 (the corresponding values for the standard normal distribution) when the number of degrees of freedom becomes larger and larger.

Since the t distribution, like the standard normal distribution, is *symmetrical* about its mean $\mu = 0$ (see Figure 9.3), we can now duplicate the argument on page 238 and thus arrive at the following $1 - \alpha$ *small sample confidence interval for μ*

$$\bar{x} - t_{\alpha/2} \frac{s}{\sqrt{n}} < \mu < \bar{x} + t_{\alpha/2} \frac{s}{\sqrt{n}} \qquad \star$$

Note that the only difference between this confidence interval formula and the one on page 239 is that $t_{\alpha/2}$ has taken the place of $z_{\alpha/2}$.

To illustrate the calculation of a small-sample confidence interval for μ, let us refer again to the illustration on page 234, where a laboratory obtained 5.1, 4.8, 5.0, 5.0, and 5.1 seconds in a random sample of five set speeds of a new adhesive. The mean and the standard deviation of these measurements are $\bar{x} = 5.0$ seconds and $s = 0.122$, and since $t_{.025}$ equals 2.776 for $5 - 1 = 4$ degrees of freedom, substitution into the formula yields

$$5.0 - 2.776 \cdot \frac{0.122}{\sqrt{5}} < \mu < 5.0 + 2.776 \cdot \frac{0.122}{\sqrt{5}}$$

or

$$4.85 < \mu < 5.15$$

This 0.95 confidence interval is precisely the interval which we gave on page 235.

The method which we used on page 237, to indicate the possible size of the error we make when using a sample mean to estimate the mean of a population, can easily be adapted to small samples (provided the population distribution has roughly the shape of a normal curve). All we have to do is substitute s for σ and $t_{\alpha/2}$ for $z_{\alpha/2}$ in the formula for the maximum error E. To give an example, suppose that in 12 test runs an experimental engine consumed on the average 12.9 gallons per minute with a standard deviation of 1.6 gallons. Since $t_{.005}$ equals 3.106 for $12 - 1 = 11$ degrees of freedom, substitution into the formula for E yields

$$E = t_{.005} \cdot \frac{s}{\sqrt{n}} = 3.106 \cdot \frac{1.6}{\sqrt{12}} = 1.43$$

Thus, if we use the mean $\bar{x} = 12.9$ gallons per minute as an estimate of the *true* average gasoline consumption of the engine, we can assert with the probability 0.99 that the error of this estimate is less than 1.43 gallons.

A BAYESIAN ESTIMATE*

In recent years there has been mounting interest in methods of inference which look upon parameters (e.g., a population mean μ) as random variables. The whole idea is not really new, but these *Bayesian methods*, as they are called, have received a considerable impetus and much wider applicability through the concept of personal, or subjective, probability. In fact, this is why supporters of the personal concept of probability refer to themselves as *Bayesians*, or *Bayesian statisticians*.

In this section we shall give a Bayesian method of estimating the mean of a population. Proponents of the subjective, or personal, point of view look upon μ as a random variable whose distribution is indicative of *how strongly a person feels about the various values which μ can take on.* In other words, they suggest that in a situation like this a person will feel most strongly about some particular value of μ, and that his enthusiasm will diminish for values of μ which are farther and farther away from the one he likes the most. Like any distribution, this kind of subjective distribution

---

* The material in this section is somewhat more advanced, and it may be omitted without loss of continuity.

for the possible values of μ has a mean which we shall denote μ_0 and a standard deviation which we shall denote σ_0.

In Bayesian estimation, these *prior feelings* about the possible values of μ are combined with *direct sample evidence* consisting of a random sample of size n, its mean $\bar{x}$, and its standard deviation s (which serves as an estimate of σ, the standard deviation of the population whose mean μ we are trying to estimate). This is accomplished by means of the formula

$$\text{Estimate} = \frac{n\sigma_0^2 \cdot \bar{x} + \sigma^2 \cdot \mu_0}{n\sigma_0^2 + \sigma^2} \qquad \star$$

which can be used under fairly general conditions. (Strictly speaking, it is usually based on the assumption that the distribution of the population from which we are sampling as well as the subjective prior distribution of μ have roughly the shape of normal distributions.)

Before we apply this formula, let us briefly examine some of its most important features. To begin with, it should be observed that it is a *weighted mean* (see Exercise 12 on page 35) of $\bar{x}$ and μ_0, whose respective weights are $n\sigma_0^2$ and σ^2. When no direct information is available and $n = 0$, the weight of $\bar{x}$ equals 0, and the estimate is based entirely on the subjective prior information. However, as more and more direct evidence becomes available (i.e., as n becomes larger and larger), the weight shifts more and more toward the direct sample evidence, namely, the sample mean $\bar{x}$. There are two other points that should be observed. When the subjective feelings about the possible values of μ are rather indefinite, that is, when σ_0 is relatively large, the estimate will be based to a greater extent on $\bar{x}$; on the other hand, when there is a great deal of variability in the population from which we are sampling, that is, when σ is relatively large, the estimate will be based to a greater extent on μ_0.

To illustrate this subjective approach, suppose that someone is planning to open a new donut shop, and that a business consultant feels most strongly that he should net on the average \$2,200 per month; also, the distribution of the feelings which he attaches to this average has a standard deviation of \$130. In other words, $\mu_0 = \$2,200$ and $\sigma_0 = \$130$. Now, if during the first nine months of operation the donut shop nets, say, \$2,410, \$2,290, \$1,950, \$2,000, \$1,920, \$1,850, \$2,030, \$2,200, and \$2,270, the problem is *how to modify the original estimate of* $\mu_0 = \$2,200$ in the light of this information. Having $n = 9$, $\bar{x} = \$2,102$, and $s = \$195$, substitution into the formula yields

$$\text{Estimate} = \frac{9(130)^2 \cdot 2,102 + (195)^2 \cdot 2,200}{9(130)^2 + (195)^2} = \$2,122$$

to the nearest dollar, and this estimate accounts for both the consultant's "feelings" as well as the direct sample information.

This introduction to Bayesian statistics has been very brief, but it should have served to bring out the following two points: (1) *In Bayesian statistics the parameter about which an inference is to be made is looked upon as having a distribution of its own, and (2) this kind of inference permits the use of direct as well as collateral information—for instance, objective information and subjective judgments.* To clarify the last point, let us add that in the donut shop example the prior judgment of the consultant could have been based on a subjective evaluation of various factors (say, business conditions in general) as well as on collateral (objective) information about the performance of other donut shops.

EXERCISES

1. A random sample (which was taken as part of a large survey) showed that 49 persons living in two-room apartments in a certain city paid an average monthly rent of $129.50 with a standard deviation of $18.75. If this sample mean is used to estimate the true average rent paid for two-room apartments in the given city, what can be said with a probability of 0.95 about the possible size of the error?

2. Use the data of Exercise 1 to construct a 0.99 confidence interval for the average monthly rent paid for two-room apartments in the given city.

3. In an automobile collision insurance study, a random sample of 120 body repair costs on a particular kind of damage had a mean of $480 with a standard deviation of $57. If this sample mean is used to estimate the true average cost of this kind of body repair, what can we assert with a probability of 0.99 about the possible size of our error?

4. Use the data of Exercise 3 to construct a 0.95 confidence interval for the actual mean cost of this kind of body repair.

5. Taking a random sample from its very extensive files, a telephone company finds that the amounts owed in 100 overdue accounts have a mean of $27.65 and a standard deviation of $7.34.
   (a) What can be said with a probability of 0.95 about the possible size of the error, if the mean amount owed on all of the company's overdue accounts is estimated to be $27.65?
   (b) Use the given data to construct a 0.99 confidence interval for the average amount owed on all of the company's overdue accounts.

6. Suppose that an airline official uses the mean $\bar{x} = 27.8$ pounds of the weights of 50 randomly selected suitcases carried by passengers on nonstop jets between Phoenix and Chicago as an estimate of the mean of the corresponding popula-

tion of suitcase weights. Making use of the fact that the standard deviation of this sample is 5.8 pounds, what can we assert with a probability of 0.90 about the possible size of his error?

7. A random sample of 50 No. 10 cans of sliced pineapple has a mean weight of 67.2 ounces and a standard deviation of 2.1 ounces. With what probability can we assert that the estimate 67.2 ounces differs from the true average weight of all the cans from which the sample was taken by less then 0.4 ounces? (*Hint:* Substitute z for $z_{\alpha/2}$ in the formula for E on page 237, solve for z, and find the normal curve area between $-z$ and z.)

8. A sample of 81 test scores on an achievement test given to thousands of twelfth grade students has a mean of 212 and a standard deviation of 36. If this mean of 212 is used to estimate the average test score obtained by all the twelfth grade students who took the test, with what probability can we assert that this estimate is "off" by less than 6.5? (*Hint:* Substitute z for $z_{\alpha/2}$ in the formula for E on page 237, solve for z, and find the normal curve area between $-z$ and z.)

9. If a sample constitutes an appreciable portion of a finite population (say, 5 per cent or more), the various formulas given in the text must be modified by using the first standard error formula of the theorem on page 222 instead of the second. For instance, the formula for E on page 237 becomes

$$E = z_{\alpha/2} \cdot \frac{\sigma}{\sqrt{n}} \sqrt{\frac{N-n}{N-1}} \qquad \bigstar$$

(a) If a sample of 40 charges for repairs on television sets randomly selected from 160 such charges has a mean of \$68.25 and a standard deviation of \$18.27, what can we assert with a probability of 0.95 about the possible size of the error, if this sample mean is used to estimate the mean of all the 160 charges?

(b) A sample of 100 scores on the Admission Test for Graduate Study in Business is randomly selected from a population of 400 such scores made by applicants for admission to a certain college in a given year. If the mean of the sample is $\bar{x} = 570$ and its standard deviation is $s = 97$, what can we assert with a probability of 0.99 about the possible size of our error if we estimate the average score for all 400 of these applicants as 570? Compare the result with that which would have been obtained if the "finite population correction factor" had been omitted.

(c) A random sample of 16 drums of a wax-base floor cleaner, drawn from among 100 such drums whose weights have a standard deviation of 12 pounds, has a mean weight of 240 pounds. Construct a 0.95 confidence interval for the actual mean weight of all 100 of these drums.

10. It is known that the standard deviation of the filled weights of a certain size carton of a detergent is 0.2 ounces. With a degree of confidence of 0.95, how large a sample is needed to determine the mean weight of all cartons of this detergent if the error of the estimate is to be less than 0.05 ounces?

11. Before bidding on a contract, a manufacturer wants to be "99 per cent sure" that he is in error by less than 4 minutes in estimating the average time it takes to perform a certain task. If the standard deviation of the time it takes to perform the task can be assumed to equal 16 minutes, on how large a sample should he base his estimate?

12. It is desired to estimate the mean lifetime of a certain kind of electric can opener. Given that $\sigma = 68$ days, how large a sample is needed to be able to assert with a probability of 0.99 that the estimate is "off" by at most 20 days?

13. A random sample of nine charges for a certain type of service has a mean of $140 and a standard deviation of $20. If we estimate the actual mean charge for this kind of service as $140, what can we assert with a probability of 0.95 about the possible size of our error?

14. In an experiment, the average amount of time it took eight fuses to blow with a 25 per cent overload was 9.2 minutes with a standard deviation of 2.3 minutes. Construct a 0.99 confidence interval for the average time it takes this kind of fuse to blow with a 25 per cent overload.

15. A major truck stop has kept extensive records on various transactions with its customers. If a random sample of 20 of these records shows average sales of 64.5 gallons of diesel fuel with a standard deviation of 2.8 gallons, construct a 0.95 confidence interval for the mean of the corresponding population.

16. In six test runs it took 12, 13, 17, 13, 15, and 14 minutes to assemble a certain mechanical device. If the mean of this sample is used to estimate the actual mean time it takes to assemble the device, what can we assert with a probability of 0.99 about the possible size of the error?

17. Five containers of a commercial solvent randomly selected from a large production lot weigh 25.5, 25.3, 26.0, 25.0, and 25.7 pounds.
   (a) What can we assert with a probability of 0.95 about the possible size of our error in estimating the population mean as 25.5 pounds?
   (b) Construct a 0.99 confidence interval for the mean weight of all the containers of the solvent from which this sample was obtained.

18. Five experimental concrete cylinders have breaking strengths of 62, 64, 58, 65, and 67 (hundreds of pounds per square inch). Construct a 0.98 confidence interval for the mean breaking strength of this kind of concrete cylinder.

19. A random sample of six daily scrap records (where scrap is expressed as a percentage of material requisitioned) shows 3.4, 4.0, 3.8, 6.0, 5.4, and 4.4 per cent scrap. If the mean of this sample is used to estimate the mean percentage of the corresponding population, what can we assert with a probability of 0.90 about the possible size of our error?

20. A distributor of soft-drink vending machines thinks that in a supermarket one of his machines should sell on the average $\mu_0 = 638$ drinks per week. He knows, of course, that this mean will vary somewhat from market to market, and he feels that this variation is measured by a standard deviation of $\sigma_0 = 12.5$ drinks.

(a) If this distributor plans to put one of his soft-drink vending machines into a brand new supermarket, what estimate would he use for the number of drinks he can expect to sell per week?

(b) How would he modify his original estimate if during 50 weeks the machine sells on the average 584 drinks with a standard deviation of 44.3 drinks?

21. A college professor is making up a final examination in banking which is to be given to a very large group of students. His feelings about the average grade they should get is expressed subjectively by a distribution which has the mean $\mu_0 = 62$ and the standard deviation $\sigma_0 = 1.2$. If, subsequently, the examination is tried on a random sample of 100 students whose grades have a mean of 74.8 and a standard deviation of 8.2, find a Bayesian estimate of the average grade all the students in the large group should get.

THE ESTIMATION OF PROPORTIONS

The information that is usually available for the estimation of a proportion (percentage or probability) is the *relative frequency* with which a given event has occurred. If an event occurs x times out of n, the relative frequency of its occurrence is $\dfrac{x}{n}$, and we generally use this *sample proportion* to estimate the true proportion p with which we are concerned. (As used here, the terms "relative frequency" and "sample proportion" are synonymous.) For example, if 624 of 800 factory workers interviewed in a study receive vacation pay from their employer, then $\dfrac{x}{n} = \dfrac{624}{800} = 0.78$, and we can use this figure as a point estimate of the true proportion of factory workers receiving vacation pay. Similarly, a large finance company might estimate the proportion of its debtors who pay their monthly installments on time as 0.85, if a sample of 500 accounts showed that 425 persons made their payment on time.

In the remainder of this section we shall assume that the situations with which we are dealing satisfy (at least approximately) the conditions of the binomial distribution (see page 167). Our information will consist of how many successes there are in a given number of trials, and it will be assumed that the trials are independent and that the probability of a success is p for each trial. In fact, p is the unknown proportion we are trying to estimate. Thus, the sampling distribution which describes the chance fluctuations of our estimates is essentially the binomial distribution, for which we indicated on pages 186 and 188 that its mean and its standard deviation are $\mu = np$ and $\sigma = \sqrt{np(1 - p)}$. An important aspect of these formulas is that they involve the "true" proportion p, and so far as σ is concerned, this causes some difficulties. In order to avoid these complications, at least for the moment, we shall begin by constructing confidence intervals for p with the use of tables designed specially for this purpose.

Tables V(a) and V(b) on pages 478 and 479 provide 0.95 and 0.99 confidence intervals for proportions; they apply to situations where the conditions underlying the binomial distribution are met at least approximately, they are easy to use, and they require practically no calculations. If a sample proportion is less than or equal to 0.50, we begin by marking the value obtained for x/n on the *bottom scale;* we then go up vertically until we reach the two contour lines (curves) which correspond to the size of the sample, and read the confidence limits for p off the *left-hand scale,* as indicated in Figure 9.4. If the sample proportion is greater than 0.50, we mark the value

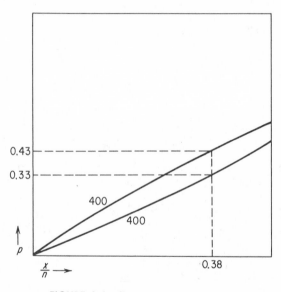

FIGURE 9.4   Confidence limits for p

obtained for x/n on the *top scale,* go down vertically until we reach the two contour lines (curves) which correspond to the size of the sample, and read the confidence limits for p off the *right-hand scale,* as indicated in Figure 9.5.

To illustrate the use of Tables V(a) and V(b) when the sample proportion is less than or equal to 0.50, suppose that in a random sample of 400 television viewers 152 were watching a certain program while 248 were not. Marking $\dfrac{x}{n} = \dfrac{152}{400} = 0.38$ on the *bottom scale* of Table V(a) and proceeding as in Figure 9.4, we find that a 0.95 confidence interval for the true propor-

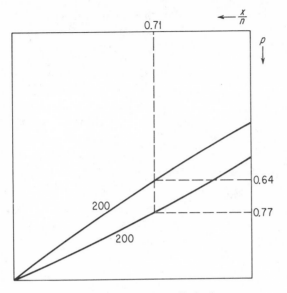

FIGURE 9.5 Confidence limits for p

tion of television viewers who watched the particular program is given by

$$0.33 < p < 0.43$$

In words, we can assert with a probability of 0.95 that the interval from 0.33 to 0.43 contains the actual proportion of television viewers who watched the particular program. Had we wanted a 0.99 confidence interval for this proportion, Table V(b) would, similarly, have yielded

$$0.32 < p < 0.44$$

Note that, as in the estimation of a mean, the interval gets wider when the degree of confidence is increased.

To illustrate the use of Tables V(a) and V(b) when the sample proportion is greater than 0.50, suppose that the owners of a large shopping center want to know what proportion of its customers live within 5 miles of the center. Suppose, also, that a sample of 200 customers reveals that 142 live within 5 miles of the center while 58 live farther away. Marking $\dfrac{142}{200} =$ 0.71 on the *top scale* of Table V(a), halfway between 0.70 and 0.72, and proceeding as in Figure 9.5, we find that the desired 0.95 confidence interval for the true proportion is

$$0.64 < p < 0.77$$

Similarly, using Table V(b), we find that a corresponding 0.99 confidence interval for the true proportion of the shopping center's customers who live within 5 miles of the center is given by

$$0.62 < p < 0.79$$

In words, we can assert with the probability 0.95 that the interval from 0.64 to 0.77 contains the true proportion of the shopping center's customers who live within 5 miles of the center, or with the probability 0.99 that the interval from 0.62 to 0.79 contains the actual proportion.

Note that in both of our examples Tables V(a) and V(b) had contour lines (curves) corresponding to the size of the samples; for values of n other than 8, 10, 12, 16, 20, 24, 30, 40, 60, 100, 200, 400, and 1,000, we literally have to read between the lines. For instance, the reader may wish to verify that, for $x = 60$ and $n = 80$, Table V(a) gives the 0.95 confidence interval

$$0.64 < p < 0.84$$

As we saw in Chapter 7, the binomial distribution can be approximated fairly well with a normal distribution having the mean $\mu = np$ and the standard deviation $\sigma = \sqrt{np(1 - p)}$, provided that n is large and p is not too close to either 0 or 1. Duplicating the argument on page 236, we can thus assert with the probability $1 - \alpha$ that a value which we obtain for x will differ from $\mu = np$ by less than $z_{\alpha/2} \cdot \sqrt{np(1 - p)}$; *dividing everything by n*, we find that this is equivalent to the assertion that the sample proportion $\frac{x}{n}$ will differ from p by less than $z_{\alpha/2} \cdot \sqrt{\dfrac{p(1 - p)}{n}}$. Since the difference between $\frac{x}{n}$ and p is the *error* which we make when using $\frac{x}{n}$ as an estimate of p, we can thus assert with the probability $1 - \alpha$ that this error will be less than the quantity

$$E = z_{\alpha/2} \cdot \sqrt{\frac{p(1 - p)}{n}}$$

An unfortunate feature of this result is that we cannot calculate the size of the *maximum error* we risk with the probability $1 - \alpha$ unless we know p, and of course, p is the quantity we are trying to estimate. However, if we substitute $\dfrac{x}{n}$ for p (as an approximation), *we can assert with a probability of (roughly) $1 - \alpha$ that the error will be less than*

$$E = z_{\alpha/2} \cdot \sqrt{\frac{\dfrac{x}{n}\left(1 - \dfrac{x}{n}\right)}{n}} \qquad \bigstar$$

In view of these various approximations (first we approximated the binomial distribution with a normal distribution, and then we approximated p with x/n), it is preferable to use this method of assaying the possible size of the error only if n is greater than or equal to 100.

To illustrate this technique, let us refer again to the example on page 248, where we had $x = 152$ and $n = 400$ in an experiment designed to estimate the true proportion of television viewers who watched a particular program. If we use the *point estimate* $\dfrac{152}{400} = 0.38$, we can now add that we are "95 per cent sure" that the error of this estimate is less than

$$1.96 \sqrt{\frac{(0.38)(0.62)}{400}} = 0.048$$

Note that $0.38 - 0.048 = 0.332$ and $0.38 + 0.048 = 0.428$ are *very close* to the 0.95 confidence limits which we obtained on page 249 with the use of Table V(a). In fact, these values constitute the *large-sample confidence limits* suggested in Exercise 5 on page 253.

As in the estimation of means, we can use the expression for the maximum error to determine how large a sample is needed to attain a desired degree of precision. If we want to be able to assert with the probability $1 - \alpha$ that a sample proportion will differ from the true proportion p by less than some quantity E, we can solve for n the expression

$$E = z_{\alpha/2} \cdot \sqrt{\frac{p(1 - p)}{n}}$$

given on page 250, and we will thus get

$$n = p(1 - p)\left[\frac{z_{\alpha/2}}{E}\right]^2 \qquad \bigstar$$

Since this formula requires knowledge of p, the quantity we are trying to estimate, it cannot be used as it stands. However, it can be shown that $p(1 - p)$ is at most equal to $\frac{1}{4}$, and that it assumes this maximum value only when $p = \frac{1}{2}$. It follows that it is always "safe" to use the above formula with $p = \frac{1}{2}$, *although this may make the resulting sample size unnecessarily large.* In case we do have some information about the possible range of values p might assume in a given example, we can take this into account in determining n. For instance, if it is reasonable to suppose that the proportion we are trying to estimate lies on the interval from 0.60 to 0.80, we substitute into the above formula whichever value is closest to 0.50; in this particular case we would substitute $p = 0.60$.

To illustrate this technique, suppose we want to estimate what proportion of household goods advertised in the classified ads of a local newspaper actually gets sold through this form of advertising. Suppose, also, that we want to be "95 per cent sure" that the error of our estimate is less than 0.05. *Having no idea what the true proportion might be,* we substitute $E = 0.05$, $p = \frac{1}{2}$, and we get

$$n = \left(\frac{1}{2}\right)\left(\frac{1}{2}\right)\left[\frac{1.96}{0.05}\right]^2 = 384.16$$

Hence, if we base our estimate on a sample of size $n = 385$, we can assert with a probability of (at least) 0.95 that the sample proportion will not be "off" by more than 0.05. (We added the words "at least" because the sample size of 385 may actually be larger than required since we substituted $p = \frac{1}{2}$.)

Had we known in this example that the quantity we are trying to estimate is in the neighborhood of, say, 0.30, the formula for n would have yielded

$$n = (0.30)(0.70)\left[\frac{1.96}{0.05}\right]^2 = 323$$

This illustrates the fact that if we do have some information about the possible size of the proportion we hope to estimate, this can appreciably reduce the size of the required sample. (Of course, if we wanted to use a probability of 0.98 or 0.99 in this kind of example, we would have only to substitute 2.33 or 2.58 for 1.96 in the formula for n.)

EXERCISES

1. A study was made of the causes of failure of a sample of 400 small business firms during the year 1970. If 264 of these firms' fixed assets exceeded 75 per cent of total worth at the time of failure, find a 0.95 confidence interval for the true proportion of small business firms that failed in that year, whose fixed assets exceeded 75 per cent of total worth at the time of failure.

2. A random sample of 100 women shoppers in a supermarket is asked which of two packages of a product is cheaper per ounce: the $4\frac{3}{8}$ ounce package priced at 19 cents or the 6-ounce package priced at 27 cents. If 34 of the shoppers chose the 6-ounce package, find a 0.99 confidence interval for the true proportion of women shoppers who would make this mistake.

3. In a health study of its employees, a large national manufacturer took a random sample of 60 employees who, at some time during the past three months, either failed to report for work or, once there, asked to be permitted to go home. If 39 of these employees gave a headache as the cause of their nonattendance or request to go home, find a 0.99 confidence interval for the true proportion of employees' absences or requests to go home that can be attributed to headaches.

4. In a random sample of 300 pedestrian fatalities in a large city-suburban area, it was found that 108 of the victims were crossing a street outside a crosswalk or in the middle of a block. Find a 0.99 confidence interval for the true proportion of pedestrian fatalities in that area in which the pedestrian crosses a street outside a crosswalk or in the middle of a block.

5. Following the suggestion on page 251, we can obtain approximate $1 - \alpha$ *large-sample confidence limits* for population proportions by means of the formula

$$\frac{x}{n} \pm z_{\alpha/2} \cdot \sqrt{\frac{\dfrac{x}{n}\left(1 - \dfrac{x}{n}\right)}{n}}$$

   (a) Use this formula to rework Exercise 1 and compare the results.
   (b) Use this formula to rework Exercise 2 and compare the results.
   (c) Use this formula to rework Exercise 4 and compare the results.

6. A television station took a random sample of 500 of its viewers, of whom 430 stated that they preferred "being informed by professional weathermen" during the evening news program to "being entertained on these programs by

pretty girls." If $\frac{430}{500} = 0.86$ is used as an estimate of the corresponding true proportion, what can we assert with a probability of 0.95 about the possible size of the error?

7. In a study conducted by a trading stamp company, it was found that among 1,000 randomly selected male shoppers who were deliberately not offered trading stamps, 850 asked for them.

   (a) Find a 0.95 confidence interval for the corresponding true proportion using Table V(a).

   (b) Find a 0.95 confidence interval for the corresponding true proportion using the formula of Exercise 5.

   (c) What can one assert with a probability of 0.99 about the possible size of the error, if the corresponding true proportion is estimated as 0.85?

8. In a study of package design, a national manufacturer wants to determine what proportion of purchases of razor blades for use by men are actually made by women. If a random sample of 500 such purchases includes 110 made by women and the manufacturer estimates the desired proportion as $\frac{110}{500} = 0.22$, what can the manufacturer assert with a probability of 0.95 about the possible size of the error?

9. In a random sample of 240 women, 144 said that for their next dishwasher they intend to buy a front-loading mobile unit. If $\frac{144}{240} = 0.60$ is used as an estimate of the corresponding true proportion for the population from which the sample came, what can one assert with a probability of 0.90 about the possible size of the error?

10. In connection with developing a new product, a food processor wants to estimate what proportion of women prefer a lightly salted cocktail cracker to a more heavily salted cracker. How large a sample must he take to be at least 95 per cent sure that the sample proportion will not be off by more than 0.02?

11. A bank wishes to estimate what proportion of all spending units with incomes of at least $7,000 before taxes have no outstanding debt on either new or used cars. How large a sample must be taken to be able to assert with a probability of at least 0.99 that the sample proportion will be off at most by 0.05?

12. A large finance company wants to estimate from a sample of its thousands of accounts what proportion of its customers plans to buy either an automobile, furniture, or major household appliances on credit during the coming year.

   (a) How large a sample will be needed to be able to assert with a probability of at least 0.95 that the sample proportion will be off by less than 0.04?

   (b) If the company has reason to believe that the actual proportion is close to 0.60, how large a sample would it need to be able to assert with a probability of 0.95 that the sample proportion it gets will be off by less than 0.04?

13. An insurance company, considering a "safe-driver plan" (with reduced rates on liability insurance for drivers who have neither had an accident nor been cited for a traffic violation in the three previous years), wants to estimate what proportion of its many policy holders would qualify for the plan.

(a) How large a sample will it have to take from its files to be able to assert with a probability of at least 0.99 that the sample proportion and the true proportion will differ by less than 0.03?

(b) If it is known from experience that the desired proportion is close to 0.20, how large a sample would the company need to be able to assert with a probability of 0.99 that the sample proportion and the true proportion will differ by less than 0.03?

14. If a sample constitutes a substantial portion (say, 5 per cent or more) of a population, the methods of this section cannot be used without appropriate modifications. If the sample itself is large, we can use the same *correction factor* as in the estimation of means, and we can write approximate $1 - \alpha$ confidence limits for p as

$$\frac{x}{n} \pm z_{\alpha/2} \sqrt{\frac{\frac{x}{n}\left(1 - \frac{x}{n}\right)}{n}} \cdot \sqrt{\frac{N - n}{N - 1}} \qquad \bigstar$$

where N is, as before, the size of the population.

(a) Among the 400 families living in a large apartment complex 200 are interviewed, and it is found that 68 of these have children of college age. Construct a 0.95 confidence interval for the actual proportion of all families living in the complex who have children of college age.

(b) A hosiery manufacturer feels that unless his employees agree to a 10 per cent wage reduction he cannot stay in business. If in a random sample of 80 of his 400 employees only 24 feel "kindly disposed" to the reduction, find a 0.99 confidence interval for the corresponding proportion for all of his employees.

THE ESTIMATION OF σ No

So far we have learned how to construct confidence intervals for means and proportions, and we have learned how to evaluate the precision of the corresponding point estimates of μ and p. Although this will take care of a great variety of situations—in fact, the vast majority of problems of estimation—similar methods can be used to estimate other parameters of populations. By studying the sampling distributions of appropriate statistics, statisticians have developed confidence intervals for population standard deviations, medians, quartiles, and the like. In principle, the ideas are always the same and the main difficulty lies in the fact that some of these sampling distributions are mathematically quite involved. Fortunately, this difficulty is resolved by the important result that for *large samples* many of these sampling distributions can be approximated with normal curves.

For instance, if we are dealing with large samples, the sampling distribution of the standard deviation s can be approximated closely with a normal distribution having the mean σ and the standard deviation

$$\sigma_s = \frac{\sigma}{\sqrt{2n}}$$

called the *standard error of s* in accordance with the terminology introduced in Chapter 8. Now, if we reason as on page 236, we arrive at the result that a sample value of s will fall on the interval from $\sigma - z_{\alpha/2} \cdot \dfrac{\sigma}{\sqrt{2n}}$ to $\sigma + z_{\alpha/2} \cdot \dfrac{\sigma}{\sqrt{2n}}$ with the probability $1 - \alpha$, and fairly straightforward algebra leads to the following *large-sample confidence interval for the population standard deviation* σ:

$$\frac{s}{1 + \dfrac{z_{\alpha/2}}{\sqrt{2n}}} < \sigma < \frac{s}{1 - \dfrac{z_{\alpha/2}}{\sqrt{2n}}} \qquad \bigstar$$

Referring again to the example on page 236, where we were interested in the average time it takes to patrol a certain size warehouse, and substituting $n = 40$, $s = 8.52$ (minutes), and $z_{.025} = 1.96$ into the above formula, we get

$$\frac{8.52}{1 + \dfrac{1.96}{\sqrt{80}}} < \sigma < \frac{8.52}{1 - \dfrac{1.96}{\sqrt{80}}}$$

$$6.98 \text{ minutes} < \sigma < 10.92 \text{ minutes}$$

Thus, we can assert with a probability of 0.95 that the interval from 6.98 minutes to 10.92 minutes contains σ, the true value of the standard deviation which measures the variability in the time it takes to patrol the given size warehouse.

No

EXERCISES

1. Referring to Exercise 1 on page 244, construct a 0.95 confidence interval for the true standard deviation of the monthly rents paid for two-room apartments in the given city.

2. Referring to Exercise 3 on page 244, construct a 0.99 confidence interval for the true standard deviation of the body repair costs on the particular kind of damage.

3. Referring to Exercise 5 on page 244, construct a 0.95 confidence interval for the true standard deviation of the amounts owed in the telephone company's overdue accounts.

4. The standard deviations of the 40 samples on page 220 are, respectively, 3.27, 3.11, 5.54, 3.39, 2.55, 4.38, 4.39, 1.78, 4.72, 6.83, 3.39, 3.51, 3.96, 6.10, 4.21, 3.71, 0.84, 4.87, 1.58, 2.24, 2.51, 4.47, 3.58, 0.84, 0.84, 2.05, 3.39, 4.53, 5.08, 4.21, 0.89, 6.91, 3.65, 4.44, 3.49, 4.30, 3.36, 4.51, 3.13, and 3.71. Calculate the standard deviation of these 40 sample standard deviations and compare the result with the value one might expect in accordance with the standard error formula $\sigma_s = \dfrac{\sigma}{\sqrt{2n}}$ with $\sigma = 4$ and $n = 5$.

A WORD OF CAUTION

The methods which we have studied in this chapter are *standard methods* which apply to a wide variety of *standard situations*. However, we often run into situations which are far from being "standard," and the methods of this chapter must, therefore, be used with a good deal of discretion. In recent years, attempts have been made to treat all problems of statistical inference (including problems of estimation) within the framework of a unified theory. Although this theory, called *decision theory*, has many conceptual and theoretical advantages, its application poses problems which are difficult to overcome. To understand these problems, one must appreciate the fact that no matter how objectively an experiment or an investigation may be planned, it is impossible to eliminate all elements of subjectivity. It is at least partially a subjective decision whether to base an experiment (e.g., the determination of an index of diffraction) on 3 measurements, on 5 measurements, or on 10 or more. Also, subjective factors invariably enter the design of equipment, the hiring of personnel, and even the precise formulation of a problem one wants to investigate. An element of subjectivity enters even when we define such terms as "good" or "best" in connection with the choice between different decision criteria (e.g., when deciding between a sample mean and a sample median in a problem of estimation), or when looking for the straight line which "best" fits a set of paired data. Above all, subjective judgments are practically unavoidable when one is asked to put "cash values" on the risks to which one is exposed. In contrast to the examples which we used in our discussion of *game theory* in Chapter 7, it is generally impossible in statistics to be completely objective in specifying rewards for being right (or close) and penalties for being wrong (or not close enough). After all, if a scientist is asked to judge the safety of a piece of equipment, how can he put a cash value on the consequences of a possible error on his part, if such an error may result in the loss of human lives?

With the exception of the section beginning on page 242, the general approach we have used in this chapter (and will use in subsequent chapters) is called the *classical approach*, which does not *formally* take into account the various subjective factors we have mentioned. In other words, in the classical approach the subjective elements do not appear as part of the formulas; rather, they will appear in the choice among formulas to be used in a given situation, in decisions concerning the size of a sample, in specifying the probabilities with which we are willing to incur certain risks, in specifying the maximum errors we consider acceptable, and so forth.

10

DECISION MAKING:
TESTS
OF HYPOTHESES

TWO KINDS OF ERRORS

In Chapter 9 we studied the problem of estimating population means, proportions, and standard deviations. The procedures we developed there enable us to make estimates of, for example, the "true" mean outside diameter of a lot of precision ball bearings, the "true" proportion of all the readers of a certain underground newspaper who are college graduates, and the "true" standard deviation of the water flows at the meters of all the homes in a city.

In this chapter we shall investigate problems of statistical inference that are of a somewhat different nature. These new problems are called problems of *testing hypotheses*. Unlike estimation problems, in which we are required to estimate an unknown population parameter, in hypothesis-testing problems we are required to decide whether or not a population parameter is in fact equal to some prescribed value. Instead of asking, for example, what the mean assessed value of all the homes in a certain district of a large city is, we shall now have to decide whether or not the mean assessed value is equal to some particular value (say, $5,400). At this point, it may seem to make little difference how we state the problem, but it will soon become clear that a number of considerations arise in connection with testing hypotheses that are not present in estimating problems.

To illustrate the nature of the situation one faces when testing a statistical hypothesis, consider the following example: A company manufactures a liquid kitchen cleaning wax which it sells in cans marked "300 grams net weight" (about 10.6 ounces). It is known from long experience that the variability of the process is stable and well established at $\sigma = 5$ grams. The cans are filled by machine, and the company makes every effort to control the mean net weight (or the mean "fill") at the 300-gram standard. However, small errors occur in the machine settings, parts wear, for instance, and the mean fill sometimes varies more or less widely from the desired 300 grams. Small departures—on the order of 1 gram or less—from standard are of no consequence, but increasingly larger departures in either direction are of increasingly more concern. Overfilling means giving away product, the value of which in time may amount to the profit on a large volume of sales. Underfilling results in a loss to consumers and also exposes the company to possible punitive action by standards enforcement agencies.

In trying to control the mean fill of the cans at 300 grams, the company has devised the following inspection procedure. Each hour during production runs, the company takes a random sample of 25 cans from the production lot, calculates the sample mean weight $\bar{x}$, and decides on the basis of this value whether or not the process is "in control" (the mean fill is 300 grams, as it is supposed to be) or "out of control" (the mean fill is not 300 grams). To make this decision, the company must have some unambiguous criterion, or rule, to follow. Accordingly, the company has specified this criterion: Consider the process to be out of control if $\bar{x}$ is either less than 297 grams or greater than 303 grams, and consider it to be in control if $\bar{x}$ is between 297 and 303 grams (both values included). So far as action is concerned, when the process is judged to be out of control, it is shut down immediately and a plant engineer is sent in to find out what (if anything) is wrong with it and put it back in control; when the process is judged to be in control, it is allowed to continue in operation without interruption.

In the language of statistics, the company wishes to test, for each submitted lot, the *hypothesis* that the mean net weight of the wax (the mean fill) is 300 grams against the *alternative* that the mean net weight is not 300 grams. And it will make this test on the basis of the following criterion: Reject the hypothesis (and accept the alternative) if $\bar{x} < 297$ grams or if $\bar{x} > 303$ grams; otherwise, accept the hypothesis.

We may call the hypothesis that the process is in control Hypothesis H and write it as H: $\mu = 300$ grams, and the alternative that the process is out of control Alternative A and write it as A: $\mu \neq 300$ grams. Clearly, the hypothesis is either true or false, and whenever it is tested the criterion will lead either to its acceptance or rejection. Unfortunately, though, on any given occasion the company may err in either one or the other of two ways (but not both). First, the company may decide that the process is out of control when in fact it is in control. This will happen if the mean (lot) can weight is actually 300 grams but the sample mean $\bar{x}$ is less than 297 grams or greater than 303 grams; the consequence of this error is that a process operating at the desired level is shut down while an engineer looks for nonexistent trouble. Second, the company may decide that the process is in control when in fact it is out of control. This will happen if the mean (lot) can weight is not 300 grams (is, say, only 290 grams) but the sample mean $\bar{x}$ is between 297 and 303 grams; the consequence of this error is that a process which is not operating at the desired standard is allowed to continue operating.

We may summarize the situation facing the company in the following table. If the hypothesis is true, the decision to accept it is the correct one;

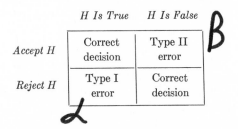

	H Is True	H Is False
Accept H	Correct decision	Type II error
Reject H	Type I error	Correct decision

conversely, if the hypothesis is false, the decision to reject it is the correct one. On the other hand, if the hypothesis is *true* and it is *rejected,* an error has been committed; the error which is committed when one rejects a true hypothesis is called a *Type I error* and the probability of committing it is designated by the Greek letter α (alpha). Conversely, if the hypothesis is *false* and it is *accepted,* an error has been committed; the error which is committed when one accepts a false hypothesis is called a *Type II error,* and the probability of committing it is designated by the Greek letter β (beta).

We shall now direct our attention to an investigation of the "goodness" of the company's criterion. Specifically, we shall calculate the probability α the company faces of rejecting H if it is true, and the probabilities the company faces of accepting H if the mean fill is some one of various possible values other than 300 grams. If these probabilities are satisfactory from an operational standpoint, the criterion may be considered to be a "good" one in the sense that it provides the company with suitable protection against committing one or the other of the two errors.

Recalling the theory of Chapter 8, we observe that the probability α of rejecting the hypothesis if it is true is just the probability of obtaining a mean of less than 297 grams or more than 303 grams in a random sample of size 25 drawn from a population whose mean is $\mu = 300$ and whose standard deviation is $\sigma = 5$ grams. This probability is represented by the shaded area of Figure 10.1, and using the normal curve approximation to the

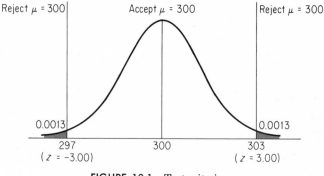

Reject $\mu = 300$ Accept $\mu = 300$ Reject $\mu = 300$

0.0013   0.0013

297   300   303
($z = -3.00$)   ($z = 3.00$)

FIGURE 10.1   Test criterion

sampling distribution of the mean, it is easily found. According to the theorem on page 222, the standard deviation of this sampling distribution (the *standard error of the mean*), $\sigma_{\bar{x}}$, is given by $\sigma/\sqrt{n}$. Hence, we calculate

$$z_1 = \frac{297 - 300}{5/\sqrt{25}} = -3.00 \quad \text{and} \quad z_2 = \frac{303 - 300}{5/\sqrt{25}} = 3.00$$

and it follows from Table I that the area in each tail of the sampling distribution shown in Figure 10.1 is $0.5000 - 0.4987 = 0.0013$. Therefore, the probability that $\bar{x}$ will either be less than 297 or greater than 303 is $0.0013 + 0.0013 = 0.0026$, and this is the probability α of rejecting the true hypothesis that the process is in control. In other words, there are about three chances in 1,000 that the company will commit the error of shutting the process down and looking for trouble which does not exist.

Suppose now that, due to some equipment malfunction, the actual mean fill has shifted and is only 296 grams. The hypothesis is false and the process is out of control. In this case, the sample mean does not come from the distribution of Figure 10.1 but instead comes from the distribution of Figure 10.2. Thus, the probability of not detecting this shift and of accepting the false hypothesis that the process is in control is represented by the shaded area of Figure 10.2. It is the probability of getting a sample mean between 297 and 303 grams from a population whose mean is $\mu = 296$ and, as before, whose standard error is $5/\sqrt{25}$. We calculate

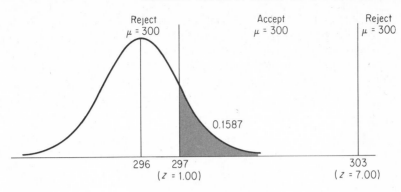

FIGURE 10.2 Test criterion

$$z_1 = \frac{303 - 296}{5/\sqrt{25}} = 7.00 \quad \text{and} \quad z_2 = \frac{297 - 296}{5/\sqrt{25}} = 1.00$$

and it follows from Table I that the shaded area of Figure 10.2 is 0.5000 − 0.3413 = 0.1587 since the area to the right of $z = 7.00$ is negligible. Therefore, the probability β of accepting the false hypothesis that the process is in control when μ is actually 296 grams is 0.1587. In other words, there are about 16 chances in 100 that the company will commit the error of letting the process run, thinking the mean fill is 300 grams when it is really only 296 grams.

In computing directly above the probability of accepting a false hypothesis, we supposed that the process mean had shifted from 300 to 296 grams. This alternative value is, after all, only one of the infinitely many possible values other than 300 grams for the actual mean weight, and for every one of them there is a probability β that the company will accept the nypothesis when it is false. We can calculate the probabilities of accepting the hypothesis that the mean weight is 300 grams for several possible alternatives and examine these probabilities to see how well, under various circumstances, the criterion controls the risks facing the company. The procedure is precisely the same as that we used in the illustration directly above. The probability of accepting the hypothesis that the mean weight is 300 grams when it is in fact, say, 305 grams is just the probability that the mean of a sample of 25 weights drawn from a population with mean 305 grams and standard deviation 5 grams will lie between 297 and 303 grams. By drawing a figure like Figure 10.2, with $\mu = 305$, and proceeding as we did above, we find that there is a probability β of 0.0228 that the company will commit the error of deciding the process is in control when the mean fill is actually 305 grams.

The third column of the table below shows, for 13 possible values of μ, the probabilities of accepting the hypothesis that $\mu = 300$ grams when μ is actually 294, 295, . . . , 305, and 306 grams. When the value of μ is 300, H is true and the probability of accepting it is the probability of *not rejecting* a true hypothesis: $1 - \alpha = 1 - 0.0026 = 0.9974$. Therefore, the probabilities shown of accepting H are the probabilities β of accepting a false

Value of μ	Probability of Type II Error	Probability of Accepting H
294	0.0013	0.0013
295	0.0228	0.0228
296	0.1587	0.1587
297	0.5000	0.5000
298	0.8413	0.8413
299	0.9772	0.9772
300	—	0.9974
301	0.9772	0.9772
302	0.8413	0.8413
303	0.5000	0.5000
304	0.1587	0.1587
305	0.0228	0.0228
306	0.0013	0.0013

hypothesis with the exception of the probability 0.9974 which is the probability of avoiding a Type I error. The probabilities of committing a Type II error, shown in the second column, are the same as the corresponding probabilities of accepting H except where $\mu = 300$ grams. In this case, H is true and there is no possibility of committing a Type II error.

The probabilities of accepting H are plotted in Figure 10.3. The curve shown there is called the *operating characteristic* curve of the test criterion, or simply its *OC-curve*. Examination of this curve provides a good understanding of the nature of the test criterion. The probability of accepting the test hypothesis is greatest when it is true. For small departures from the 300-gram standard there is a high probability of accepting H; that is, when the process deviates only slightly from standard, it will most likely be allowed to continue to operate. But for larger and larger departures from standard in either direction, the probabilities of failing to detect them and accepting H in error become smaller and smaller. This is precisely what the company wants, and any test procedure which did not behave in this way would not be at all suited to its needs.

The OC-curve of Figure 10.3 applies to the case where the hypothesis that $\mu = 300$ grams is accepted if the mean of a random sample of 25 weights

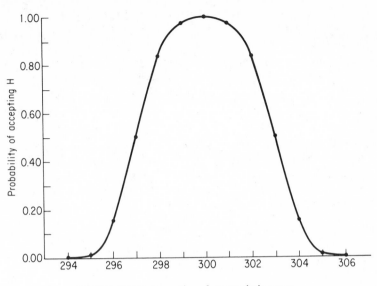

FIGURE 10.3 Operating characteristic curve

falls between 297 and 303 grams. If it wishes, the company can change the shape of the OC-curve (and hence, change the amount of protection it is getting against committing Type I and Type II errors) by changing the test criterion or the sample size. In fact, OC-curves can often be made to assume a particular desired shape by an appropriate choice of the sample size and/or the dividing lines of the test criterion.

A detailed study of OC-curves would go considerably beyond the scope of this text, and the purpose of our illustration was mainly to show how statistical methods can be used to analyze and measure the risks to which one is exposed in testing hypotheses. Of course, these methods are not limited to problems connected with quality control in industrial plants. To name but a few of countless possibilities, Hypothesis H might well be the hypothesis that a shipment will arrive on time, the hypothesis that the demand for a product will increase in the next year, the hypothesis that the price of tin will go up faster than the price of copper, or the hypothesis that a new business will succeed.

EXERCISES

1. Suppose that on the basis of a sample we want to test the hypothesis that private passenger cars in the United States are driven on the average 12,000

miles per year. Explain under what conditions we would commit a Type I error and under what conditions we would commit a Type II error.

2. Suppose that on the basis of a sample we want to test the hypothesis that the average assessed value of all the homes in one area of a large city is $4,300. Explain under what conditions we would commit a Type I error and under what conditions we would commit a Type II error.

3. Verify the values of the OC-curve given in the column entitled "Probability of Accepting H" on page 264.

4. (*Power function*) In hypothesis testing, a function whose values are the probabilities of *rejecting* a given hypothesis (and accepting the alternative) for various values of the parameter under consideration is called a *power function*. Thus, in the filling example, for all values of μ other than the hypothetical 300 grams, the power function gives the probabilities $1 - \beta$ of *not* committing a Type II error; for $\mu = 300$, however, it gives the probability α of committing a Type I error. Obviously, the values of the power function are 1 *minus* the corresponding values of the OC-curve. Find the power function for the filling example, and plot and examine its curve.

5. Suppose that in the filling example the criterion is changed so that the hypothesis that $\mu = 300$ grams is accepted if the sample mean falls between 297.5 and 302.5 grams; otherwise, the hypothesis is rejected.

   (a) Find the OC-curve of this test, calculating the probability β for the same 12 alternative mean fills shown on page 264.

   (b) Plot the OC-curve.

   (c) Find the power function and plot its curve.

6. An automatic machine in a food-processing plant is supposed to set the lids on pint jars of mayonnaise so that the average "twist" required for a person to loosen the lids ("break the sets") is 30 inch-pounds. It is known from long experience that the variability of the sets is stable and given by $\sigma = 2.0$ inch-pounds. The processor does not want the lids set too loosely since this may cause discoloration and spoilage of the mayonnaise, and he does not want them set too tightly since people resent having to struggle with stubborn lids. Consequently, the processor sets up a hypothesis that the lids are set at 30 inch-pounds on the average (the process is in control) and an alternative that the lids are not set at 30 inch-pounds (the process is out of control). The hypothesis is tested periodically by taking from production lots of sealed jars random samples of 16 jars, determining the mean twist $\bar{x}$ required to break the sets, accepting the hypothesis if $\bar{x}$ is between 28.8 inch-pounds and 31.2 inch-pounds and rejecting it if $\bar{x}$ is either less than 28.8 inch-pounds or greater than 31.2 inch-pounds.

   (a) Find the OC-curve of this test criterion, calculating the probability β of deciding the process is in control when it is not for alternative mean sets of 27.5, 28.0, 28.5, 29.0, 29.5, 30.5, 31.0, 31.5, 32.0 and 32.5 inch-pounds.

   (b) Plot the OC-curve.

7. Referring to Exercise 6, suppose that the criterion is changed so that the hypothesis is accepted if $\bar{x}$ falls between 29.0 and 31.0 and rejected otherwise.
   (a) Find the OC-curve of this test, calculating the probability β for the same 10 alternative means given in Exercise 6.
   (b) Plot the OC-curve and compare it with the one plotted in Exercise 6.

8. Referring to Exercise 6, suppose that the criterion is changed so that the hypothesis is accepted if $\bar{x}$ falls between 28.5 and 31.5 and rejected otherwise.
   (a) Find the OC-curve of this test, calculating the probability β for the same 10 alternative means given in Exercise 6.
   (b) Plot the OC-curve and compare it with those plotted in Exercises 6 and 7.

NULL HYPOTHESES AND SIGNIFICANCE TESTS

In the example of the preceding section we formulated a hypothesis that the filling process was in control and stated it as H: $\mu = 300$ grams. The reason we stated the hypothesis this way, instead of hypothesizing that the process was not in control or that μ was not 300 grams, is so that we could calculate the probability of committing a Type I error. In the example, it was known that $\sigma = 5$ grams and, using the theory of Chapter 8, we readily calculated the probability α that the mean weight of a random sample of 25 fills was less than 297 grams or more than 303 grams when the process mean fill was actually 300 grams. We shall follow the general rule of stating hypotheses in such a way that we can always calculate α or, in other words, so that we know what to expect if they are true. Following this rule we shall sometimes find ourselves hypothesizing just the opposite of what we may want to prove. For instance, if we want to show that a new copper-bearing steel has a higher yield strength than ordinary steel, we formulate the hypothesis that the two yield strengths are the same. Similarly, if we want to show that one method of teaching Morse code is more effective than another, we hypothesize that the two methods are equally effective; and if we want to show that the proportion of sales slips incorrectly written up in one department of a store is greater than that in another department, we formulate the hypothesis that the two proportions are identical. Since we hypothesize that there is *no difference* in the yield strengths, *no difference* in the effectiveness of the two methods, and *no difference* in the two proportions, we call hypotheses such as these *null hypotheses*.

We have shown above how we can easily calculate the probability of committing a Type I error when we are given a specific hypothesis, an unambiguous criterion, and enough information about the situation to apply the theory of Chapter 8. We have also shown that we can easily calculate the probability of committing a Type II error when some *specific value* other than the one prescribed by the hypothesis is the true value. Indeed, we can calculate the probabilities of accepting a null hypothesis if various other possible *specific values* are the true values, and then draw an OC-curve and examine it to see whether the test criterion satisfactorily controls the risks of committing Type I and Type II errors.

Although a positive probability β of accepting a false hypothesis exists for all values of μ alternative to the test value, we can sometimes avoid a Type II error altogether. To illustrate how this might be done, suppose that a large title-search company knows from experience that the mean number of typing errors made on submitted copies of Form A by typists preparing these forms is 2.3 per day with a standard deviation of 0.75. In trying to confirm its suspicion that one particular typist makes more errors on the average than the others, the company takes a random sample of nine forms prepared by this typist and tests the null hypothesis that there is no difference between her performance and that of the others (that $\mu = 2.3$ applies to this girl also) by using the following criterion:

Reject the null hypothesis that $\mu = 2.3$ (and accept the alternative that $\mu > 2.3$) if the typist averages 2.7 or more errors per form; otherwise, reserve judgment (pending further performance checks).

With this criterion there is no possibility of committing a Type II error; when the hypothesis is tested it may be rejected outright (and the typist presumed to be less proficient than average), but otherwise, it will not actually be accepted.

The procedure we have just outlined is referred to as a *test of significance*. If the difference between what we expect under the hypothesis and what we observe in a sample is too large to be reasonably attributed to chance, we reject the null hypothesis. If the difference between what we expect and what we observe is so small that it might well be attributed to chance, we say that the result is *not (statistically) significant*. We then either accept the null hypothesis or reserve judgment depending on whether a definite decision one way or the other is required. In the example above, the company's suspicion that the typist is worse than average is confirmed if she averages 2.7 or more errors; in that case it is felt that the difference between the sample mean and $\mu = 2.3$ is too large to be attributed to chance. (See Exercise 9 on page 273.) If the typist averages less than 2.7 errors, the company feels that the test did not confirm its suspicion (but nevertheless, it may remain suspicious). In tests of this sort, as in most criminal proceedings, the burden of guilt is put on the "prosecution," and the "defendant" is found not guilty unless his guilt is proven beyond a reasonable doubt.

Referring again to the filling example of the previous section, we note that the company is actually performing a significance test each hour. The criterion for the tests may be written in the following form:

> Reject the Hypothesis: $\mu = 300$ grams (and accept the Alternative: $\mu \neq 300$ grams) if the mean of the 25 sample fills is less than 297 grams or greater than 303 grams; accept the hypothesis if it falls on the interval from 297 to 303 grams.

The reason this criterion specifies that the company is either to reject the hypothesis or to accept it is because one cannot really reserve judgment in these circumstances. This is because once the process is in operation, it must either be allowed to continue in operation or be shut down. Any time the company rejects the hypothesis (and stops the filling operation) there is a probability $\alpha = 0.0026$ of stopping it in error. And any time it lets the process continue in operation, it has actually accepted the hypothesis that the process is in control. No matter how the criterion is phrased, so long as the filling process continues, there is no way possible to avoid the consequences of committing a Type II error. Whether or not one can afford the luxury of reserving judgment in any given situation depends entirely on the nature of the situation. In general, whenever a decision must be reached one way or the other when a test is made, we are required either to reject or accept a hypothesis, and we have no way to avoid the risk of accepting a false one.

The general problem of constructing statistical decision criteria and testing hypotheses often seems confusing, and it helps to proceed in the systematic way outlined below:

1. *We formulate a (null) hypothesis in such a way that the probability of a Type I error can be calculated; we also formulate an alternative hypothesis so that the rejection of the null hypothesis is equivalent to the acceptance of the alternative hypothesis.*

In the filling example the null hypothesis was $\mu = 300$ grams, and the alternative was $\mu \neq 300$. We refer to this kind of alternative as a *two-sided alternative*. The company has specified this alternative because it wants protection against thinking the filling process is in control when it is either underfilling or overfilling the cans. In the typist example, the null hypothesis is $\mu = 2.3$ mistakes, and the alternative is $\mu > 2.3$. This is called a *one-sided alternative*, and the firm has chosen this alternative because it feels that the burden of proof is on it to show that the typist is poorer (makes more mistakes) than the average. (The firm could hardly argue that the girl is poorer than average if she averages less than 2.3 mistakes per form.) We can also write a one-sided alternative with the inequality sign pointing in the other direction. For instance, if we wanted to determine whether the average time required for secretarial-pool employees to do a certain job is less than 40 minutes, we might take a random sample of, say, 20 of these times and test the null hypothesis that $\mu = 40$ against the alternative that $\mu < 40$. In writing an alternative hypothesis we usually specify that the population mean is not equal to, is less than, or is greater than the value assumed under the null hypothesis. In connection with choosing the correct alternative in a particular situation, we want to stress the fact that the choice depends entirely on the nature of the problem itself.

2. *We specify the probability α of committing a Type I error; if possible, desired, or necessary, we may also make some specifications about the probabilities β of Type II errors for specific alternatives.*

The probability of committing a Type I error is usually referred to as the *level of significance* at which a test is performed. Usually, tests are performed at a level of significance of 0.05 or 0.01. Testing a hypothesis at a level of significance of, say, $\alpha = 0.05$ simply means that we are fixing the probability of rejecting the hypothesis if it is true at 0.05.

Once we have set up a hypothesis, the probability of committing a Type I error is absolutely under our control and can be made as small as we like. How small we actually make it in a particular case depends on the consequences (cost, inconvenience, embarrassment, etc.) of rejecting a true hypothesis. Ordinarily, the more serious the consequences which result from committing a Type I error, the smaller the risk we are willing to take of committing it. However, we are restrained in practice from setting very low probabilities α by the fact that, other things being equal, the smaller we make the probability of rejecting a true hypothesis, the larger the probability β of accepting a false one becomes. If, for instance, in the filling example the probability of shutting down the process when it is in control were reduced from $\alpha = 0.0026$ to $\alpha = 0.0001$, the probability of not detecting that the process was out of control when the mean fill was actually 304 grams would be increased from $\beta = 0.1587$ to $\beta = 0.4602$. What we usually need in practice is some reasonable balance between the probabilities of committing the two errors, and (except in industrial quality control) levels of significance of 0.05 or 0.01, or thereabouts, have been found to provide this balance.

3. *We use suitable statistical theory to construct a criterion for testing the (null) hypothesis formulated in Step 1 against the alternative specified in Step 1 at the level of significance specified in Step 2.*

In the filling example, we based the criterion on the normal curve approximation to the sampling distribution of $\bar{x}$; in general, it depends on the *statistic* upon which we want to base the decision and on its sampling distribution. A good portion of the remainder of this book will be devoted to the construction of such criteria. As we shall see later, this usually involves choosing an appropriate statistic, specifying the sample size, and then determining the dividing lines (critical values) of the criterion. In the previous examples, we used a *two-sided test* (or *two-tail test*) with the two-sided alternative $\mu \neq 300$ in the filling example, rejecting the null hypothesis for either *small* or *large* values of $\bar{x}$; in the example of the number of mistakes made by a typist, we used a *one-sided test* (or *one-tail test*) with the one-sided alternative $\mu > 2.3$, rejecting the null hypothesis only for *large* values of $\bar{x}$; and in the example of the speed required to do a certain job, we used a one-sided test (or one-tail test) with the one-sided alternative $\mu < 40$, rejecting the null hypothesis only for *small* values of $\bar{x}$. In general, a test is said to be one-sided or two-sided (one-tailed or two-tailed), depending on whether the null hypothesis is rejected for values of the statistic falling into *either* tail or *both* tails of its sampling distribution.

4. *We specify whether the alternative to rejecting the hypothesis formulated in Step 1 is to accept it or to reserve judgment.*

This, as we have said, depends on whether we must make a decision one way or the other on the basis of the test, or whether the circumstances of the problem are such that we can delay a decision pending further study. Sometimes we may accept a null hypothesis with the hope that we are not exposing ourselves to excessively high risks of committing serious Type II errors. Of course, if it is necessary and we have enough information, we can calculate the probabilities needed to get an overall picture from the OC-curve of the test criterion.

Before we discuss various special tests in the next few sections, let us point out that the discussion of this and the preceding section is not limited to tests concerning means. The concepts we have introduced apply equally well to hypotheses concerning proportions, standard deviations, randomness of samples, relationships among several variables, trends of time series, and so on.

EXERCISES

1. Whether an error is a Type I error or a Type II error depends on how a hypothesis is formulated. For instance, suppose that a large insurance firm is

concerned about the extent to which its policy holders are interested in a proposed new type of homeowner's policy.

(a) If the company formulates the hypothesis "at least 65 per cent of our policy holders will be interested in the new policy," what is the alternative and what are the consequences of a Type I error and of a Type II error?

(b) If the company formulates the hypothesis "less than 65 per cent of our policy holders will be interested in the new policy," what is the alternative and what are the consequences of a Type I error and of a Type II error?

2. Suppose that a high school principal is being tested by a psychological testing service to determine whether or not he is emotionally fit to take the position of superintendent of schools in a large city.

(a) If the testing service hypothesizes that the man is fit, what is the alternative and what are the consequences of a Type I error and of a Type II error?

(b) If the testing service hypothesizes that the man is not fit, what is the alternative and what are the consequences of a Type I error and of a Type II error?

3. Suppose that a study is being made to test manufacturer claims that lead-weighted, leather reducing belts are effective in reducing the weights and waistlines of wearers.

(a) If it is hypothesized that the belts are effective, what is the alternative and what are the consequences of a Type I error and of a Type II error?

(b) If it is hypothesized that the belts are not effective, what are the consequences of a Type I error and of a Type II error?

4. Suppose that a cotton textile manufacturer has received a large lot of buttons from a supplier and wants to test the hypothesis that the lot meets the supplier's guaranteed quality specifications against the alternative that the lot does not meet the specifications.

(a) What are the consequences of a Type I error and of a Type II error?

(b) Explain why, in cases like this, the probability of committing a Type I error is often called the "producer's risk," and the probability of committing a Type II error is often called the "consumer's risk."

5. Using standard drug treatment, the mortality rate of a certain disease is known to be 0.04. A drug manufacturer has developed a new drug to treat this disease and wants to test the null hypothesis $r = 0.04$ against a suitable alternative, where r is the mortality rate of patients treated with the new drug.

(a) What alternative hypothesis should the manufacturer use if he wants to introduce the new drug only if it proves definitely superior to the old drug in tests?

(b) What alternative hypothesis should the manufacturer use if he wants to introduce the new drug, provided only that it does not prove definitely inferior to the old drug in tests?

6. A shoe manufacturer is considering the purchase of a new machine for stamping out uppers. If μ_1 is the average number of good uppers stamped out by the old

machine per hour and μ_2 is the corresponding average for the new machine, the manufacturer wants to test the null hypothesis $\mu_1 = \mu_2$ against a suitable alternative.

(a) What should the alternative be if the manufacturer does not want to buy the new machine unless tests prove it to be definitely superior to the old one?

(b) What should the alternative be if the manufacturer wants to buy the new machine unless tests prove it to be definitely inferior to the old one?

7. Suppose that a large supermarket has one checkout clerk whom it suspects of making more mistakes than the average of all its clerks.

(a) If the market decides that it will let the clerk go, provided this suspicion is confirmed on the basis of observations made on the clerk's performance, what hypothesis and alternative should the market set up?

(b) If the market decides to let the clerk go unless he can prove himself significantly better than the average of all clerks, what hypothesis and alternative should the market set up?

8. Referring to Exercise 7, suppose it is known that the average number of register mistakes per day per clerk is 18 and the standard deviation is 4, and the market decides to fire the clerk only if in a random sample of 10 days' work he averages more than 20 mistakes.

(a) What is the probability of firing the clerk when his work is in fact of average quality?

(b) What is the probability of keeping the clerk on if he averages 23 mistakes per day?

9. Verify for the typist example on page 268 that the probability of committing a Type I error is 0.055.

TESTS CONCERNING MEANS

To illustrate the procedure outlined in the preceding section, suppose that we want to determine whether or not the actual mean weekly food expenditure of families of three persons within a certain income range in a large city is $40. From information gathered in other pertinent studies, we assume that the variability of these expenditures is given by a standard deviation of $\sigma = \$12.20$. Our determination is to be made on the basis of the mean $\bar{x}$ of a random sample of 100 family food expenditures.

Beginning with Step 1 above, we formulate the null hypothesis to be tested and the two-sided alternative as

$$Hypothesis: \quad \mu = \$40$$
$$Alternative: \quad \mu \neq \$40$$

That is, we will consider as evidence against the hypothesized $40 mean expenditure an observed mean expenditure $\bar{x}$ which is either significantly less than or significantly greater than $40. We are willing to take a risk of 0.05 of rejecting the hypothesis if it is true, so we specify a level of significance of $\alpha = 0.05$.

We can write a criterion for this test in which the dividing lines between acceptance and rejection (the critical values) are stated in terms of dollars, but instead we shall formulate the criterion in terms of *standard units* because, stated in these units, a criterion is applicable to many problems, not just one. Approximating the sampling distribution of $\bar{x}$, as before, with a normal distribution, the criterion to be used for this test is shown in Figure 10.4. The dividing lines of the criterion are $z = -1.96$ and $z = 1.96$,

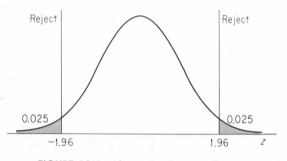

FIGURE 10.4 Alternative hypothesis $\mu \neq \mu_0$

so that the areas in the two tails of the distribution (the shaded regions of Figure 10.4) are both equal to 0.025. Since the standard deviation of this sampling distribution is given by $\sigma/\sqrt{n}$ the test criterion of Figure 10.4 may be expressed as follows:

Reject the hypothesis (and accept the alternative) if $z < -1.96$ or $z > 1.96$; accept the hypothesis (or reserve judgment) if $-1.96 \leq z \leq 1.96$, where

$$z = \frac{\bar{x} - \mu_0}{\sigma/\sqrt{n}} \qquad \qquad ★$$

and where μ_0 is the value of μ assumed under the given hypothesis.

Suppose now that we find the sample mean expenditure to be $\bar{x} = \$42.80$. We make the test by substituting this value, together with $\mu = 40$, $\sigma = 12.20$, and $n = 100$ into the formula for z, getting

$$z = \frac{42.80 - 40.00}{12.20/\sqrt{100}} = 2.30$$

Since z exceeds 1.96, we reject the hypothesis and conclude that the mean weekly food expenditure in the population from which the sample came is *not* \$40. A difference of \$2.80 between the observed mean expenditure and the \$40 expected under the hypothesis is too large to be accounted for by chance, we feel. It appears from the data that the mean is actually greater than \$40.

In this example, we can reject the hypothesis at a level of significance of $\alpha = 0.05$, but we could not have rejected the hypothesis if we had specified a level of significance of $\alpha = 0.01$ (the critical values at this level are $z = -2.58$ and $z = 2.58$). *This illustrates the important point that the level of significance should always be specified before a significance test is actually performed. This will spare one the temptation of later choosing a level of significance which happens to suit one's particular objectives.*

In the example above, we used the two-sided alternative $\mu \neq \$40$ because large departures in either direction from the hypothetical value would constitute evidence that the actual mean expenditure was not \$40. Suppose that in another study, however, we want to determine whether the actual mean wage in one type of employment in a given region is \$190 or whether it is, in fact, *greater than* \$190. A moment's reflection suggests that a two-sided alternative is inappropriate in this case, since a sample mean of anything *less than* \$190 would hardly constitute evidence that the mean population wage is *greater than* \$190. Clearly, a one-sided alternative with the region of rejection of the hypothesis in the right tail of the sampling distribution is the appropriate one. Therefore, we set up the following hypothesis and alternative:

Hypothesis: $\mu = \$190$
Alternative: $\mu > \$190$

Choosing again a level of significance of 0.05, the normal curve approximation to the sampling distribution of $\bar{x}$ leads to the test criterion shown in Figure 10.5, where the dividing line, $z = 1.64$, is such that the area in the right-hand tail equals 0.05. Formally, this criterion can be stated as follows:

Reject the hypothesis (and accept the alternative) if $z > 1.64$; accept the hypothesis (or reserve judgment) if $z \leq 1.64$, where

$$z = \frac{\bar{x} - \mu_0}{\sigma/\sqrt{n}} \qquad \bigstar$$

and where μ_0 is the value of μ assumed under the given hypothesis.

Suppose we take a random sample of 200 weekly wages from the population under study and calculate $\bar{x} = \$192.01$, and we assume on the basis of corollary studies that the standard deviation of the population is $\sigma = \$18.85$. Substituting in the formula for z, we find

$$z = \frac{192.01 - 190.00}{18.85/\sqrt{200}} = 1.51$$

and the hypothesis cannot be rejected. The difference of $2.01 between the observed mean wage and the $190 expected under the hypothesis is not large enough to provide evidence that the true mean wage is greater than $190. In this instance, if it is not urgent to reach an immediate decision, rather than accept the hypothesis outright and possibly commit a Type II error (if the true mean were $194, for example, this probability is $\beta = 0.09$), we might decide to reserve judgment.

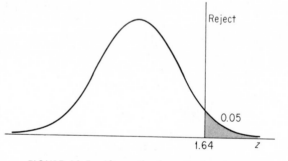

FIGURE 10.5 Alternative hypothesis $\mu > \mu_0$

Let us consider one further study to illustrate the need for a one-sided alternative and a criterion in which the region of rejection is all in the left tail of the sampling distribution. Suppose now that we want to determine whether the mean number of hours worked during a particular week by part-time employees in retailing in a metropolitan area is 20 hours or whether it is less than 20 hours. The standard deviation of the population is assumed to be $\sigma = 4.1$. Since evidence that the mean is actually less than 20 hours would be provided by a sample mean which is significantly less

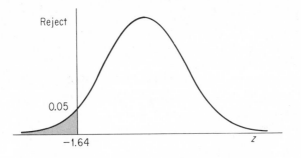

FIGURE 10.6   Alternative hypothesis $\mu < \mu_0$

than 20 hours, we shall use the test criterion shown in Figure 10.6 where the entire region of rejection is in the left tail of the curve.

With a level of significance of 0.05, this test is simply the mirror image of the test shown in Figure 10.5. We want to test the Hypothesis: $\mu = 20$ against the Alternative: $\mu < 20$, and by analogy with the right-hand counterpart of this test, we will reject the hypothesis (and accept the alternative) if $z < -1.64$; otherwise, we will accept the hypothesis (or perhaps reserve judgment).

Suppose that, in conducting the investigation, we took a random sample of 60 part-time employee records for the period and calculated $\bar{x} = 18.7$. From this we find that

$$z = \frac{18.7 - 20.0}{4.1/\sqrt{60}} = -2.46$$

which is less than -1.64, and the hypothesis is rejected. At the 5 per cent level of significance, there is evidence that, on the average, part-time employees were used less than 20 hours during the week.

In the three examples discussed above, the level of significance was set at 0.05, but as we have said, this level is directly under the control of the experimenter. If we choose $\alpha = 0.01$, for example, the critical values for a two-tail test are $z = -2.58$ and $z = 2.58$; for a right-hand one-tail test the critical value is $z = 2.33$; and for a left-hand one-tail test the critical value is $z = -2.33$. In general, when we perform a two-tail test like the one of Figure 10.4 at the level of significance α, we reject the hypothesis if $z < -z_{\alpha/2}$ or if $z > z_{\alpha/2}$; for the right-hand one-tail test of Figure 10.5 we reject the hypothesis if $z > z_{\alpha}$; and for the left-hand one-tail test of Figure 10.6 we reject the hypothesis if $z < -z_{\alpha}$. (As before, the value $z_{\alpha/2}$ is defined as the value of the standard normal distribution for which the area under the curve to its right equals $\alpha/2$, and z_α is the value for which the area to its right equals α.)

In most practical cases, the population standard deviation is not known. Provided we have a large sample ($n \geq 30$), we simply replace the unknown σ by its best estimate, the sample standard deviation s, and proceed exactly as we have above. The test hypothesis is rejected according to some specified criterion if

$$z = \frac{\bar{x} - \mu_0}{s/\sqrt{n}} \qquad \qquad \bigstar$$

falls in the region of rejection; otherwise, it is accepted (or judgment is withheld). These procedures, which are based on normal curve theory, are restricted to large samples, however.

In cases where we do not know the population standard deviation *and* we have a small sample ($n < 30$), we must proceed in a different way, basing our test on the Student-t distribution (see page 240). We must now assume that the distribution of the population which we are sampling has roughly the shape of a normal distribution, and with this assumption, we can use the statistic

$$t = \frac{\bar{x} - \mu_0}{s/\sqrt{n}} \qquad \qquad \bigstar$$

whose sampling distribution is the Student-t distribution with $n - 1$ degrees of freedom. This formula for t is identical with the formula for z given in the preceding paragraph, but in small sample tests, the critical values of a criterion will be read from the t distribution of Table II instead of from the normal distribution of Table I. At a level of significance α, the critical values for the two-tail test are $-t_{\alpha/2}$ and $t_{\alpha/2}$, and the critical value for the one-tail test is either $-t_\alpha$ or t_α, depending on whether the alternative hypothesis is $\mu < \mu_0$ or $\mu > \mu_0$ (see Figures 10.7, 10.8, and 10.9); in any case, the critical values are read from Table II with $n - 1$ degrees of freedom. (Here the value $t_{\alpha/2}$ is defined as the value of the t distribution for which the area under the curve to its right equals $\alpha/2$, and t_α is the value for which the area to its right equals α.)

To illustrate such a small sample test, suppose that we want to decide whether, under prescribed operating conditions in a shop, the average tool life per sharpening of a multitoothed metal cutting tool (called a "broach") is 3,000 pieces or whether the average life is less than 3,000 pieces. Our

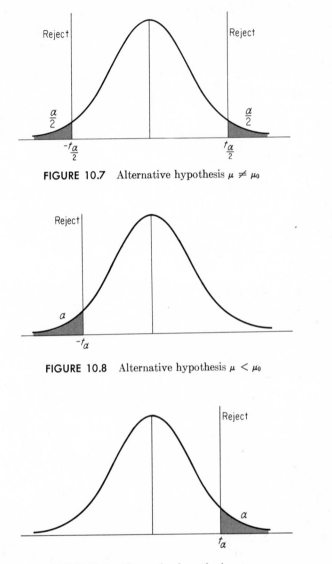

FIGURE 10.7 Alternative hypothesis $\mu \neq \mu_0$

FIGURE 10.8 Alternative hypothesis $\mu < \mu_0$

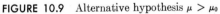

FIGURE 10.9 Alternative hypothesis $\mu > \mu_0$

decision is to be based on tests made on a random sample of six broaches, and we do not know the population standard deviation. Choosing a level of significance of $\alpha = 0.05$, we set up the following hypothesis and one-sided alternative

$$\begin{aligned} \textit{Hypothesis:} \quad & \mu = 3,000 \\ \textit{Alternative:} \quad & \mu < 3,000 \end{aligned}$$

Using a left-hand one-tail test, we will reject the hypothesis (and accept the alternative) if $t < -t_{.05} = -2.015$ for $n - 1 = 5$ degrees of freedom; otherwise, we will accept the hypothesis (or perhaps reserve judgment).

Suppose that in the six tests the broaches showed tool lives of 2,970, 3,020, 3,005, 2,900, 2,940, and 2,925 pieces. The mean of this sample is $\bar{x} = 2,960$ pieces and the standard deviation is $s = 46.8$. Substituting in the formula for t, we have

$$t = \frac{2,960 - 3,000}{\dfrac{46.8}{\sqrt{6}}} = -2.09$$

which is less than -2.015, so we reject the hypothesis. Evidently, the average tool life of the broach is less than 3,000 pieces.

DIFFERENCES BETWEEN MEANS

There are many statistical problems in which we must decide whether an observed difference between two sample means can be attributed to chance. We may want to decide, for instance, whether there is really a difference between two kinds of steel, if a sample of one kind had an average strength of 56,800 psi while a sample of another kind had an average strength of 55,600 psi. Similarly, we may want to decide on the basis of samples whether there actually is a difference in the size of delinquent charge accounts in two branches of a department store, whether men can perform a given task faster than women, whether one kind of television tube is apt to last longer than another, and so on.

The method which we shall employ to test whether an observed difference between two sample means may be attributed to chance is based on the following theory: *If $\bar{x}_1$ and $\bar{x}_2$ are the means of two large independent random samples of size n_1 and n_2, the sampling distribution of the statistic $\bar{x}_1 - \bar{x}_2$ can be approximated closely with a normal curve having the mean $\mu_1 - \mu_2$ and the standard deviation*

$$\sqrt{\frac{\sigma_1^2}{n_1} + \frac{\sigma_2^2}{n_2}}$$

where μ_1, μ_2, σ_1, and σ_2 are, respectively, the means and the standard deviations of the two populations from which the two samples were obtained.

By "independent" samples we mean that the selection of one sample is in no way affected by the selection of the other. Thus, the theory does *not* apply to "before-and-after" kinds of comparisons, nor does it apply, say, to the comparison of the I.Q.'s of husbands and wives. A special method for handling this kind of problem is referred to in Exercise 22 on page 288.

One difficulty in applying the above theory is that in most practical applications σ_1 and σ_2 are not known. When this is the case, provided both our samples are large (both n_1 and n_2 should be at least 30), we can substitute the sample standard deviations s_1 and s_2 for σ_1 and σ_2, and write the formula for the *standard error of the difference between two means as*

$$\sqrt{\frac{s_1^2}{n_1} + \frac{s_2^2}{n_2}}$$

To illustrate how this theory is applied, suppose that a metropolitan department store operating two suburban branch stores wants to determine whether or not the mean balance outstanding on 30-day charge accounts is the same in Branches 1 and 2. After taking random samples of size 80 and 100 from the accounts of Branch 1 and Branch 2, respectively, the store summarizes its findings as follows:

$$n_1 = 80 \qquad \bar{x}_1 = \$64.20 \qquad s_1 = \$16.00$$
$$n_2 = 100 \qquad \bar{x}_2 = \$71.41 \qquad s_2 = \$22.13$$

Letting μ_1 and μ_2 represent the actual average balances outstanding in Branch 1 and Branch 2, the null hypothesis the store wants to test and the two-sided alternative are

Hypothesis: $\mu_1 - \mu_2 = 0$
Alternative: $\mu_1 - \mu_2 \neq 0$

Specifying that the level of significance is to be $\alpha = 0.05$, we shall again use the criterion of Figure 10.4, although z must, of course, be calculated in a different way. Formally, this criterion may be stated as follows:

Reject the null hypothesis (and accept the alternative) if $z < -1.96$ or $z > 1.96$; accept the null hypothesis (or reserve judgment) if $-1.96 \leq z \leq 1.96$, where

$$z = \frac{\bar{x}_1 - \bar{x}_2}{\sqrt{\dfrac{s_1^2}{n_1} + \dfrac{s_2^2}{n_2}}} \qquad \bigstar$$

Note that this formula for z is obtained by subtracting from $\bar{x}_1 - \bar{x}_2$ the mean of its sampling distribution (which under the null hypothesis is $\mu_1 - \mu_2 = 0$) and then dividing by the standard deviation of its sampling distribution with s_1 and s_2 substituted for σ_1 and σ_2.

Substituting the numerical values given above into the formula for z, the store finds that

$$z = \frac{64.20 - 71.41}{\sqrt{\frac{(16.00)^2}{80} + \frac{(22.13)^2}{100}}} = -2.53$$

and since this value is less than -1.96, it concludes that the observed difference of \$7.21 between the average outstanding balances in the two branches is *significant*. That is, the store feels that this difference is too large to be accounted for by chance, and there must be a real difference in the two population means.

The significance test for the difference between two means which we have described above applies only to large samples. However, a small-sample criterion for tests concerning the difference between two means is provided by the Student-t distribution. To use this criterion we must assume that the two samples come from populations which can be approximated closely with normal distributions and which, furthermore, have *equal variances*. Specifically, we test the null hypothesis $\mu_1 = \mu_2$ against an appropriate one-sided or two-sided alternative with the statistic

$$t = \frac{\bar{x}_1 - \bar{x}_2}{\sqrt{\frac{\Sigma (x_1 - \bar{x}_1)^2 + \Sigma (x_2 - \bar{x}_2)^2}{n_1 + n_2 - 2} \cdot \left(\frac{1}{n_1} + \frac{1}{n_2}\right)}} \qquad \bigstar$$

where $\Sigma (x_1 - \bar{x}_1)^2$ is the sum of the squared deviations from the mean of the first sample while $\Sigma (x_2 - \bar{x}_2)^2$ is the sum of the squared deviations from the mean of the second sample. Note that since, by definition, $\Sigma (x_1 - \bar{x}_1)^2 = (n_1 - 1) \cdot s_1^2$ and $\Sigma (x_2 - \bar{x}_2)^2 = (n_2 - 1) \cdot s_2^2$, the above formula can be simplified somewhat when the two sample variances have already been calculated from the data.

Under these assumptions, it can be shown that the sampling distribution of this t statistic is the Student-t distribution with $n_1 + n_2 - 2$ degrees of freedom. At the level of significance α, the criterion for this test is either that of Figure 10.7, 10.8, or 10.9.

To illustrate this small-sample test for the difference between two means, let us suppose that a large institutional buyer has four packages of frozen raspberries (supposedly of the same grade) submitted to it by two different packers. The buyer grades the fruit (on a scale with 100 for perfection) with the following results:

$$Brand\ 1: \quad 72,\ 68,\ 76,\ 64$$
$$Brand\ 2: \quad 75,\ 74,\ 80,\ 83$$

The means of these two samples are 70 and 78, and we want to test at the 5 per cent level of significance whether this difference of eight points in the ratings is significant. In order to calculate t according to the above formula, we first determine

$$\Sigma\ (x_1 - \bar{x}_1)^2 = 2^2 + (-2)^2 + 6^2 + (-6)^2 = 80$$
$$\Sigma\ (x_2 - \bar{x}_2)^2 = (-3)^2 + (-4)^2 + 2^2 + 5^2 = 54$$

Substituting these values together with $n_1 = n_2 = 4$, $\bar{x}_1 = 70$, and $\bar{x}_2 = 78$ into the formula for t, we obtain

$$t = \frac{70 - 78}{\sqrt{\dfrac{80 + 54}{4 + 4 - 2} \cdot \left(\dfrac{1}{4} + \dfrac{1}{4}\right)}} = -2.39$$

Since -2.39 falls between -2.447 and 2.447, where 2.447 is the value of $t_{.025}$ for $4 + 4 - 2 = 6$ degrees of freedom, it follows that the difference between the two sample means is *not significant*.

EXERCISES

1. It is desired to test the hypothesis that the average performance level for a certain classification operation is 150 correct classifications per time period. A sample of 49 individuals is tested, and the average performance level found to be 140 correct classifications with a standard deviation of 15. Is this evidence at the 0.05 level of significance that the actual performance level in the population sampled is not 150?

2. We wish to test, at the 5 per cent level of significance, the hypothesis that the true mean gasoline consumption of a certain kind of engine is 14 gallons per minute against the alternative that it is not 14 gallons. If tests on a random sample of four engines showed mean consumption of 17 gallons per minute with standard deviation of 4 gallons, can the hypothesis be rejected?

3. A sample of four spools of lithium metal ribbons of a certain width and thickness is taken from a large population of such spools. If the lengths of the ribbons are 30.1, 30.1, 30.0, and 30.2 feet, is this evidence at the 5 per cent level of significance that the true mean length of the ribbons is greater than 30.0 feet?

4. A manufacturer guarantees a certain ball bearing to have mean outside diameter of 0.8525 inches with standard deviation of 0.0003. If a random sample of nine bearings from a large lot of these bearings has a mean outside diameter of 0.8529 inches, at the 1 per cent level of significance does this lot meet the manufacturer's guarantee on the mean outside diameter?

5. Investigating an alleged unfair trade practice, the Federal Trade Commission takes a random sample of 49 "9-ounce" candy bars from a large shipment. The mean of the sample weights is 8.94 ounces and the standard deviation is 0.12 ounces.

   (a) Show that, at a level of significance of 0.01, the commission has grounds upon which to proceed against the manufacturer on the unfair practice of short-weight selling.

   (b) If the level of significance were set at, say, 0.0001, the commission would have no grounds for such a proceeding (the critical value of z is -3.72). What would be the advantages or disadvantages to anyone in setting the level of significance at such an extremely low figure?

6. Two random samples of 12 men each from the over-45 age group are chosen. The cholesterol level of Group 1 was lowered by at least 10 per cent for a year, but the cholesterol level of Group 2 was not lowered. To test the effect of blood cholesterol level on mental acuity in men of this age group, the sample subjects were given a battery of psychological tests covering a wide range of abilities (such as problem solving, reasoning, and reaction time). The mean composite score for Group 1 was 130 points with standard deviation of 10 points; the mean composite score for Group 2 was 118 points with standard deviation of 12 points. Does this constitute evidence at the 0.05 level of significance that men in the over-45 age group with lower cholesterol levels are mentally sharper than those with higher levels?

7. A random sample of 400 adults showed an average weight of 16 pounds above the weight corresponding from the best health records. If the sample standard deviation was 5 pounds, at the 2 per cent level of significance does this evidence contradict a claim that the actual average overweight in the population sampled is 14 pounds?

8. An agricultural experiment showed that six test plots planted with one variety of corn yielded on the average 89.5 bushels per acre with a standard deviation of 5.8 bushels, and six test plots planted with another variety of corn yielded 82.1 bushels per acre with standard deviation of 7.1 bushels. At the 0.05 level of significance is there a real difference between the two average yields?

9. A manufacturer packs a chemical solvent in drums guaranteed to contain 250 pounds; the standard deviation of the fills is presumed from experience to be 4 pounds. A prospective buyer wishes to test, at the 0.05 level of significance, the

hypothesis that the mean weight of a large lot of such drums is in fact 250 pounds with the decision whether or not to reject the hypothesis to be made on the basis of 16 drum weights randomly selected from the lot.

(a) Compute to three decimal places the probabilities of rejecting the hypothesis for possible mean weights of 246, 247, . . . , 253, and 254 pounds, using (i) the right-hand one-tail test, (ii) the left-hand one-tail test, and (iii) the two-tail test. Enter these probabilities in a suitable table.

(b) Plot the power curves of the three tests of part (a) on the same graph.

(c) What is the best test to use (and why) against a specific alternative mean of 247 pounds? Of 252 pounds? What is the best test to use if the mean weight is in fact 250 pounds?

(d) What general conclusions can you draw from an examination of the three curves? What seems to be the best procedure to follow when testing a hypothesis against an alternative that specifies the true value is (i) greater than the hypothetical value, (ii) less than the hypothetical value and (iii) not equal to the hypothetical value?

10. A production process is designed to fill No. $1\frac{1}{4}$ cans with 14.5 ounces net weight of sliced pineapple. The mean weight varies from time to time, but the standard deviation is considered to be stable and well established at 0.64 ounces. In order to test incoming lots for weight, a large institutional buyer takes a random sample of 30 cans from each lot and determines the mean net weight.

(a) For what values of $\bar{x}$ should the buyer reject the null hypothesis $\mu = 14.5$ ounces, if he uses a two-sided alternative and a level of significance of 0.05?

(b) Using the criterion established in part (a), what is the probability that the buyer will fail to detect a lot whose mean is only 14.3 ounces?

11. Repeat Exercise 10 using the one-sided alternative $\mu < 14.5$ ounces, which seems more appropriate for a buyer who feels he needs protection only against accepting lots which are underfilled. ("Buyers" often feel they need protection against getting either less than or more than a guaranteed standard. Can you think of an example of this?)

12. We wish to test at a level of significance of 0.10 the hypothesis that the true mean weekly wage in a certain type of employment in a large city is $105 against the alternative that it is not $105. Standard deviation is assumed to be $21 and the sample size is 49. Using the best test of this hypothesis against the alternative, what is the probability of accepting $105 as the true value when the true value in fact deviates from $105 by $7?

13. Specifications call for one dimension of a mechanical part to be 0.5485 inches. Six parts randomly drawn from a large production lot of such parts showed a mean dimension of 0.5479 inches and a standard deviation of 0.0006 inches. Can it be concluded at the 0.05 level of significance that the lot of parts meets the specification on this dimension?

14. A standardized test of the ability to think scientifically has been administered to thousands of students with mean score of 84 and standard deviation of 10.4 points. A group of 50 students is randomly selected for instruction in a course in which

special emphasis is placed on problem formulation, deductive and inductive logic, interpretation of data, and other aspects of scientific thinking. Following the instruction period, the students are given the standard test and their average score is 86.1. Does this constitute evidence at the 0.05 level of significance that the special instruction raises the achievement level on the test?

15. In order to evaluate the clinical effects of a new steroid in treating chronically underweight persons, 40 such persons were given 25-milligram dosages of the drug over a 12-week period, while another 40 such persons were given 50-milligram dosages of the drug over the same period. After 6 months, it was observed that the persons in the first group had gained on the average 9 pounds with a standard deviation of 7 pounds, while those in the second group had gained on the average 11 pounds with a standard deviation of 6 pounds.

   (a) Treating the two groups as random samples from conceptually large populations, test at the 0.05 level of significance whether there is a real difference in the average weight gains of persons receiving the two dosages.

   (b) If the 40 subjects given the 50-milligram dosages are observed for six months after the drug use was discontinued and 30 of them either maintained their weight or continued to gain, within what limits may we be 95 per cent confident that the actual population proportion lies?

16. A testing laboratory tests five biochemistry kits intended for use in junior high school science courses to determine how many times the basic experiments included may be performed using the materials furnished with the kits. The following are the results: 25, 28, 30, 29, and 30 times. At the 10 per cent level of significance, does this support the manufacturer's claim that on the average the kits contain enough materials for more than 26 experiments?

17. One important consideration in selecting magnetic tape for computer use is the amount of wear the tape produces in the read-write heads. In laboratory wear tests continuous 20-meter loops of tape are driven at a constant velocity of 1 meter per second with tension accurately controlled within ± 8 grams. The measure of head wear is the weight loss (in milligrams) produced during one hour of continuous run on a precision metal disk used to simulate the recording head. The weight losses observed in a random sample of five loops of Brand A tape were 0.04, 0.07, 0.06, 0.05, and 0.08; the weight losses observed in a random sample of five loops of Brand B tape were 0.08, 0.10, 0.09, 0.10, and 0.08. Is this evidence at the 0.05 level of significance that head wear produced by Brand A is less than that produced by Brand B?

18. A soft-drink vending machine is set to dispense 6 ounces per cup. The machine is tested nine times, and the mean cup fill is 6.4 ounces with standard deviation 0.5 ounces. At a level of significance of 0.10, is this evidence that the machine is overfilling the cups?

19. (*Sample size*) If we wish to test a hypothesis that the mean of some population is μ_0 in such a way that the probability of rejecting it in error is α, and the probability of accepting it is β if some specific alternative value μ_A is the true value, we must take a random sample of size n where

$$n = \frac{\sigma^2(z_\alpha + z_\beta)^2}{(\mu_A - \mu_0)^2}$$ ★

if the alternative is one-sided; and

$$n = \frac{\sigma^2(z_{\alpha/2} + z_\beta)^2}{(\mu_A - \mu_0)^2}$$ ★

if the alternative is two-sided.

Suppose, for example, that for a population with $\sigma = 5$ we wish to test the hypothesis that the true mean is 200 pounds against the alternative that the true mean is less than 200 pounds. The probability of rejecting the hypothesis if it is true is to be fixed at $\alpha = 0.05$, and the probability of accepting the hypothesis if the true mean is actually 198 pounds is to be fixed at $\beta = 0.20$. Substituting into the first formula above, we find

$$n = \frac{(5)^2(1.64 + 0.84)^2}{(198 - 200)^2} = 38.4$$

so we must take a sample of either 38 or 39 weights.

We wish to test the hypothesis that the true mean of a population is \$500; the population standard deviation is assumed to be \$10. If this hypothesis is true, we want to be 95 per cent sure of accepting it, and if the true mean deviates from \$500 by \$5 in either direction, we want to be 80 per cent sure of rejecting the hypothesis. Verify that the desired control over the two errors requires that we take a sample of either 31 or 32 items from the population.

20. A company wants to compare the lifetimes of two stones used in an abrasive process (called "superfinishing"), which produces rapidly a fine microfinish on machined surfaces. In laboratory tests, hot rolled steel driveshafts of the same degree of surface roughness are processed for two minutes each under specified conditions using both stones. If the average lifetime of 10 stones of the first kind was 60 pieces with a standard deviation of 5 pieces, while the average lifetime of 10 stones of the second kind was 64 pieces with a standard deviation of 3 pieces, is the difference between these two means significant at a level of significance of 0.05?

21. For a very large population of lengths of metal strips whose standard deviation is assumed to be 0.10 inches, a buyer wants to test (on the basis of a random sample of strips drawn from the lot) the hypothesis that the true mean length of the strips is 4 inches against the hypothesis that it is less than 4 inches. Consequences of rejecting the hypothesis if it is true and of accepting it if the mean length is actually 3.95 inches are considered to be equally serious, and their risks both set at 0.02. For what values of $\bar{x}$ should the buyer reject the hypothesis?

22. If we want to study the effectiveness of a new diet on the basis of weights "before and after," or if we want to study whatever differences there may be between the I.Q.'s of husbands and wives, the methods introduced in this chapter cannot be used. The samples are not independent; in fact, in each case the data are *paired*. To handle data of this kind, we work with the (signed) differences of the paired data and test whether these differences may be looked upon as a sample from a population for which $\mu = 0$. If the sample is small, we use the t test on page 278; otherwise, we use a large sample test. Apply this technique to the following data designed to test whether there is a systematic difference in the blood-pressure readings obtained with two different instruments:

Patient	Reading Obtained with Instrument A	Reading Obtained with Instrument B
1	147	151
2	118	120
3	144	145
4	165	168
5	140	146
6	125	124
7	132	135
8	151	152

Use a level of significance of 0.05 to test whether there is a difference in the true average readings obtained with the two instruments.

23. In a study of the effectiveness of a reducing diet, the following "before-and-after" figures were obtained for a random sample of 10 adult married females in the age group from 30 to 40 years (data in pounds):

Dieter	Before	After
Mrs. A	134	131
Mrs. B	147	140
Mrs. C	165	164
Mrs. D	152	153
Mrs. E	139	133
Mrs. F	122	122
Mrs. G	138	135
Mrs. H	147	148
Mrs. I	153	147
Mrs. J	178	165

Use a level of significance of 0.01 to test the null hypothesis (that the diet is not effective) against a suitable one-sided alternative.

24. To investigate the claim that women are better at a certain job than men, the personnel department of a large cotton textile mill gives an appropriate per-

formance test to a random sample of 40 male employees and to a random sample of 50 female employees. If the women averaged 88.9 points with a standard deviation of 11.4, and the men averaged 81.3 points with a standard deviation of 10.2, is this evidence at the 0.01 level of significance that women are actually better at the job than men?

25. (*Control charts for means*) In industrial quality control it is often necessary to test the same hypothesis over and over again at regular intervals of time. Suppose, for example, that a process for making compression springs is in control if the free lengths of the springs have a mean of $\mu = 1.5$ inches; it is known from experience that the standard deviation of the free lengths of these springs is $\sigma = 0.02$ inches. In order to test whether the process is in control, random samples of size n are taken, say, every hour, and it is decided in each case on the basis of $\bar{x}$ whether to accept or reject the null hypothesis $\mu = 1.5$. To simplify this task, quality control engineers use *control charts* like that of Figure 10.10, where the vertical scale is the scale of measurement of $\bar{x}$,

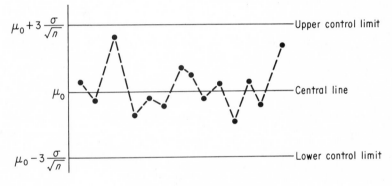

FIGURE 10.10   Control chart

the central line is at $\mu = \mu_0$, and the *upper and lower control limits* are at $\mu_0 + 3 \dfrac{\sigma}{\sqrt{n}}$ and $\mu_0 - 3 \dfrac{\sigma}{\sqrt{n}}$. Each sample mean is plotted on this chart, and the process is considered to be in control so long as the $\bar{x}$'s fall between the control limits. Assuming that the data constitute random samples from a normal population, the probability of a Type I error is less than 0.003. Using $\mu_0 = 1.5$ and $\sigma = 0.02$, construct a control chart for the means of samples of size $n = 5$. Also plot on this chart the following data, constituting the means of such random samples taken at hourly intervals during the operation of the process: 1.510, 1.495, 1.521, 1.505, 1.524, 1.520, 1.488, 1.465, 1.529, 1.444, 1.531, 1.502, 1.490, 1.531, 1.475, 1.478, 1.522, 1.491, 1.491, 1.482. Was the process ever out of control?

26. In the production of certain shafts, a process is said to be in control if the outside diameters have a mean of $\mu = 2.500$ inches and a standard deviation of $\sigma = 0.002$ inches.

(a) Construct a control chart for the means of random samples of size 4.

(b) Plot on the chart obtained in part (a) the following figures, representing the means of 25 such random samples taken at regular intervals of time: 2.5014, 2.5022, 2.4995, 2.5006, 2.4985, 2.5002, 2.5016, 2.4985, 2.4992, 2.5028, 2.5016, 2.5020, 2.5001, 2.4995, 2.5003, 2.5012, 2.4988, 2.4975, 2.5011, 2.4992, 2.5002, 2.5014, 2.4976, 2.5003, 2.5015. Was the process ever out of control?

27. Under heavy pressure, a small boy agrees to give his younger sister three pieces of candy from his candy bag containing 20 cherry cordials and 10 coconut candies; he also agrees to draw the pieces at random. Knowing that her brother prefers cherries to coconuts, the girl decides to test the hypothesis that the boy actually does choose the pieces at random, finally deciding to reject the hypothesis if she gets at most one cherry cordial and to suggest that her brother deliberately selected the pieces of candy to suit himself. What is the probability that the girl will commit a Type I error if her brother does in fact select the candy at random?

TESTS CONCERNING PROPORTIONS

Once the reader has grasped the fundamental ideas underlying tests of hypotheses, the various tests we shall study in this and in later sections should not present any difficulties. The tests we shall discuss in this particular section apply to problems in which we must decide, on the basis of sample data, whether the true value of a proportion (percentage, or probability) equals a certain given value p. For instance, a management consulting firm studying cost control in a large laundry may want to know whether the true proportion of "cleaned" shirts rejected at final inspection because of faulty work is $p = 0.10$. Similarly, the personnel manager of a large textile mill may want to know whether 0.40 is the true proportion of employees recruited through newspaper ads who will pass a standard mechanical aptitude test without company training, and a manufacturer may want to know whether it is true that 60 per cent of all housewives (in a certain market area) prefer a new sweetened fruit soup to an unsweetened one.

Questions of this kind are usually decided on the basis of a sample proportion or the observed number of "successes" in n "trials." In the above examples, the management consulting firm would base its decision on the *number* (or proportion) of "cleaned" shirts rejected in a sample of, say, 400; the personnel manager would base his decision on the *number* of employees recruited by newspaper ads who pass the test, say, among 200 employees thus recruited; and the manufacturer would base his decision on the *number* of housewives preferring the sweetened soup, say, in a sample of 600 housewives selected at random.

The sampling distribution of the number of successes in n trials is the binomial distribution first studied in Chapter 7. In fact, when n is small, tests concerning a "true" proportion p (i.e., tests concerning the parameter p of a binomial distribution) are usually based directly on tables of binomial probabilities, such as those referred to in the Bibliography at the end of this book. When n is large, however, we approximate the binomial distribution with a normal curve and (as was indicated on page 206) this approximation is usually satisfactory so long as np and $n(1 - p)$ both exceed 5. The mean and the standard deviation are taken as np and $\sqrt{np(1 - p)}$, respectively, in accordance with the formulas given on pages 186 and 188 in Chapter 7.

To illustrate this kind of test, let us suppose that a large national brokerage firm wants to test, at a level of significance of 0.05, a claim that 80 per cent of its customers who sell stock to establish tax losses immediately reinvest the proceeds in other stocks. The test is to be made on the basis of a random sample of 320 sales drawn from a large population of sales known to have been made to establish losses. The hypothesis the firm wants to test and the two-sided alternative are

Hypothesis: $p = 0.80$

Alternative: $p \neq 0.80$

and since both np and $np(1 - p)$ exceed 5, it is reasonable to use the normal curve approximation to the distribution of the number of successes in n trials. The appropriate test criterion is again the two-tail test of Figure 10.4:

Reject the hypothesis (and accept the alternative) if $z < -1.96$ or $z > 1.96$; accept the hypothesis (or reserve judgment) if $-1.96 \leq z \leq 1.96$ where

$$z = \frac{x - np_0}{\sqrt{np_0(1 - p_0)}} \qquad \bigstar$$

and where p_0 is the value assumed for p under the given hypothesis.

In making the test, the firm found in the sample 245 cases in which proceeds of the sale were immediately reinvested in other stocks. Substituting $x = 245$, $n = 320$, and $p_0 = 0.80$ into the above formula for z, the firm gets

$$z = \frac{245 - 320(0.80)}{\sqrt{320(0.80)(0.20)}} = -1.54$$

and it follows that there is no evidence upon which to reject the claim. (If it accepts the hypothesis, the firm may be committing a Type II error, a possibility which can be avoided only by reserving judgment on the claim, if that is possible.)

To give an example where a one-sided alternative is the appropriate one, let us suppose that an automobile manufacturer wants to decide whether or not more than 60 per cent of the car owners in a large city would prefer to buy for their next car one with high seatbacks (to help prevent whiplash) than one with headrests. The decision is to be based on a random sample of 250 car owners.

The hypothesis and the alternative are

$$\textit{Hypothesis:} \quad p = 0.60$$
$$\textit{Alternative:} \quad p > 0.60$$

The level of significance is set at 0.01, and the criterion for this right-hand, one-tail test calls for the hypothesis to be rejected if $z > 2.33$, where z is again calculated from the formula

$$z = \frac{x - np_0}{\sqrt{np_0(1 - p_0)}} \qquad\qquad \bigstar$$

in which p_0 is the value specified for p under the hypothesis. [We may again use the normal curve approximation to the binomial distribution because both np and $np(1 - p)$ are greater than 5.] If, in the sample of 250 car owners, 169 say they prefer to buy a car with high seatbacks, we have

$$z = \frac{169 - 250(0.60)}{\sqrt{250(0.60)(0.40)}} = 2.45$$

Since this value exceeds the critical value for the test, the manufacturer rejects the hypothesis. There is strong evidence that, in the population sampled, the actual value of p is greater than 0.60.

DIFFERENCES AMONG PROPORTIONS

There are many applications in which we must decide whether observed differences among two or more sample proportions (or percentages) are significant or whether they can be attributed to chance. For instance, if one mail-order solicitation yields a 10 per cent response while another, more expensive, one yields a 12 per cent response, we may want to decide whether the observed 2 per cent difference is significant or whether it is merely due to chance. To illustrate how problems of this sort are handled, let us suppose that, in connection with developing a new frozen orange juice concentrate, a processor wants to determine whether the proportions of people who prefer (in taste tests) an acid-type juice to a nonacid-type juice are the same in three distinct geographical regions. The processor has taken random samples of size 200, 150, and 200 in Regions A, B, and C, respectively, and found 124, 75, and 120 people in these three samples who prefer the acid-type juice to the nonacid type. The results of the market tests are shown in the table below:

	Region A	Region B	Region C	
Prefer Acid Type	124	75	120	319
Prefer Nonacid Type	76	75	80	231
	200	150	200	550

and in order to decide whether the actual proportions of people who prefer the acid-type juice are the same in the three regions, the processor must determine whether the observed differences among the three sample proportions, 0.62, 0.50, and 0.60, for Regions A, B, and C are too large to be accounted for by chance or whether they may reasonably be attributed to chance.

In the following analysis, we shall denote the true proportions of people preferring the acid-type juice in the three regions p_1, p_2, and p_3. Thus, the null hypothesis we want to test and the alternative hypothesis are

Hypothesis: $p_1 = p_2 = p_3$
Alternative: p_1, p_2, and p_3 are not all equal

If the null hypothesis is true, we can combine the three samples and estimate the common proportion of people who prefer the acid-type juice as

$$\frac{124 + 75 + 120}{200 + 150 + 200} = \frac{319}{550} = 0.58$$

With this estimate of the common population proportion, we would expect to find $200(0.58) = 116$ people who prefer the acid-type juice in Region A (the first sample), $150(0.58) = 87$ in Region B, and $200(0.58) = 116$ in Region C. Subtracting these totals from the respective sample totals, we would expect $200 - 116 = 84$ people who prefer the nonacid-type juice in Region A, $150 - 87 = 63$ in Region B, and $200 - 116 = 84$ in Region C. These results are summarized in the following table, where the *expected frequencies* are shown in parentheses below the (actually) *observed frequencies:*

	Region A	Region B	Region C	
Prefer Acid Type	124 (116)	75 (87)	120 (116)	319
Prefer Nonacid Type	76 (84)	75 (63)	80 (84)	231
	200	150	200	550

To test the null hypothesis $p_1 = p_2 = p_3$, we now compare the frequencies that were actually observed with the ones we could expect if the null hypothesis were true. It stands to reason that the null hypothesis should be accepted if the two sets of frequencies are very much alike; after all, we would then have obtained almost exactly what we could have expected if the null hypothesis were true. On the other hand, if the discrepancies between the two sets of frequencies are large, the observed frequencies do not agree with what we could expect, and this suggests that the null hypothesis must be false.

Using the letter f for the observed frequencies and the letter e for the expected frequencies, we shall base their comparison on the following χ^2 (*chi-square*) statistic

$$\chi^2 = \Sigma \frac{(f - e)^2}{e} \qquad \bigstar$$

In words, χ^2 is the sum of the quantities obtained by dividing $(f - e)^2$ by e separately for each "cell" of the table. Calculating χ^2 for our example we have

$$\chi^2 = \frac{(124 - 116)^2}{116} + \frac{(75 - 87)^2}{87} + \frac{(120 - 116)^2}{116} + \frac{(76 - 84)^2}{84}$$

$$+ \frac{(75 - 63)^2}{63} + \frac{(80 - 84)^2}{84} = 5.583$$

If there is a close agreement between the f's and e's, the differences $f - e$ and hence χ^2 will be small; if the agreement is poor, some of the differences $f - e$ and hence χ^2 will be large. Consequently, we reject the null hypothesis if χ^2 is large, and we accept it or reserve judgment if χ^2 is small. The exact criterion for this decision is based on the sampling distribution of the χ^2 statistic, which can be approximated very closely by a theoretical distribution called the *chi-square distribution*. The criterion which is based on this sampling distribution is shown in Figure 10.11; using a level of signifi-

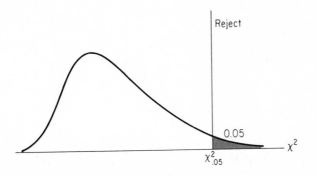

FIGURE 10.11   Chi-square distribution

cance of 0.05, the dividing line of the criterion, $\chi^2_{.05}$, is such that the shaded area in Figure 10.11 is 0.05. Its value may be obtained from Table III. The chi-square distribution, like the Student-t distribution, depends on a parameter called the number of degrees of freedom. In our example the number of degrees of freedom is 2, and in general, in the comparison of k sample proportions it is $k - 1$. (Intuitively, we can justify this formula for the number of degrees of freedom with the argument that if we calculate $k - 1$ of the expected frequencies in either row of the table, all the other expected frequencies may be obtained by subtraction from the totals of appropriate rows or columns.)

Referring to Table III, we find that the value of $\chi^2_{.05}$ for $3 - 1 = 2$ degrees of freedom is 5.991. Since 5.583, the value we actually obtained for χ^2, is less than 5.991, the null hypothesis cannot be rejected. We conclude that the observed differences among the sample proportions are not so large that they cannot reasonably be accounted for by chance. In reaching this conclusion, we note that there may be real differences between p_1, p_2, and p_3, but if there are this study has failed to reveal them.

In general, if we want to compare k sample proportions, we first calculate the expected frequencies as we did on page 294. Combining the data, we estimate p as

$$\frac{x_1 + x_2 + \cdots + x_k}{n_1 + n_2 + \cdots + n_k} \qquad \bigstar$$

where the n's are the sizes of the respective samples and the x's are the corresponding numbers of successes. Multiplying the n's by this estimate of p, we obtain the expected frequencies for the first row of the table, and subtracting these values from the totals of the corresponding samples, we obtain the expected frequencies for the second row of the table. (Note that the expected frequency for any one of the cells may also be obtained by multiplying the total of the row to which it belongs by the total of the column to which it belongs, and dividing by the grand total, $n_1 + n_2 + \cdots + n_k$, for the entire table.)

Next, χ^2 is calculated according to the formula

$$\chi^2 = \Sigma \frac{(f - e)^2}{e} \qquad \bigstar$$

with $(f - e)^2/e$ determined separately for each of the $2k$ cells of the table. The criterion for testing at the level of significance α, the null hypothesis $p_1 = p_2 = \cdots = p_k$ against the alternative that these p's are not all equal then reads as follows:

Reject the null hypothesis (and accept the alternative) if $\chi^2 > \chi^2_\alpha$; *accept the null hypothesis (or reserve judgment) if* $\chi^2 \leq \chi^2_\alpha$, *where the value of* χ^2_α *is based on* $k - 1$ *degrees of freedom.*

When calculating the expected frequencies it is customary to round to the nearest integer or to one decimal, as we did in our example. The entries in Table III are given to three decimals, but there is seldom any need to carry more than two decimals in calculating the value of the χ^2 statistic itself. It should also be noted that since the above test is only approximate, it is best not to use it when one (or more) of the expected frequencies is less than 5.

The method which we have discussed can be used only to test the null hypothesis against the alternative hypothesis that the p's are not all equal. However, in the special case where $k = 2$ and we are interested in the significance of the difference between two sample proportions, we can test against either of the alternatives $p_1 < p_2$ or $p_1 > p_2$ by using an *equivalent test* based on the normal curve rather than the chi-square distribution. This alternative procedure is discussed in Exercise 8 below.

EXERCISES

1. A doctor wishes to determine whether a new skeletal muscle relaxant will produce beneficial results in a significantly higher proportion of patients suffering from various severe neurological disorders than the 0.80 proportion receiving beneficial results from standard treatment. If, in a random sample of 100 patients, 85 received beneficial results from the new muscle relaxant, how should the doctor interpret this at the 0.10 level of significance?

2. An experiment is conducted to determine the effect on the germination time of soaking loblolly pine seeds in hydrogen peroxide (usually used as a germ killer). If 55 of 100 treated seeds germinated in a day and a half or less, does it seem reasonable at the 5 per cent level of significance to suppose that half of all such seeds treated will germinate in at most a day and a half under the conditions of the experiment?

3. The manager of the cafeteria of a large manufacturing company claims that more than three fourths of the employees who eat in the cafeteria would prefer to have smaller portions on the $1.25 "Special Plate" and a corresponding reduction in price to $1.10. What can the company conclude at a level of significance of 0.05 if, in a random sample of 300 employees who eat in the cafeteria, 238 are in favor of the change?

4. A large city park and recreation department plants 1,000 tulip bulbs each of the highest-grade bulbs supplied by seed firms A, B, and C. Of the bulbs bought from these firms, 105, 70, and 95, respectively, failed to germinate. At the 0.05 level of significance, does this constitute evidence that the true proportion of bulbs of the given grade that fail to germinate is not the same for the three firms?

5. A study of 10,000 young schoolchildren showed that 400 children suffered from migraine headaches believed to be caused by such stroboscopic effects as flickering television. At the 5 per cent level of significance, does this support a claim that more than 3 per cent of the children in the population sampled suffer from such headaches?

6. In a random sample of 400 television viewers, 300 felt that television station policy of running three consecutive commercials followed by a station break and two more commercials (or five commercials) in less than three minutes was extremely poor policy and gave the appearance of "cluttering" the air with commercials. At the 1 per cent level of significance, is this finding consistent with a claim that more than two thirds of the viewers in the population sampled feel that this is an extremely poor policy?

7. Tests are made on the proportion of defective castings produced by two molds. If among 100 castings from Mold I there were 14 defectives and among 200

castings from Mold II there were 36 defectives, test at a level of significance of 0.01 the null hypothesis that there is no difference between the true proportions of defective castings produced by the two molds.

8. Repeat Exercise 7 basing your decision on the statistic

$$z = \frac{\dfrac{x_1}{n_1} - \dfrac{x_2}{n_2}}{\sqrt{p(1 - p)\left(\dfrac{1}{n_1} + \dfrac{1}{n_2}\right)}} \qquad \text{with} \qquad p = \frac{x_1 + x_2}{n_1 + n_2} \qquad \bigstar$$

whose sampling distribution is approximately the standard normal distribution provided that both samples are large. The criterion for this two-tail test is again the one shown in Figure 10.4, with 2.58 substituted for 1.96 to change the level of significance to 0.01. Also verify that the *square* of the value obtained for z in this exercise equals the value obtained for χ^2 in Exercise 7.

9. In a random sample of 100 individuals who skipped breakfast 45 reportedly experienced midmorning fatigue, and in a random sample of 400 individuals who ate breakfast 120 reportedly experienced midmorning fatigue. Use first the χ^2 statistic, then the z statistic of Exercise 8 to determine whether, at a level of significance of 0.01, there is a real difference between the two proportions of individuals reporting midmorning fatigue in the populations sampled. Also, verify that $z^2 = \chi^2$.

10. In a random sample of 400 housewives who had been cooking for less than one year, 300 felt that cup content (in addition to weight content) should be shown on various canned and packaged food items which are measured (instead of weighed) into other dishes; of 400 randomly selected housewives who had been cooking for at least five years, 180 felt that the cup content should also be shown on such food items. At the 5 per cent level of significance, does this suggest that less-experienced cooks feel the need of such help more than more-experienced cooks do?

11. A study is made of a large population of work connected electrical accidents.

(a) In a random sample of 200 of these accidents where the condition of the equipment could be determined, unsafe conditions were identified in 158 accidents. At the 5 per cent level of significance, is this finding consistent with a claim that in 80 per cent of such accidents unsafe conditions are involved?

(b) In a random sample of 100 accidents drawn from the population of those in which it was possible to determine what the worker was doing at the time of the accident, unsafe acts were involved in 68 instances. Estimate with probability 0.95 the actual proportion of such accidents in which unsafe acts were involved.

(c) There was a large number of accidents in which contact with conductors caused injuries. In a random sample of 300 of these conductor injuries, it was found that in 100 cases conductors carrying more than 600 volts were involved. At a level of significance of 0.01, does this finding support a claim that, in more than one fourth of the conductor caused injuries, conductors carrying more than 600 volts were involved?

12. To determine the attitude of its field sales personnel toward whether the company should follow its present policy of maintaining profit levels on unit sales or adopt a proposed new high-volume, low-price policy, a national manu- facturer took a sample of its salesmen in four broad geographic areas with the following results:

	North	South	East	West
Maintain Old Policy	42	22	54	52
Adopt New Policy	48	38	146	78

Find the expected frequencies under the null hypothesis that the true propor- tions of field men who favor the present policy are the same, calculate χ^2, and test the null hypothesis at a level of significance of 0.05.

13. A study is made of the supply of part-time and casual workers as well as the demand for these workers in a large city.

(a) In a random sample of 100 women over 45 years of age seeking work, typing was the primary skill of 80 women. Test, at the 0.05 level of signifi- cance, a claim that typing was the primary skill of more than 70 per cent of the women in the population sampled.

(b) In a random sample of 100 homeowners who sought help for household services, the service most badly needed by 78 homeowners was general cleaning. Test, at the 0.05 level of significance, the hypothesis that two thirds of the homeowners in the population sampled needed general cleaning help most badly.

14. In studying its baggage-handling problems, a major commercial airline takes a random sample of 100 lost bags each from flights originating in City A and City B. In 46 of the City A losses and in 54 of the City B losses, the cause of the loss was found to be tagging errors at check-in time. At the 0.01 level of significance, does it appear that the City B crew is doing a poorer job of baggage tagging at check-in than the City A crew?

15. In a random sample of 100 farmers each from Regions A and B, it was found that 38 farmers in Region A and 42 farmers in Region B used no fertilizer at all

on their farms. Test, at the 0.05 level of significance, the hypothesis that the same proportion of farmers in the two regions use no fertilizer.

16. In a grape-planting investigation a random sample of 100 cuttings of root stock Saltcreek (which root with difficulty) was totally immersed in water for one day prior to planting, and another random sample of 100 cuttings was completely soaked for five days prior to planting. At the end of a given time after planting, it was found that 67 of the cuttings soaked five days and 63 of the cuttings soaked one day had rooted successfully. Is this evidence, at the 0.01 level of significance, that soaking the stock for five days increases the proportion of successful rootings?

17. (*Control charts for attributes*) In order to control the proportion of defectives or other characteristics (attributes) of mass-produced items, quality control engineers take random samples of size n at regular intervals of time and plot their results (the sample proportions) on a *control chart* like that of Figure 10.12.

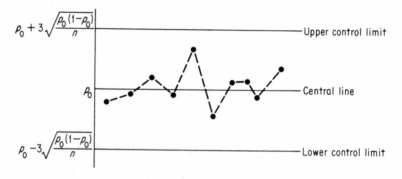

FIGURE 10.12 Control chart

If the production process is considered to be under control when the true proportion of defectives is p_0, the *central line* of the control chart for the proportion of defectives is at p_0, and the *3-sigma upper and lower control limits* are at

$$p_0 + 3\sqrt{\frac{p_0(1 - p_0)}{n}} \quad \text{and} \quad p_0 - 3\sqrt{\frac{p_0(1 - p_0)}{n}}$$

As was explained in connection with Figure 10.10, a process is assumed to be under control so long as the sample proportions, plotted on the control chart, remain between the two control limits.

Construct a control chart for the proportion of defectives obtained in repeated random samples of size 100 from a process which is considered to be under control when $p_0 = 0.20$. Given that 24 consecutive samples of size 100 contained, respectively, 23, 14, 19, 25, 20, 18, 13, 27, 24, 25, 21, 20, 30, 24, 17,

23, 19, 16, 13, 19, 21, 20, 25, and 34 defectives, plot the corresponding proportions on the control chart and comment on the performance of the process.

18. Construct a control chart (see Exercise 17) for the proportion of defectives
obtained in repeated random samples of size 400 from a process which is
considered to be under control when $p_0 = 0.10$. Given that 25 consecutive
samples of size 400 contained, respectively, 41, 32, 55, 49, 37, 45, 40, 23, 35, 27,
47, 45, 38, 40, 42, 36, 48, 50, 36, 25, 20, 34, 28, 25, and 24 defectives, plot the
corresponding proportions on the control chart and comment on the performance of the process.

CONTINGENCY TABLES No

The χ^2 statistic plays an important role in many other problems
dealing with *count data*, that is, in problems where information is obtained
by counting rather than by measuring. The method of analyzing such data
that we shall discuss in this section is an extension of the method of the
preceding section, and it applies to problems in which, unlike binomial
problems, each trial permits *more than two* outcomes. For example, in a
poll to determine voter attitude toward a proposed piece of legislation,
people in five different age groups may be asked to state whether they favor
the legislation, oppose it, or have no opinion; and in the example of the
preceding section, people tasting the orange juice in the three regions might
have been given four juices with distinctly different levels of acidity and
asked to express a preference for one of the four. In the first case, we would
have presented the results in a 3 × 5 table, that is, a table having 3 rows
and 5 columns; in the second case the results would be summarized in a
4 × 3 table. More generally, when there are k samples and each trial permits r alternatives, we refer to the resulting table as an $r \times k$ table. Tables
of this sort are of two kinds: tables, like those of the two examples above,
in which the column totals are fixed (because they are the respective sample sizes) and the row totals vary as the result of chance, and tables in
which both the column totals and the row totals depend on chance. These
latter tables, in which items are classified according to two characteristics
and chance determines both the row and column in which an item is entered, are called *contingency tables*.

To illustrate the analysis of count data summarized in an $r \times k$ table,
suppose that we want to determine whether or not there is a real relationship between the incomes of salesmen of hotel supplies and their formal
educational level. For this purpose we have taken a random sample of 400
such salesmen and constructed the following 3 × 3 contingency table:

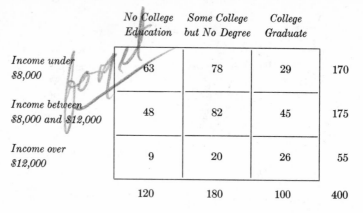

	No College Education	Some College but No Degree	College Graduate	
Income under $8,000	63	78	29	170
Income between $8,000 and $12,000	48	82	45	175
Income over $12,000	9	20	26	55
	120	180	100	400

The null hypothesis we want to test, with, say $\alpha = 0.05$, is that the income and educational level of salesmen of hotel supplies are *independent* (or, in other words, that there is no relationship between the income of salesmen and their formal education). If this null hypothesis is true, the probability of randomly selecting a salesman of hotel supplies for our sample who has an income under $8,000 and no college education equals the *product* of the probability that he has an income under $8,000 and the probability that he has no college education (see page 129). Using the row and column totals of the above table to *estimate* these two probabilities, we obtain 170/400 for the probability that a salesman of hotel supplies has an income under $8,000 and 120/400 for the probability that he has no college education. Hence, we estimate the "true proportion" of salesmen of hotel supplies with incomes under $8,000 and no college education as $\frac{170}{400} \cdot \frac{120}{400}$, and in a sample of size 400 we would *expect* to find $\frac{170}{400} \cdot \frac{120}{400} \cdot 400 = \frac{170 \cdot 120}{400} = 51$ salesmen who fit this description.

Having thus obtained the *expected frequency* for the first cell of the first row of our table, it should be noted that after canceling 400's, we obtained this value by multiplying the total of the first row by the total of the first column and then dividing by the grand total of 400. In general we can find the expected frequency for any one cell of a contingency table by *multiplying the total of the row to which it belongs by the total of the column to which it belongs and then dividing by the grand total for the whole table.* Using this procedure, we obtain an expected frequency of $\frac{170 \cdot 180}{400} = 76.5$ for the second cell of the first row, and $\frac{175 \cdot 120}{400} = 52.5$ and $\frac{175 \cdot 180}{400} = 78.75$ for the first two cells of the second row.

As it can be shown that the sum of the expected frequencies for each row or column must equal the sum of the corresponding observed frequencies, we obtain an expected frequency of

$$170 - (51 + 76.5) = 42.5$$

for the third cell of the first row,

$$175 - (52.5 + 78.75) = 43.75$$

for the third cell of the second row, and

$$120 - (51 + 52.5) = 16.5$$
$$180 - (76.5 + 78.75) = 24.75$$

and

$$100 - (42.5 + 43.75) = 13.75$$

for the three cells of the third row. Rounding the expected frequencies to one decimal, these results are summarized in the following table, where the expected frequencies are shown in parentheses below the frequencies that were actually observed:

	No College Education	Some College but No Degree	College Graduate	
Income under $8,000	63 (51)	78 (76.5)	29 (42.5)	170
Income between $8,000 and $12,000	48 (52.5)	82 (78.8)	45 (43.8)	175
Income over $12,000	9 (16.5)	20 (24.8)	26 (13.8)	55
	120	180	100	400

From here on, the work is like that of the preceding section; we calculate the χ^2 statistic according to the formula

$$\chi^2 = \Sigma \frac{(f - e)^2}{e}$$

★

with $(f - e)^2/e$ calculated separately for each cell of the table. Then we use the criterion of Figure 10.11 as it is formulated on page 296, with the one exception that the number of degrees of freedom is given by a different formula. If a contingency table has r rows and k columns, the number of degrees of freedom for χ^2 is $(r - 1)(k - 1)$. In our numerical example it is $(3 - 1)(3 - 1) = 4$, and it should be noted that, after we had calculated four of the expected frequencies, all the others were obtained by subtraction from the totals of appropriate rows or columns.

Returning now to the observed and expected frequencies of the table on page 303, we get

$$
\chi^2 = \frac{(63 - 51)^2}{51} + \frac{(78 - 76.5)^2}{76.5} + \frac{(29 - 42.5)^2}{42.5} + \frac{(48 - 52.5)^2}{52.5}
$$
$$
+ \frac{(82 - 78.8)^2}{78.8} + \frac{(45 - 43.8)^2}{43.8} + \frac{(9 - 16.5)^2}{16.5} + \frac{(20 - 24.8)^2}{24.8}
$$
$$
+ \frac{(26 - 13.8)^2}{13.8}
$$
$$
= 22.81
$$

and since this value exceeds 9.488, the value of $\chi^2_{.05}$ for 4 degrees of freedom, the null hypothesis must be rejected. In other words, we have shown that there exists a significant dependence (or relationship) between the income and educational level of salesmen of hotel supplies.

Since the χ^2 statistic we are using here has only approximately a chi-square distribution, it should not be used in cases where any of the expected cell frequencies are less than 5. When there are expected frequencies smaller than 5, it is often possible to combine some of the cells, subtract one degree of freedom for each cell eliminated, and then perform the test just as we have described it here.

GOODNESS OF FIT

In this section we shall treat a further application of the χ^2 criterion, in which we compare an observed frequency distribution with a distribution which we might expect according to theory or assumptions. To illustrate this application, suppose that we want to decide whether or not a coin is actually balanced and equally likely to fall heads and tails. In order to decide, we perform an experiment which consists of tossing the coin three times and recording the number of heads showing in the three tosses, then repeating the process until we have completed 160 sets of three tosses each. All the tosses are made by a mechanical coin tosser which is assumed to ensure randomness in the tossing. Suppose that we obtain the results shown in the "observed frequency" column of the following table:

Number of Heads	Observed Frequency	Expected Frequency
0	14	20
1	71	60
2	65	60
3	10	20

If the coin is actually balanced and the tosses are independent, the respective probabilities of 0, 1, 2, and 3 heads in any one of the 160 sets are given by the binomial distribution $f(x) = \dfrac{\binom{3}{x}}{8}$ for $x = 0, 1, 2,$ and 3. These probabilities are 1/8, 3/8, 3/8, and 1/8, and the expected frequencies in the right-hand column are obtained by multiplying each of these probabilities by 160.

We now formulate the null hypothesis that the coin is balanced (i.e., the probability of heads on any single toss is 0.50), and the alternative that the coin is not balanced, and set a level of significance of $\alpha = 0.05$. To determine whether the discrepancies between the observed frequencies and those we would expect (if the binomial distribution is the proper model in this experiment) are due to chance, we calculate χ^2 by means of the formula

$$\chi^2 = \Sigma \, \frac{(f - e)^2}{e}$$

and again use the criterion of Figure 10.11, rejecting the hypothesis (and stating that the binomial distribution is a poor fit to the data) if $\chi^2 > \chi^2_{.05}$, and accepting the hypothesis (or reserving judgment) if $\chi^2 \leq \chi^2_{.05}$. *In goodness of fit problems, the number of degrees of freedom is given by the number of terms $(f - e)^2/e$ added to obtain χ^2, minus the number of quantities, obtained from the observed data, that are used in calculating the expected frequencies.*

In our numerical example we shall sum four terms to get χ^2, and the only quantity needed from the observed data to calculate the expected frequencies is the total frequency of 160. Hence, the number of degrees of freedom is $4 - 1 = 3$ and with a level of significance of 0.05 we find from Table III the critical value $\chi^2_{.05} = 7.815$. Substituting the frequencies from the table above into the formula for χ^2, we get

$$\chi^2 = \frac{(14 - 20)^2}{20} + \frac{(71 - 60)^2}{60} + \frac{(65 - 60)^2}{60} + \frac{(10 - 20)^2}{20} = 9.24$$

and since this is greater than the critical value, we reject the null hypothesis that the coin is balanced. Evidently it is not, and the binomial distribution with $p = \frac{1}{2}$ provides a poor fit to the data.

The method we have illustrated in this section is used quite generally to test how well theoretical distributions fit (or describe) observed data. In the exercises which follow, we shall ask the reader to test whether it is reasonable to treat an observed distribution as if it had (at least approximately) the shape of a normal distribution, and also whether an observed set of data might be satisfactorily described by a Poisson distribution.

EXERCISES

1. In a study devoted to stockholder attitude toward its community relations programs, a large national manufacturer takes a random sample of 340 of its stockholders, classifies their holdings as either small, medium sized, or large, and asks each person to rate the company programs as either good, fair, or poor. The following are the results:

	Small	Medium	Large
Good	35	50	20
Fair	50	45	25
Poor	45	55	15

At the 0.01 level of significance, is there a relationship between the size of stockholders' holdings and their attitude toward the company programs?

2. In an occupational survey a large university takes a random sample of 385 male graduates of 10 years earlier who had been working continuously in business since their graduation. Each person was asked to state whether, in his opinion, his career progress had been better than was expected at graduation, about what was expected, or poorer than expected, and each person was ranked as having stood in the bottom third, middle third, or top third of his class. The results are shown in the table at the top of page 307.

	Bottom Third	Middle Third	Top Third
Better than Expected	80	40	70
About what Was Expected	40	20	50
Poorer than Expected	30	15	40

At the 0.05 level of significance, is there a relationship between the class standing and the career progress of the men?

3. In a study to determine whether there is any relationship between temperament and speed of professional advancement, a random sample of 420 men is taken and the following table is constructed:

	Speed of Advancement			
	Slow	Average	Fast	Very Fast
Extreme Self-control	42	31	56	28
Average Self-control	16	82	47	21
Little Self-control	13	26	39	19

Test, at the 0.05 level of significance, the hypothesis that there is no real relationship between temperament and speed of advancement.

4. A large opinion research organization wants to determine whether there is any relationship between the quality of interviewers' work and their scores on an introvert-extrovert personality test. Each interviewer is rated by his superior as being above average, average, or below average on the basis of such factors as persistence, need for supervision, complaints from alleged respondents, neatness in completing schedules, and so on. The results are shown in the following table:

	Work above Average	Work Is Average	Work below Average
Introvert	18	28	14
Average	37	63	30
Extrovert	15	29	16

Test, at a level of significance of 0.05, the hypothesis that there is no relationship between personality as measured by the test and work performance quality. What can one conclude about the effectiveness of the test in predicting whether a prospective interviewer will do good-quality work?

5. If the analysis of a contingency table shows that there is a relationship between the two variables under consideration, the strength of the relationship can be measured by the *contingency coefficient*

$$C = \sqrt{\frac{\chi^2}{\chi^2 + n}} \qquad \bigstar$$

where n is the total frequency for the table. This coefficient assumes values between 0 (corresponding to independence) and a maximum value of less than 1 depending on the size of the table. For example, for a $k \times k$ table the maximum value of C is $\sqrt{(k-1)/k}$. The larger C is, the stronger the relationship is between the variables. Calculate C for Exercise 3 above in which there was a significant relationship.

6. Five coins are tossed 320 times, and 0, 1, 2, 3, 4, and 5 heads showed 13, 49, 87, 109, 56, and 6 times, respectively. At the 0.05 level of significance, is it reasonable to suppose that the coins are balanced?

7. We want to test whether the three dice in a chuck-a-luck cage will actually fall with the *six face* showing in accord with what might be expected if the dice are balanced and the tossing is random. If the cage is spun 1,080 times and 3, 2, 1, and 0 sixes show 2, 33, 390, and 655 times, respectively, what should we conclude at the 0.01 level of significance?

8. A tetrahedral (four-sided) die lettered A, B, C, and D and intended for use in a monopoly-type game is supposed to be loaded in such a way that the B face falls twice as often as the A face, the C face twice as often as the B face, and the D face twice as often as the C face. In order to test the accuracy of the loading, the die is randomly tossed 225 times. If the A, B, C, and D faces fall 11, 24, 64, and 126 times, respectively, is this evidence at the 0.05 level of significance that the die is not properly loaded?

9. A pair of tetrahedral dice numbered 1, 2, 3, and 4 is tossed 256 times. A total of 2, 3, 4, 5, 6, 7, and 8 points showed on the dice in 12, 33, 52, 59, 50, 28, and 22 of the tosses, respectively. Does it appear, at the 0.05 level of significance, that the dice are not balanced?

10. At the 0.01 level of significance, is it reasonable to suppose that the hourly demand for a particular kind of service at a large bank is appropriately described by a Poisson distribution with $\lambda = 1$ if in a random sample of 1,000 hours there were 355, 362, 190, 65, 22, and 6 hours in which there were 0, 1, 2, 3, 4, and at least 5 of these demands, respectively? (Combine the two smallest classes in the distribution so that all the expected frequencies are greater than 5.)

11. A company observed that in a 200-day period there were 24, 50, 56, 32, 20, 10, 6, and 2 days on which 0, 1, 2, 3, 4, 5, 6, and 7 or more record changes, respectively, were required in a certain record file. Is it reasonable to suppose, at the 0.05 level of significance, that the distribution of the daily number of record changes is actually a Poisson distribution with $\lambda = 2$? (Combine some adjacent classes of the distribution so that no expected frequency is less than 5.)

12. The following is a distribution of the times required by a random sample of 200 men to complete a screening test required of applicants by a firm specializing in executive placement:

Time (minutes)	Number of Men
24 or less	15
25–29	50
30–34	75
35–39	40
40–44	15
45 or over	5

The mean and standard deviation of these times, calculated before they were grouped, are $\bar{x} = 32.1$ minutes and $s = 5.6$ minutes.

(a) Find the area under a normal curve with $\mu = 32.1$ and $\sigma = 5.6$ which lies to the left of 24.5, between 24.5 and 29.5, between 29.5 and 34.5, between 34.5 and 39.5, between 39.5 and 44.5, and to the right of 44.5.

(b) Calculate the expected normal curve frequencies for the six classes of the distribution by multiplying each of the areas found in part (a) by 200.

(c) Test, at the 0.05 level of significance, the null hypothesis that this sample might reasonably have come from a population having approximately the shape of a normal distribution. In making the χ^2 test of the goodness of fit of a normal distribution, the number of degrees of freedom is $k - 3$, where k is the number of classes, and three degrees of freedom are lost since the sums of the expected and observed frequencies must agree and the mean and standard deviation of the normal curve must equal the mean and standard deviation of the observed distribution. (Combine the two highest classes in this example so that no expected frequency is less than 5.)

A WORD OF CAUTION

There is an interesting analogy between significance testing and the principle of criminal law that a person is presumed innocent until he is proven guilty. In significance tests the burden of proof is on us to show that observed differences (discrepancies among sets of data, or discrepancies between theory and practice) are too large to be reasonably attributed to chance; in criminal proceedings the burden of proof is on the prosecution, which must show beyond any reasonable doubt that the defendant is guilty. Thus, it may sometimes happen that a guilty party is found "not guilty" simply because the prosecution did not have enough evidence to "get a conviction." However, the fact that a defendant is found not guilty in a court of law does not imply that he must be considered innocent in the court of public opinion. Similarly, we may fail to reject a false hypothesis because of chance factors, or simply because we do not have enough data to "get a conviction." But if we are unable to reject a null hypothesis, this does not mean that we must conclude that it is true and accept it; this is why we suggested the possibility of reserving judgment on the truth of a hypothesis whenever it is appropriate and feasible to do so.

We should also note that in statistics the term "significant" is used in a technical sense. If we say that something is "statistically significant" we do not mean to imply that it is necessarily of any practical significance or importance. For instance, a direct-mail retailer might find from sample mailings that a new, expensive appeal significantly increases (statistically) the percentage of returns over the percentage produced by the old, relatively inexpensive appeal. Yet the expected increase in returns and profitability could be so small that there is no possible economic justification for switching to the new appeal.

Finally, if we look carefully again at the table on page 261, we note a marked resemblance to the games of strategy which we discussed in Chapter 7. In the two-person games of that chapter, each player has the choice of two or more strategies, and in testing a hypothesis H we must decide whether to accept H or reject it. We can call our opponent Nature and note that in this game Nature has control over the truth or falsity of H; that is, Nature controls the statistical parameters we are concerned with. So, conceptually, it is not unreasonable to look on decision problems such as those we have discussed in this chapter as two-person games against Nature. However, the trouble with this approach in practice comes when we try to assign cash values to the various outcomes and construct a payoff table.

In the situation discussed in Exercise 6 on page 266, for example, a Type I error is committed when we erroneously reject the hypothesis that the machine is setting the lids correctly; the cost consequences of this error might be just the cost of unnecessary work on a machine which is operating properly, plus the cost of lost production. However, if we commit a Type II error and erroneously accept the hypothesis that the machine is setting the lids correctly, it becomes very difficult to evaluate the costs. "How great is the loss due to discoloration if the lids are set too loosely?" "How great is the loss if some people die as a result of eating spoiled mayonnaise?" "How great is the loss in good will and future sales if buyers are discouraged by lids that are set too tightly?" In general, the whole problem of assigning cash values to the consequences of all possible outcomes is so hard to solve that it limits the application of game theory in statistical decision problems.

In the classical methods discussed in this chapter the specification of "suitable" probabilities α and β is undesirably subjective and perhaps often debatable; nevertheless, these methods have been tested in the acid of use and they have proved themselves extremely valuable in business and industry, and in virtually all of the natural and social sciences.

11

DECISION MAKING:
ANALYSIS
OF VARIANCE*

DIFFERENCES AMONG k MEANS

Let us now generalize the work on pages 280 to 283 and consider the problem of deciding whether observed differences among more than two sample means can be attributed to chance, or whether they are indicative of actual differences among the means of the corresponding populations. For instance, we may want to decide on the basis of sample data whether there really is a difference in the effectiveness of three methods of teaching computer programming, we may want to compare the average mileage yield of four kinds of gasoline, or we may want to compare the performance of the graduates of six different technical schools. To give a concrete example, suppose that the branch manager of a bank can drive to work along four different routes, and that the following are the number of minutes it took him on five different occasions for each route:

---

*The material in this chapter is somewhat more advanced and may be omitted without loss of continuity.

Route 1	Route 2	Route 3	Route 4
22	25	25	27
26	27	29	29
25	28	33	28
25	26	30	30
32	29	33	31

The means of these four samples are, respectively, 26, 27, 30, and 29 minutes, and we would like to know whether the differences among these means are significant or whether they can be attributed to chance.

If μ_1, μ_2, μ_3, and μ_4 are the true average times it would take him to get to work along the four different routes, we want to test the null hypothesis $\mu_1 = \mu_2 = \mu_3 = \mu_4$ against the alternative that μ_1, μ_2, μ_3, and μ_4 are *not all equal*. Evidently, this null hypothesis would be supported if the sample means were all nearly the same size, and the alternative hypothesis would be supported if the differences among the sample means were large. Thus, we need a precise measure of the discrepancies among the $\bar{x}$'s, and the most obvious, perhaps, is their standard deviation or their variance, which we can calculate according to the formula on page 43. Since $\dfrac{26 + 27 + 30 + 29}{4} = \dfrac{112}{4} = 28$, we get

$$s_{\bar{x}}^2 = \frac{(26 - 28)^2 + (27 - 28)^2 + (30 - 28)^2 + (29 - 28)^2}{4 - 1}$$

$$= \frac{10}{3}$$

where we used the subscript $\bar{x}$ to indicate that this variance measures the variability of the sample means.

Let us now make an assumption which is critical to the method of analysis we shall employ: *It will be assumed that the populations from which we are sampling can be approximated closely with normal distributions having the same standard deviation* σ. If the null hypothesis $\mu_1 = \mu_2 = \mu_3 = \mu_4$ is true, we can then look upon our four samples as samples from *one and the same population* and, hence, upon $s_{\bar{x}}^2 = \dfrac{10}{3}$ as an estimate of $\sigma_{\bar{x}}^2$, the square of the standard error of the mean. Now, if we make use of the theorem on page 222, according to which $\sigma_{\bar{x}} = \dfrac{\sigma}{\sqrt{n}}$ for samples from infinite populations, we can look upon $s_{\bar{x}}^2 = \dfrac{10}{3}$ as an estimate of $\sigma_{\bar{x}}^2 = \left(\dfrac{\sigma}{\sqrt{n}}\right)^2 = \dfrac{\sigma^2}{n}$ and, therefore,

upon $n \cdot s_{\bar{x}}^2 = 5 \cdot \dfrac{10}{3} = \dfrac{50}{3}$ as an estimate of σ^2, the common variance of the four populations.

If σ^2 were known, we could compare $n \cdot s_{\bar{x}}^2 = \dfrac{50}{3}$ with σ^2 and *reject* the null hypothesis if $\dfrac{50}{3}$ were much larger than σ^2. However, in our example (as in most practical problems) σ^2 is unknown, and we have no choice but to estimate it on the basis of the given data. Having assumed, in fact, that the four samples come from identical populations, we can use any one of these sample variances (namely, s_1^2, s_2^2, s_3^2, or s_4^2) as an estimate of σ^2, and hence, we can also use their mean

$$\frac{s_1^2 + s_2^2 + s_3^2 + s_4^2}{4}$$

$$= \frac{1}{4}\left[\frac{(22-26)^2 + (26-26)^2 + (25-26)^2 + (25-26)^2 + (32-26)^2}{5-1} \right.$$

$$+ \frac{(25-27)^2 + (27-27)^2 + (28-27)^2 + (26-27)^2 + (29-27)^2}{5-1}$$

$$+ \frac{(25-30)^2 + (29-30)^2 + (33-30)^2 + (30-30)^2 + (33-30)^2}{5-1}$$

$$+ \left. \frac{(27-29)^2 + (29-29)^2 + (28-29)^2 + (30-29)^2 + (31-29)^2}{5-1} \right]$$

$$= \frac{118}{16}$$

which utilizes *all* the information we have about the population variance σ^2. We now have two estimates of σ^2,

$$n \cdot s_{\bar{x}}^2 = \frac{50}{3} = 16\frac{2}{3} \qquad \text{and} \qquad \frac{s_1^2 + s_2^2 + s_3^2 + s_4^2}{4} = \frac{118}{16} = 7\frac{3}{8}$$

and if the *first estimate* (which is based on the variation among the sample means) is much larger than the *second estimate* (which is based on the variation within the samples and, hence, measures variation that is due to chance), it stands to reason that the null hypothesis should be rejected. After all, *in that case the variation among the sample means would be greater than it should be if it were due only to chance.*

To put the comparison of these two estimates of σ^2 on a rigorous basis, we use the statistic

$$F = \frac{\text{Estimate of } \sigma^2 \text{ based on the variation among the } \bar{x}\text{'s}}{\text{Estimate of } \sigma^2 \text{ based on the variation within the samples}}$$

which is appropriately called a *variance ratio*. If the null hypothesis and the assumption which we made are true, the sampling distribution of this statistic is the so-called *F distribution*, an example of which is shown in Figure 11.1. Since the null hypothesis will be rejected only when F is *large*

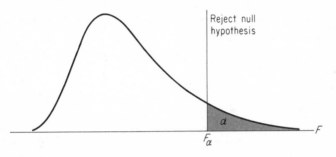

FIGURE 11.1 F distribution

(i.e., when the variability of the $\bar{x}$'s is too great to be attributed to chance), we ultimately base our decision on the criterion of Figure 11.1. Here F_α is such that the area under the curve to its right equals α, the level of significance, and it depends on two quantities (parameters) referred to, respectively, as the *numerator and denominator degrees of freedom*. When we compare the means of k samples of size n, the numerator degrees of freedom equals $k - 1$ and the denominator degrees of freedom equals $k(n - 1)$.* The values of F_α, for various degrees of freedom, are given in Table IVa for $\alpha = 0.05$ and in Table IVb for $\alpha = 0.01$ at the end of the book.

Returning now to our numerical example, we find that for $k = 4$ and $n = 5$ the numerator and denominator degrees of freedom are, respectively, 3 and 16, and that $F_{.05}$ equals 3.24 according to Table IVa. Since this is *not* exceeded by the value which we obtain for F for the given data, namely,

---

* So far as the formula for the numerator degrees of freedom is concerned, note that the numerator is $n \cdot s_{\bar{x}}^2$ where $s_{\bar{x}}^2$ is the variance of k means and, hence, has $k - 1$ degrees of freedom in accordance with the terminology introduced in the footnote to page 240. So far as the formula for the denominator degrees of freedom is concerned, note that the denominator is the mean of k sample variances with each having $n - 1$ degrees of freedom.

$$F = \frac{16\frac{2}{3}}{7\frac{3}{8}} = 2.26$$

it follows that *the null hypothesis cannot be rejected*. Even though the first route seems to be quite a bit faster than the third, this (as well as the other differences among the means) may very well be attributed to chance.

The technique we have just described is the simplest form of a powerful statistical tool called the *analysis of variance*, ANOVA for short. Although we could go ahead and perform F tests for differences among k means without further discussion, it will be instructive to look at the problem from an analysis-of-variance point of view, and we shall do so in the next section.

ONE-WAY ANALYSIS OF VARIANCE

The basic idea of analysis of variance is to express a measure of the total variability of a set of data as a *sum* of terms, each of which can be attributed to a specific source, or cause, of variation. With reference to the example of the preceding section, two such sources of variation might be (1) actual differences among the *true* average times it would take the branch manager of the bank to drive to work along the different routes and (2) chance, which in problems of this kind is usually referred to as *experimental error*. The measure of the total variation of a set of data which we use in analysis of variance is the *total sum of squares**

$$SST = \sum_{i=1}^{k} \sum_{j=1}^{n} (x_{ij} - \bar{x}_{..})^2$$

where x_{ij} is the jth observation of the ith sample ($i = 1, 2, \ldots, k$ and $j = 1, 2, \ldots, n$), and $\bar{x}_{..}$ is the *grand mean*, namely, the mean of all the $n \cdot k$ measurements or observations. Note that if we divided the total sum of squares SST by $nk - 1$, we would obtain the *variance* of the combined data; hence, the total sum of squares of a set of data is interpreted in much the same way as its variance.

---

* The use of double subscripts and double summations is explained briefly in Technical Note 1 at the end of Chapter 3.

Letting $\bar{x}_{i.}$ denote the mean of the ith sample, we can now write the following *identity*, which forms the basis of a *one-way analysis of variance:*

$$SST = n \cdot \sum_{i=1}^{k} (\bar{x}_{i.} - \bar{x}_{..})^2 + \sum_{i=1}^{k} \sum_{j=1}^{n} (x_{ij} - \bar{x}_{i.})^2$$

If we look closely at the two terms into which the total sum of squares SST has thus been partitioned, we find that the first term is a measure of the variation *among the sample means;* in fact, if we divided it by $k - 1$, we would get the quantity which, on page 314, was denoted $n \cdot s_{\bar{x}}^2$. Similarly, the second term is a measure of the variation *within the individual samples,* and if we divided it by $k(n - 1)$ we would get the mean of the variances of the individual samples, namely, the quantity which we put into the denominator of F on page 316.

It is customary to refer to the first term, the quantity which measures the variation *among the sample means,* as the *treatment sum of squares* $SS(Tr)$; the second term, which measures the variation *within the individual samples,* is usually referred to as the *error sum of squares* SSE. The term "treatment" is accounted for by the fact that most analysis-of-variance techniques were originally developed in connection with agricultural experiments, where different fertilizers, for example, were regarded as different *treatments* applied to the soil. The "error" in "error sum of squares" refers to the experimental error, that is, to chance. Although this may sound confusing at first, we shall refer to the four routes in our example as four different treatments, and in other experiments we may refer to four different kinds of packaging as four different treatments, to five different advertising campaigns as five different treatments, and so on.

Before we go any further, let us verify the identity $SST = SS(Tr) + SSE$ with reference to the example of the preceding section. Substituting into the formulas for the various sums of squares, we get

SST
$$\begin{aligned}
=\ & (22 - 28)^2 + (26 - 28)^2 + (25 - 28)^2 + (25 - 28)^2 + (32 - 28)^2 \\
& + (25 - 28)^2 + (27 - 28)^2 + (28 - 28)^2 + (26 - 28)^2 + (29 - 28)^2 \\
& + (25 - 28)^2 + (29 - 28)^2 + (33 - 28)^2 + (30 - 28)^2 + (33 - 28)^2 \\
& + (27 - 28)^2 + (29 - 28)^2 + (28 - 28)^2 + (30 - 28)^2 + (31 - 28)^2 \\
=\ & 168
\end{aligned}$$

$SS(Tr)$
$$\begin{aligned}
=\ & 5[(26 - 28)^2 + (27 - 28)^2 + (30 - 28)^2 + (29 - 28)^2] \\
=\ & 50
\end{aligned}$$

SSE

$$
\begin{aligned}
= &(22 - 26)^2 + (26 - 26)^2 + (25 - 26)^2 + (25 - 26)^2 + (32 - 26)^2 \\
&+ (25 - 27)^2 + (27 - 27)^2 + (28 - 27)^2 + (26 - 27)^2 + (29 - 27)^2 \\
&+ (25 - 30)^2 + (29 - 30)^2 + (33 - 30)^2 + (30 - 30)^2 + (33 - 30)^2 \\
&+ (27 - 29)^2 + (29 - 29)^2 + (28 - 29)^2 + (30 - 29)^2 + (31 - 29)^2 \\
= &\ 118
\end{aligned}
$$

and it follows that

$$ SS(Tr) + SSE = 50 + 118 = 168 = SST $$

To test the null hypothesis $\mu_1 = \mu_2 = \mu_3 = \mu_4$ against the alternative hypothesis that these *treatment means* are not all equal, we now proceed as in the preceding section and compare $SS(Tr)$ and SSE by means of an appropriate F statistic. In actual practice, it has become the custom to exhibit the necessary work as follows in a so-called *analysis-of-variance table*:

Source of Variation	Degrees of Freedom	Sum of Squares	Mean Square	F
Treatments	$k - 1$	$SS(Tr)$	$MS(Tr) = \dfrac{SS(Tr)}{k - 1}$	$\dfrac{MS(Tr)}{MSE}$
Error	$k(n - 1)$	SSE	$MSE = \dfrac{SSE}{k(n - 1)}$	
Total	$nk - 1$	SST		

Here the second column lists the degrees of freedom (the number of *independent* deviations from the mean on which the respective sums of squares are based), the fourth column lists the *mean squares* which are obtained by dividing the sums of squares by their respective degrees of freedom, and the right-hand column gives the F statistic as the ratio of the two mean squares. *Note that the two mean squares are, in fact, the two estimates of σ^2 referred to on page 314, and that the numerator and denominator degrees of freedom for the F test are as before $k - 1$ and $k(n - 1)$, namely, the figures corresponding to treatments and error in the "degrees of freedom" column.*

Referring again to our numerical example, we now get the following analysis-of-variance table:

Source of Variation	Degrees of Freedom	Sum of Squares	Mean Square	F
Treatments	3	50	$\dfrac{50}{3}$	$\dfrac{50/3}{118/16} = 2.26$
Error	16	118	$\dfrac{118}{16}$	
Total	19	168		

Finally, the test of significance is performed (as before) by comparing the value obtained for F with F_α, which equals 3.24 for $\alpha = 0.05$ and 3 and 16 degrees of freedom. The conclusion is the same as before—since $F = 2.26$ is less than 3.24, the null hypothesis that *the four routes take on the average an equal amount of time* cannot be rejected.

The numbers which we used in our illustration were intentionally chosen so that the calculations would be relatively simple. In actual practice, the calculations of the sums of squares can be quite tedious unless we use the following short-cut formulas, in which $T_{i.}$ denotes the total of the observations corresponding to the ith treatment (i.e., the sum of the values in the ith sample), and $T_{..}$ denotes the *grand total* of all the data:

$$SST = \sum_{i=1}^{k} \sum_{j=1}^{n} x_{ij}^2 - \frac{1}{kn} \cdot T_{..}^2 \qquad \bigstar$$

$$SS(Tr) = \frac{1}{n} \cdot \sum_{i=1}^{k} T_{i.}^2 - \frac{1}{kn} \cdot T_{..}^2 \qquad \bigstar$$

and by subtraction

$$SSE = SST - SS(Tr) \qquad \bigstar$$

It will be left to the reader to verify in Exercise 3 below that for the example of the preceding section these formulas yield

$$SST = 15{,}848 - \frac{1}{20} (560)^2 = 168$$

$$SS(Tr) = \frac{1}{5} (130^2 + 135^2 + 150^2 + 145^2) - \frac{1}{20} (560)^2 = 50$$

and

$$SSE = 168 - 50 = 118$$

EXERCISES

1. An experiment was conducted to compare three methods of teaching the programming of a certain digital computer. Random samples of size 4 were taken from each of three groups of students taught, respectively, by Method A (straight teaching-machine instruction), Method B (personal instruction and some direct experience working with the computer), and Method C (personal instruction but no work with the computer itself), and the following are the grades obtained by these students on an appropriate achievement test:

$$
\begin{array}{llllll}
Method\ A: & 76, & 80, & 70, & 74 \\
Method\ B: & 95, & 85, & 91, & 89 \\
Method\ C: & 77, & 82, & 81, & 84
\end{array}
$$

Use the short-cut formulas to calculate the required sums of squares, construct an analysis-of-variance table, and test at the level of significance $\alpha = 0.05$ whether the differences among the three sample means are significant.

2. Random samples of five brands of tires required the following braking distances (in feet) at a speed of 30 miles per hour:

Brand A	Brand B	Brand C	Brand D	Brand E
26	28	28	27	30
25	28	25	32	24
29	31	28	31	29
24	33	27	30	25

Use the short-cut formulas to calculate the required sums of squares, construct an analysis-of-variance table, and test at the level of significance $\alpha = 0.01$ whether the differences among the five sample means can be attributed to chance.

3. Use the short-cut formulas to verify the values of SST, $SS(Tr)$, and SSE given on page 319 for the illustration used in the text.

4. The following are eight consecutive weeks' earnings (in dollars) of three door-to-door vacuum cleaner salesmen employed by a given firm:

Mr. Jones:	153,	192,	169,	176,	212,	185,	178,	200
Mr. Brown:	176,	182,	173,	187,	199,	188,	169,	184
Mr. North:	165,	201,	177,	195,	189,	173,	182,	198

Use the short-cut formulas to calculate the necessary sums of squares, construct an analysis-of-variance table, and test at the level of significance $\alpha = 0.05$ whether the differences among the average weekly earnings of these salesmen are significant.

5. The following are the scores obtained by 10 students randomly selected from each of four different schools in a test designed to measure their knowledge of current events:

School 1:	75,	88,	62,	97,	62,	81,	93,	76,	53,	99
School 2:	80,	82,	64,	45,	67,	84,	55,	39,	60,	57
School 3:	64,	90,	58,	64,	82,	71,	59,	66,	87,	63
School 4:	94,	69,	85,	57,	93,	88,	69,	78,	56,	78

Use the short-cut formulas to calculate the necessary sums of squares, construct an analysis-of-variance table, and test at the level of significance $\alpha = 0.05$ whether the differences among the average grades obtained by the students from the four schools are significant.

6. Referring to the illustration in the Word of Caution at the end of Chapter 3, perform an analysis of variance to test whether the differences among the sample means (of the area covered by the three paints) are significant. Use the level of significance $\alpha = 0.05$.

7. The method discussed in the text applies only when each sample is of the same size. When the sample sizes are *unequal* and the ith sample contains n_i observations, the computing formulas for the sums of squares become

$$SST = \sum_{i=1}^{k} \sum_{j=1}^{n_i} x_{ij}^2 - \frac{1}{N} \cdot T_{..}^2. \qquad \bigstar$$

$$SS(Tr) = \sum_{i=1}^{k} \frac{T_i^2}{n_i} - \frac{1}{N} \cdot T_{..}^2. \qquad \bigstar$$

$$SSE = SST - SS(Tr) \qquad \bigstar$$

where $N = n_1 + n_2 + \cdots + n_k$. The numerator and denominator degrees of freedom for the F test, that is, for treatments and error, are $k - 1$ and $N - k$, and the total number of degrees of freedom is $N - 1$.

(a) A consumer testing service, wishing to test the accuracy of the thermostats of three different kinds of electric irons, set them at 480°F and obtained the following actual temperature readings by means of a thermocouple:

$$Iron\ X:\quad 474,\quad 496,\quad 467,\quad 471$$
$$Iron\ Y:\quad 492,\quad 498$$
$$Iron\ Z:\quad 460,\quad 495,\quad 490$$

Use $\alpha = 0.05$ to test the null hypothesis that there is no difference in the true average settings at 480°F of the thermostats of the three kinds of irons.

(b) Three groups of rats were injected, respectively, with 0.5 milligrams, 1.0 milligram, and 1.5 milligrams of a new tranquilizer; the following are the number of minutes it took them to fall asleep:

The 0.5-milligram group:	11,	13,	9,	14,	15,	13	
The 1.0-milligram group:	9,	11,	10,	8,	12		
The 1.5-milligram group:	10,	5,	8,	6,	10,	9,	6

Test at the level of significance $\alpha = 0.05$ whether we can reject the null hypothesis that differences in dosage have no effect on the time taken to fall asleep.

TWO-WAY ANALYSIS OF VARIANCE

In the example which we used as an illustration throughout this chapter, we were unable to show that there really is a difference in the average time it takes the branch manager of the bank to drive to work along the different routes even though the sample means varied from 26 minutes to 30 minutes. Of course, there may actually be no difference, but it is also possible that *SSE*, which served as an estimate of chance variation, may actually have been "inflated" by identifiable sources of variation. For instance, it is possible that every time he took Route 3 it rained, or that every time he took Route 1 he left for work late and, hence, missed rush-hour congestion in traffic. Things like this cannot be told from an inspection of the data (that would require more careful planning in setting up the experiment), but if we look at the figures on page 313, careful scrutiny shows that *in each case the first sample value was the smallest while the last sample value was the largest.* This seems strange, so let us suppose we discover that in each case the first observation was taken on a Monday, the second on a Tuesday, the third on a Wednesday, the fourth on a Thursday, and the fifth on a Friday. Thus, the variability which we have been ascribing to chance (or at least part of it) may well be due to differences in traffic conditions on different days of the week.

This suggests that we should have performed a *two-way analysis of variance*, in which the total variability of the data is partitioned into one component which we ascribe to possible differences due to one variable (the different *treatments*), a second component which we ascribe to possible differences due to a second variable (referred to as *blocks*, thanks again to the origin of this method in agricultural research), while the remainder of the variability is ascribed to chance. With reference to our illustration, we might thus refer to the different routes as *different treatments* and to the different days of the week as *different blocks*.

Thus, let us see what part of the total variation of the data can be attributed to traffic conditions on different days of the week. If we let $\bar{x}._j$ stand for the mean of all the values obtained for the jth day, the answer is given by the *block sum of squares*

$$SSB = k \cdot \sum_{j=1}^{n} (\bar{x}._j - \bar{x}..)^2$$

which measures the variability of the average times obtained for the different days. This formula is very similar to the one for $SS(Tr)$—all we have to do is substitute the means obtained for the different days for the means obtained for the different routes, and correspondingly sum on j instead of i and interchange n and k. Corresponding to the short-cut formula for computing $SS(Tr)$, we now have analogously

$$SSB = \frac{1}{k} \cdot \sum_{j=1}^{n} T^2._j - \frac{1}{kn} \cdot T^2.. \qquad \star$$

where $T._j$ is the total of the observations for the jth block (in our example, the total of the values obtained for the jth day).

In a *two-way analysis of variance* we compute SST and $SS(Tr)$ according to the short-cut formulas on page 319, SSB according to the formula of the preceding paragraph, and we then obtain SSE by subtraction, namely, by means of the formula

$$SSE = SST - SS(Tr) - SSB \qquad \star$$

Note that the error sum of squares for a two-way analysis of variance does *not* equal the error sum of squares for the corresponding one-way analysis, even though we denote it with the same symbol. In fact, what we have been doing in this section has been to divide the error sum of squares of the one-way analysis into *two terms:* the block sum of squares and the error sum of squares for the two-way analysis.

Analogous to the analysis-of-variance table for a one-way analysis, that for a two-way analysis is usually presented in the following fashion:

Source of Variation	Degrees of Freedom	Sum of Squares	Mean Square	F
Treatments	$k - 1$	$SS(Tr)$	$MS(Tr) = \dfrac{SS(Tr)}{k - 1}$	$\dfrac{MS(Tr)}{MSE}$
Blocks	$n - 1$	SSB	$MSB = \dfrac{SSB}{n - 1}$	$\dfrac{MSB}{MSE}$
Error	$(n - 1)(k - 1)$	SSE	$MSE = \dfrac{SSE}{(n - 1)(k - 1)}$	
Total	$nk - 1$	SST		

The mean squares are again given by the corresponding sums of squares divided by their degrees of freedom, and the two F values are obtained by dividing, respectively, the mean square for treatments and the mean square for blocks by the mean square for error. Note also that the degrees of freedom for SSE can be obtained by subtracting the degrees of freedom for treatments and blocks from $nk - 1$, the total number of degrees of freedom. A two-way analysis of variance leads to two tests of significance: The first F statistic, $\dfrac{MS(Tr)}{MSE}$, serves to test the null hypothesis that the effects of the k treatments are *all equal* (against the alternative that they are *not all equal*), and the second F statistic, $\dfrac{MSB}{MSE}$, serves to test the null hypothesis that the effects of the n blocks are *all equal* (against the alternative that they are *not all equal*). The numerator and denominator degrees of freedom for the corresponding F tests are $k - 1$ and $(n - 1)(k - 1)$ for treatments, and $n - 1$ and $(n - 1)(k - 1)$ for blocks.

Returning now to our numerical example, we find that the totals for the five days are, respectively, 99, 111, 114, 111, and 125, so that

$$SSB = \frac{1}{4}(99^2 + 111^2 + 114^2 + 111^2 + 125^2) - \frac{1}{20}(560)^2$$

$$= 86$$

and hence, the *new* error sum of squares is

$$SSE = 168 - 50 - 86 = 32$$

since we know from page 319 that $SST = 168$ and $SS(Tr) = 50$. Thus, the table for the two-way analysis of variance of the given data becomes

Source of Variation	Degrees of Freedom	Sum of Squares	Mean Square	F
Treatments	3	50	$\dfrac{50}{3}$	$\dfrac{50/3}{32/12} = 6.25$
Blocks	4	86	$\dfrac{86}{4}$	$\dfrac{86/4}{32/12} = 8.06$
Error	12	32	$\dfrac{32}{12}$	
Total	19	168		

Using a level of significance of 0.05, we find that for 3 and 12 degrees of freedom $F_{.05}$ equals 3.49, and that for 4 and 12 degrees of freedom $F_{.05}$ equals 3.26. Since the first of these two values is exceeded by $F = 6.25$, we can reject the null hypothesis concerning the treatments; and since the second is exceeded by $F = 8.06$, we can also reject the null hypothesis concerning the blocks. In other words, *the differences between the sample means obtained for the four routes are significant, and so are the differences obtained for the different days of the week.* An interesting aspect of this example is that the *proper* analysis led to significant results, whereas the analysis which failed to account for the differences in traffic conditions on different days of the week did not. Of course, which kind of analysis is "proper" for a given set of data will have to depend on how the entire experiment or study was planned and conducted.

The methods we have described in this chapter are only two examples of an analysis of variance. If an experiment is appropriately designed (or planned), it may be possible to express the total sum of squares as a sum of many terms which can be attributed to various sources of variation—that is, to many different variables and even to the joint effects (*interactions*) of two or more variables which may be of relevance in a given investigation.

EXERCISES

1. To study the performance of three detergents at three different water temperatures, the following "whiteness" readings were obtained with specially designed equipment:

	Detergent A	Detergent B	Detergent C
Cold Water	57	55	67
Warm Water	49	52	68
Hot Water	54	46	58

Perform a two-way analysis of variance, using the level of significance $\alpha = 0.05$.

2. The following are the number of defective pieces produced by four operators working, in turn, four different machines:

		Operator			
		B_1	B_2	B_3	B_4
	A_1	34	28	33	29
	A_2	30	24	35	22
Machine	A_3	27	20	40	27
	A_4	28	29	29	26

Use a level of significance of 0.05 to test for significant differences among the operators and also for significant differences among the machines.

3. Analyze the data of Exercise 4 on page 320 by means of a two-way analysis of variance; that is, regard the salesmen as "treatments" and the successive weeks as "blocks." Use $\alpha = 0.01$ for each of the tests of significance.

4. Three different, though supposedly equivalent, forms of a standardized test are given to each of six students. The following are their scores:

	Student					
	1	2	3	4	5	6
Form A	83	72	85	43	78	81
Form B	91	70	88	51	68	87
Form C	85	79	76	49	75	74

Test at a level of significance of 0.05 whether it is reasonable to treat the three forms as equivalent.

A WORD OF CAUTION

Although the analysis of variance is a very powerful statistical tool, we saw from our example what can happen when we use the wrong kind of analysis, in this case, a one-way analysis when we should have used a two-way analysis. In addition to this there are many other pitfalls, for it is often difficult to determine whether the necessary assumptions are met. In the two-way analysis we had *only one observation* from each population, that is, one observation for each combination of routes and days. Thus, it is virtually impossible in this example to determine whether the populations from which we are sampling actually have roughly the shape of normal distributions which all have the *same variance*. To take care of this we could have taken several observations from each population, but there are many situations in which this is neither feasible nor practical.

Let us also point out that in some instances analysis of variance techniques cannot be used because experiments are improperly planned. If, in our example, the branch manager of the bank had failed to try Route 4 on a Friday, the analysis of the experiment would have been a good deal more complicated. This ties in closely with problems of *experimental design,* which we shall touch upon briefly in Chapter 15.

12

DECISION MAKING: SHORT-CUT STATISTICS AND NONPARAMETRIC METHODS

SHORT-CUT STATISTICS

When sample sizes are small and the numbers themselves not unwieldy, the calculations necessary to find such statistics as a mean or standard deviation can be made easily on a desk calculator, or even by hand. Often in practice, however, the sheer volume of routine arithmetic operations required to reduce large sets of data to a few relevant summary measures is so great that hand or simple machine calculation is out of the question. In many cases, though, it is possible to estimate various quantities of interest by measures called "short-cut statistics" which may be arrived at with comparatively little arithmetic. These "quick-and-easy" estimates are often used even though there may be a loss of *efficiency*, that is, even though we may expose ourselves to greater chance fluctuations.

There are several ways, for example, of estimating the mean μ of a population other than by using the sample mean $\bar{x}$. The sample *median*, which for fairly small sets of data may be found more easily than the mean, is one such estimate; when the data are already ordered, the median can be located quickly. Another easily found estimate of the population mean (certainly for ordered data) is the sample *midrange* which is the average of the largest and the smallest observations in a sample. For grouped sample data it may sometimes be convenient to estimate the population mean by using some formula based on the *fractiles* of the distribution; for instance, we might use the *midquartile* $\frac{Q_1 + Q_3}{2}$ where Q_1 and Q_3 are the first and the third quartiles of the distribution (see page 54).

There are several short-cut techniques for estimating the standard deviation σ of a population. When dealing with a fairly large set of data, we can calculate the mean of the largest 5 per cent of the data, the mean of the smallest 5 per cent, and then make use of the fact that the difference between these two means estimates, roughly, four times the standard deviation. Also, if a distribution follows closely the pattern of a normal curve, we can make use of the fact that approximately 68 per cent of the data falls within one standard deviation on either side of the mean, and estimate σ by means of the quantity $\frac{1}{2}(P_{84} - P_{16})$ or by means of the difference $P_{84} - P_{50}$. These P's are the corresponding *percentiles* as defined on page 54, and by using the special probability graph paper shown on page 202, they can sometimes even be read directly off the plot of a cumulative distribution.

When the sample size is quite small, the most widely used short-cut method for estimating the population standard deviation is based on the *sample range*. To obtain an estimate of σ from a sample drawn from a population whose distribution is roughly that of a normal curve, we divide the sample range by a constant, usually denoted d_2 in industrial applications, which depends on the size of the sample. Values of d_2 for samples of size 2 through 12 are shown below:

n	2	3	4	5	6	7	8	9	10	11	12
d_2	1.13	1.69	2.06	2.33	2.53	2.70	2.85	2.97	3.08	3.18	3.26

To illustrate this technique, suppose that a random sample of nine springs is drawn from a large population of such springs. For each spring the number of pounds of force required to compress the spring a specified distance is measured with the following results:

$$10.1, \quad 10.3, \quad 9.9, \quad 10.2, \quad 9.8, \quad 10.0, \quad 9.7, \quad 10.4, \quad 10.1$$

The range is $10.4 - 9.7 = 0.7$, and since $d_2 = 2.97$ for $n = 9$, we have $\frac{0.7}{2.97} = 0.24$ pounds. The median of this sample is 10.1, so we can estimate the mean and standard deviation of the population of forces from which the sample came to be 10.1 pounds and 0.24 pounds, respectively. (See Exercise 6 below.)

EXERCISES

1. Estimate the mean of the 40 times listed on page 236 by using the median, and estimate the standard deviation by taking one fourth of the difference between the mean of the highest 5 per cent of the data and the mean of the lowest 5 per cent of the data. Compare these estimates with the calculated values $\bar{x} = 38.2$ minutes and $s = 8.52$ minutes.

2. Estimate the mean of the 100 seat occupancy times on page 11 by using the median, and estimate the standard deviation by taking one fourth of the difference between the mean of the highest 5 per cent of the data and the mean of the lowest 5 per cent of the data. Compare these values with the calculated values $\bar{x} = 44.38$ and $s = 16.9$.

3. A random sample of 60 lengths is drawn from a large population of such lengths. The five largest items are 15.5, 15.4, 15.4, 15.2, and 15.1 feet, and the five shortest items are 14.2, 14.2, 14.4, 14.4, and 14.5 feet. Estimate the standard deviation of the population from which this sample came.

4. A sample of 80 weights (in ounces) is put into a frequency distribution having the intervals 15.80–15.99, 16.00–16.19, . . . , and 17.00–17.19, with corresponding frequencies 1, 10, 18, 22, 16, 11, and 2.
   (a) Calculate the mean and standard deviation of this distribution.
   (b) Estimate the mean of the distribution by using the midquartile, and estimate the standard deviation by taking one half the difference between P_{84} and P_{16}. Compare these estimates with the values calculated in part (a).

5. A sample of temperature readings is put in a frequency distribution which has $Q_3 = 26.3°C$ and a coefficient of quartile variation of 25.5 per cent (see Exercise 11 on page 56). Estimate the mean of the population from which the sample came.

6. Calculate $\bar{x}$ and s for the sample of nine forces given on page 329, and compare these estimates of the population mean and standard deviation with the estimates based on the median (10.1 pounds) and the sample range (0.24 pounds).

7. Refer to the two samples given on page 334 and verify that, based on the sample ranges, the estimates of the standard deviations of the populations of Brand A and Brand B lifetimes are 4.79 and 2.85 hours, respectively.

8. A random sample of 10 large banks is taken, and their net profits as a percentage of total capital accounts for a given year are determined as follows: 6.3, 6.6, 8.4, 8.7, 7.8, 7.7, 8.0, 7.9, 9.4, and 7.9. Use the median and an estimate based on

the sample range to estimate the mean and standard deviation of the population from which the sample came. Calculate $\bar{x}$ and s and compare the short-cut estimates with these values.

9. Eight cans of cleansing powder are randomly selected from a large production lot, and their net weights are determined with the following results: 13.9, 14.1, 14.0, 14.1, 14.2, 14.1, 14.0, and 13.9 ounces. Use the median and an estimate based on the sample range to estimate the mean and standard deviation of the population from which the sample came. Calculate $\bar{x}$ and s and compare the short-cut estimates with these values.

10. Given the following measurements on tensile strength (in 1,000 psi) in a random sample of nine castings, use the median and an estimate based on the sample range to estimate the mean and standard deviation of the population from which the sample came: 37, 32, 34, 31, 26, 39, 33, 36, and 30. Calculate the mean and standard deviation of the sample and compare the short-cut estimates with these values.

NONPARAMETRIC TESTS

Most of the tests of hypotheses we have discussed in our work require that we make certain assumptions about the population(s) from which the samples are obtained. Since there are many situations in which necessary assumptions (e.g., that a population has roughly the shape of a normal curve) cannot be made, statisticians have developed various alternate techniques of analysis which have become known as *nonparametric* or *distribution-free* tests. Strictly speaking, these terms were not intended to be synonymous; the first was meant to apply to tests in which we make no hypothesis about population parameters, and the second was meant to apply to tests in which we, furthermore, make no assumptions about the nature, shape, or form of the populations. It has now become the custom to refer to either kind of test simply as a *nonparametric* test.

Aside from the fact that they may be used under very general conditions, nonparametric tests are often easier to explain and understand than the standard tests they replace. Moreover, in many of these tests the computational burden is relatively light. For these reasons, nonparametric methods have become quite popular, and there is an extensive literature devoted to them. Of the many nonparametric tests which now exist, we shall discuss below only the so-called *sign test* and some tests based on *rank sums.**

---

* We note, however, that the test for a trend in a time series based on runs above and below the median is a nonparametric test (see Technical Note 8 on page 409); so is the test of the relationship between two variables as measured by the rank correlation coefficient (see page 370).

THE SIGN TEST

In order to make the standard small-sample t test for the difference between two means (i.e., to test the Hypothesis: $\mu_1 - \mu_2 = 0$ against the Alternative: $\mu_1 - \mu_2 \neq 0$, where σ_1 and σ_2 are not known and the sample sizes are both small) it is necessary to assume that the two samples are drawn from (approximately) normal populations having equal standard deviations. If either of these assumptions cannot be met, we can test the hypothesis of no difference between the population means by using a non-parametric test called the *sign test*. The name of this test follows from the fact that, although we work with the differences between pairs of observations, we use only the *signs* (positive or negative) of these differences and not their actual magnitudes. The sign test also provides a convenient alternative for the method (see Exercise 22 on page 288) we have suggested to test the difference between two means when the two samples are *not independent*.

To illustrate the application of the sign test to a problem in which the two samples are in fact not independent, let us consider the following. The average number of days required to convert receivables into cash, called the "average collection period" (calculated by dividing average receivables on the books during a year by average daily sales), is often used as a measure of the efficiency of credit departments. A random sample of 20 wholesalers of drugs and drug sundries is chosen, and the average collection period is calculated for each of them for a given year and also for the fourth year following. We wish to test at the 1 per cent level of significance the hypothesis that, in the population from which the sample came, collection departments are no more efficient now than before (that the average collection period is no shorter for the recent year than for the early year, in other words). The data are given on page 333 in the second and third columns of the table. The fourth column shows by a plus sign the companies for which there was an increase in the average collection period and by a minus sign the ones for which there was a decrease in the average collection period. Counting the signs, we find that there are 16 minus signs and 4 plus signs.

The null hypothesis that the average collection period is no shorter now than it was formerly is equivalent to the null hypothesis that we are as likely to observe a minus sign as we are to observe a plus sign. Therefore, we can test the original hypothesis by testing the null hypothesis that the probability of getting a minus sign is $p = 0.50$ against the alternative that $p > 0.50$. We use this one-sided alternative because $p > 0.50$ means that there is better than a fifty-fifty chance that the average collection period is shorter; if p were less than 0.50, the average collection period would have

Average Collection Period

Wholesaler	Early	Recent	Sign of change
A	50	38	−
B	54	40	−
C	44	28	−
D	42	30	−
E	46	32	−
F	48	33	−
G	55	40	−
H	40	28	−
I	41	43	+
J	47	50	+.
K	44	28	−
L	48	33	−
M	43	30	−
N	50	51	+
O	51	37	−
P	48	50	+
Q	45	30	−
R	50	35	−
S	40	28	−
T	34	26	−

been longer than it formerly was. There are $n = 20$ "trials" and $p = 0.50$, and since the mean and standard deviation of the binomial distribution are np and $\sqrt{np(1 - p)}$, respectively, we have

$$\mu = 20(0.50) = 10$$

and

$$\sigma = \sqrt{10(0.50)} = 2.236$$

Both np and $n(1 - p)$ are greater than 5 here, so we can base the test on the normal approximation to the binomial distribution. Hence, we calculate

$$z = \frac{15.5 - 10}{2.236} = 2.46$$

and since this exceeds $z_{.01} = 2.33$, we reject the null hypothesis and accept the alternative, concluding that the collection departments are more efficient now than formerly.

When n is very small, it may be preferable to base a test like this on a table of binomial probabilities rather than on the normal curve approximation. Reference to such a table shows that the probability of 16 or more successes in 20 trials with $p = 0.50$ is 0.005 which is less than 0.01, so the conclusion is the same. Because of its great simplicity, especially when binomial probability tables can be used, the sign test is sometimes used as a short cut even though standard techniques are applicable.

RANK-SUM TESTS: THE MANN–WHITNEY TEST

Another nonparametric test, called the *Mann–Whitney test* (or the *Wilcoxon test* or the *U test*), which is somewhat less wasteful of information than the sign test, can also be used as an alternative to the standard small sample t test for the difference between two means. As with other nonparametric tests, the Mann–Whitney test applies under very general conditions. In fact, in using this test it is necessary only to assume that the populations sampled are continuous, and in practice even the violation of this assumption is usually not very serious.

To illustrate how the Mann–Whitney test is performed, let us suppose that we want to test, at a level of significance of 0.05, the hypothesis that the mean lifetimes of two brands of 9-volt batteries used mainly in small transistor radios are equal against the alternative that the lifetimes are not equal. Tests made on a random sample of 12 batteries drawn from the population of Brand A batteries and on a random sample of 12 batteries drawn from the population of Brand B batteries showed the following lifetimes (in hours):

Brand A: 6.9, 11.2, 14.0, 13.2, 9.1, 13.9, 16.1, 9.3, 2.4, 6.4, 18.0, 11.5
Brand B: 15.5, 11.1, 16.0, 15.8, 18.2, 13.7, 18.3, 9.0, 17.2, 17.8, 13.0, 15.1

The means of these two samples are 11.0 and 15.1 hours for the Brand A and Brand B batteries, respectively. Short-cut methods suggest that the variance in the Brand A batteries is almost three times the variance in the Brand B batteries, so that it may be unreasonable to assume that the samples came from populations with equal variability.

We begin the Mann–Whitney test by ranking the observations (as though they comprised a single sample) in either an increasing or decreasing order of magnitude. The low-to-high ranking is shown at the top of the next page, and underneath each item the letter A or B is written to indicate which brand the item belonged to.

2.4	6.4	6.9	9.0	9.1	9.3	11.1	11.2	11.5	13.0	13.2	13.7
A	A	A	B	A	A	B	A	A	B	A	B
13.9	14.0	15.1	15.5	15.8	16.0	16.1	17.2	17.8	18.0	18.2	18.3
A	A	B	B	B	B	A	B	B	A	B	B

Assigning the data *in this order* the ranks 1, 2, 3, . . . , and 24 we find that the lifetimes of the Brand A batteries occupy ranks 1, 2, 3, 5, 6, 8, 9, 11, 13, 14, 19, and 22, while those of Brand B occupy ranks 4, 7, 10, 12, 15, 16, 17, 18, 20, 21, 23, and 24. There are no ties here between values belonging to different samples, but if there were, we would assign each of the tied observations the mean of the ranks which they jointly occupied.

The null hypothesis we are testing is that the two samples come from populations having equal means. If this hypothesis is true, it seems reasonable to suppose that the means of the ranks assigned to the values of the two samples should be more or less the same. The alternative hypothesis is that the means of the populations are not equal, and if this is the case and the difference is pronounced, most of the smaller ranks will go to the values of one sample, while most of the higher ranks will go to those of the other sample.

Using *rank sums*, rather than average ranks, we base the test of this null hypothesis on the statistic

$$U = n_1 n_2 + \frac{n_1(n_1 + 1)}{2} - R_1 \qquad \bigstar$$

where n_1 and n_2 are the first and second sample sizes and R_1 is the sum of the ranks assigned to the values of the first sample. In practice we find whichever rank sum can be most easily obtained, since it is immaterial which sample is called the "first" one.

Under the null hypothesis that the $n_1 + n_2$ observations came from identical populations (or, what is the same thing, from one population), it can be shown that the sampling distribution of U has the mean

$$\mu_U = \frac{n_1 n_2}{2} \qquad \bigstar$$

and the standard deviation

$$\sigma_U = \sqrt{\frac{n_1 n_2 (n_1 + n_2 + 1)}{12}} \qquad \bigstar$$

Furthermore, if n_1 and n_2 are both greater than 8 (some statisticians prefer that both be greater than 10), the sampling distribution of U can be approximated closely with a normal curve. Hence, we can reject the Hypothesis: $\mu_1 - \mu_2 = 0$ (and accept the Alternative: $\mu_1 - \mu_2 \neq 0$) at a level of significance α if

$$z = \frac{U - \mu_U}{\sigma_U} \qquad \star$$

is either less than $-z_{\alpha/2}$ or greater than $z_{\alpha/2}$. (If either n_1 or n_2 is so small that the normal curve approximation cannot be used, the test must be based on special tables, referred to in the Bibliography at the end of this book.)

For the example we are discussing let us call the sample of Brand A batteries the first sample; we then have $n_1 = 12$, $n_2 = 12$, $R_1 = 1 + 2 + 3 + 5 + 6 + 8 + 9 + 11 + 13 + 14 + 19 + 22 = 113$, and

$$U = 12 \cdot 12 + \frac{12 \cdot 13}{2} - 113 = 109$$

Also we find that

$$\mu_U = \frac{12 \cdot 12}{2} = 72$$

$$\sigma_U = \sqrt{\frac{12 \cdot 12 \cdot 25}{12}} = 17.3$$

and hence that

$$z = \frac{109 - 72}{17.3} = 2.14$$

Since $z = 2.14 > z_{.025} = 1.96$, we reject the test hypothesis at a level of significance of 0.05 and conclude that the difference between the two sample mean lifetimes is too great to be accounted for by chance.

An interesting feature of the Mann–Whitney test is that, when there is no difference in the means of two populations, it can also be used to test the hypothesis that two samples come from identical populations (or the same population) against the alternative that the two populations have *unequal dispersions.* The test proceeds in the same way we have described it above except for a modification in the way in which ranks are assigned. As before, the values of the two samples are arranged jointly in an increasing (or decreasing) order of magnitude, but now they are *ranked from both ends toward the middle.* We assign Rank 1 to the smallest value, Ranks 2 and 3 to the largest and second largest, Ranks 4 and 5 to the second and third smallest, Ranks 6 and 7 to the third and fourth largest, and so on. Subsequently, U is calculated, and the test performed, exactly as before. With this kind of ranking, however, a *small rank sum* tends to indicate that the population from which the sample was obtained has a *greater variation* than the other, since its values occupy the more extreme positions.

To illustrate how the Mann–Whitney test is used to test the null hypothesis that two samples come from identical populations against the alternative that the populations have unequal dispersions, consider the following: Nine electronics students are taught sound-pattern recognition by Method P and 10 other electronics students are taught sound-pattern recognition by Method Q. After the first instruction period the students are tested to determine how quickly they can identify a certain distinctive sound pattern occurring amidst a general noise background. The nine students taught by Method P identified the pattern in 10.3, 12.7, 13.1, 15.5, 10.6, 9.9, 12.2, 10.1, and 17.9 seconds; and the 10 students taught by Method Q identified the pattern in 11.3, 11.8, 12.1, 14.1, 11.5, 12.4, 10.8, 12.9, 13.4, and 11.7 seconds. Treating these two groups as random samples from two conceptually large populations, we can apply the U test to determine that it is *not* possible to reject the hypothesis that the *mean* recognition times of the two populations are equal. (See Exercise 12 on page 342.)

Working now with the 19 observations ordered from low to high, we rank them from the ends toward the middle. Calling the students taught by Method P the first sample, we find that these sample values have Ranks 1, 4, 5, 8, 18, 14, 10, 3, and 2. Therefore, $R_1 = 65$ and $U = 70$; also $\mu_U = 45$, $\sigma_U = 12.2$, and

$$z = \frac{70 - 45}{12.2} = 2.05$$

Since this z value exceeds 1.96, the null hypothesis is rejected at the 0.05 level of significance, and we conclude that there is a real difference in the variability in the pattern identification times of the students taught by the two methods. (Since U is greater than expected, R_1 is smaller than expected, and this means that there is greater variability in the first sample than in the second.)

RANK-SUM TESTS: THE KRUSKAL–WALLIS TEST

The Kruskal–Wallis (or H) test is a nonparametric alternative to a one-way analysis of variance inasmuch as it is used to test the null hypothesis that k independent samples come from identical populations against the alternative that the means of these populations are not all equal. Unlike the standard one-way analysis of variance, however, the Kruskal–Wallis test does not require the assumption that the samples come from (approximately) normal populations having the same variance σ^2.

In the Kruskal–Wallis test (as in the U test) the k samples are ranked from low to high, or vice versa, as though they constituted a single sample. Letting R_i be the sum of the ranks assigned to the n_i observations in the ith sample, and $n = n_1 + n_2 + \cdots + n_k$, the test is based on the statistic

$$H = \frac{12}{n(n+1)} \sum_{i=1}^{k} \frac{R_i^2}{n_i} - 3(n+1) \qquad \bigstar$$

If the null hypothesis is true and each sample has at least five observations, the sampling distribution of H can be approximated closely with a chi-square distribution with $k - 1$ degrees of freedom. Consequently, we can reject the null hypothesis that $\mu_1 = \mu_2 \cdots = \mu_k$ and accept the alternative that the μ's are not all equal at the level of significance α, if $H > \chi_\alpha^2$ for $k - 1$ degrees of freedom. If any sample has less than five items, the χ^2 approximation cannot be used, and the test must be based on special tables.

To illustrate the Kruskal–Wallis test, suppose that in a large company three groups of technical sales trainees are taught a certain trouble-shooting procedure by different methods, A, B, and C. At the end of instruction the subjects are given performance tests, and we want to decide whether in general the mean performance quality for trainees taught by the three methods is the same. The following are the performance scores made by the trainees in the three groups:

$$\begin{aligned}
Method\ A: &\quad 33, \quad 30, \quad 32, \quad 25, \quad 30, \quad 34 \\
Method\ B: &\quad 29, \quad 28, \quad 27, \quad 28, \quad 21, \quad 24, \quad 28 \\
Method\ C: &\quad 31, \quad 22, \quad 24, \quad 26, \quad 23
\end{aligned}$$

After first ranking the 18 observations from low to high and assigning ranks to the observations in the three samples, we find that $R_1 = 6 + 13 + 14 + 16 + 17 + 18 = 84.0$, $R_2 = 1 + 4.5 + 8 + 9 + 10 + 11 + 12 = 55.5$, and $R_3 = 2 + 3 + 4.5 + 7 + 15 = 31.5$ (there is one tie and two observations are ranked 4.5). Substituting these values together with $n_1 = 6$, $n_2 = 7$, and $n_3 = 5$ into the formula for H we get

$$H = \frac{12}{18 \cdot 19} \left(\frac{84.0^2}{6} + \frac{55.5^2}{7} + \frac{31.5^2}{5} \right) - 3(19) = 6.507$$

Since this value is greater than 5.991, the $\chi^2_{.05}$ value for 2 degrees of freedom, we reject the null hypothesis and conclude that the means of the three populations from which the samples came are not all equal.

EXERCISES

1. In order to pretest a program aimed at making employees cost conscious, a large textile mill randomly selects 12 experienced cutters (who are presumably not learning with practice) and gives them a short cost course. To measure the effectiveness of the course in creating cost consciousness, the company analyzes the following data showing the amount of cloth waste (in pounds) out of a given amount of work, left on the cutting-room floor by each cutter in a preinstruction period, and the amount of waste in an identical amount of work in a postinstruction period.

					Cutter							
	A	B	C	D	E	F	G	H	I	J	K	L
Before	13.8	15.5	17.2	14.6	15.4	12.0	15.1	14.2	13.0	16.4	14.3	15.7
After	13.1	14.4	13.0	14.7	14.5	12.2	14.3	13.6	12.5	13.2	14.0	13.9

At the 5 per cent level of significance, is there evidence that the amount of waste is less in the postinstruction period than it was in the preinstruction period?

2. The following data show, for a random sample of 14 large manufacturing companies, the expenditures in research and development (in cents per common share) for the years 1962 and 1970:

						Company								
Year	A	B	C	D	E	F	G	H	I	J	K	L	M	N
1962	14	9	17	8	20	22	11	11	15	13	10	10	17	27
1970	16	12	14	6	25	27	15	9	19	10	16	18	12	31

At a level of significance of 0.05, is there actually an increase in research and development expenditures between the two time periods in the population from which the sample came?

3. On the first trial of a practice period, 15 experienced code clerks scored 85, 83, 81, 82, 77, 83, 88, 84, 87, 86, 83, 81, 72, 82, and 76 points on a newly constructed digit-symbol learning test, and on the tenth trial they scored 92, 89, 84, 82, 92, 90, 90, 86, 91, 92, 94, 78, 81, 90, and 81 points, respectively. Test at the 0.05 level whether the apparent gain in score is significant.

4. In its quality control section, a large food manufacturer tests the consistency of a salad dressing by dropping a plummet from a standard height into 12 jars of the dressing randomly selected from a large production lot, and measuring the distance the plummet penetrates (which is proportional to the consistency of the mix). These jars are set aside for 10 days and then are tested again for consistency. If the following are the penetration distances (in inches) on the two tests, is there any evidence of a change in consistency (as reflected by a change in the distance)?

| | | | | | | Jar | | | | | | |
Test	1	2	3	4	5	6	7	8	9	10	11	12
First	3.4	3.2	3.7	3.1	3.5	3.9	3.4	3.4	3.6	3.5	3.8	3.8
Second	3.5	3.1	3.5	3.2	3.5	3.7	3.2	3.5	3.3	3.3	3.9	3.1

5. The sign test can also be used to test the hypothesis that the mean of a symmetical population equals some constant μ_0 against a suitable one-sided or two-sided alternative. We discard all sample values equal to μ_0, record a *plus sign* for each sample value greater than μ_0 and a *minus sign* for each value less than μ_0, and then test the hypothesis that $p = \dfrac{1}{2}$ on the basis of the number of plus (or minus) signs corresponding to the n sample values.

(a) Use this method to test, on the basis of the following sample data, whether in a given year the average number of days wholesalers of drugs and drug sundries require to convert receivables into cash is $\mu_0 = 32$ days: 51, 27, 29, 36, 31, 38, 34, 52, 31, 41, 44, 29, 51, 29, 39, 31, 33, 34, 29, and 41 days.

(b) In order to test the hypothesis that the mean batch volume of Portland cement concrete prepared with certain fixed amounts of cement, sand, gravel, and water is 1.345 cubic feet, 20 experimental batches are prepared. If the batch volume was greater than 1.345 in 14 batches and less than this in the other 6 batches, can we reject the hypothesis at the 5 per cent level of significance?

6. In addition to its use in credit analysis, the ratio analysis of working capital is important in determining how efficiently working capital is being used. One important measure is the ratio of inventory to net working capital, since it shows the extent to which owners have invested in the least liquid of all working-capital components. The following figures show the inventory-to-net-working-capital ratios (times 100) for the same 20 wholesalers referred to in Exercise 5(a) in a survey conducted in 1958: 77.4, 79.5, 70.4, 73.2, 68.9, 80.1, 74.4, 82.4, 81.0, 78.5, 79.0, 75.0, 82.3, 81.2, 75.4, 77.8, 79.3, 80.2, 75.0, and

75.7. Use this information to test, at the 0.05 level, whether the corresponding mean for all wholesalers of drugs and drug sundries is $\mu_0 = 79.3$.

7. Referring to Exercise 6, suppose that in a survey conducted in 1963 the corresponding ratios (times 100) for the 20 wholesalers (listed in the same order) were 76.3, 82.5, 75.9, 75.6, 73.0, 79.2, 77.2, 82.4, 80.3, 81.2, 82.1, 74.9, 80.8, 82.4, 80.5, 75.0, 78.5, 80.0, 80.8, and 81.3. Use the sign test and a level of significance of 0.05 to test whether there has been a change from 1958 to 1963 in the mean ratio of inventory to net working capital in the population of wholesalers from which the samples were obtained.

8. Tests are made on n specimens of a certain kind of wood to measure its radial shrinkage in passing from green to oven-dry condition. In two cases, or 2.5 per cent of the tests, the shrinkage was 5.5 per cent; twice as many of the remaining tests showed shrinkage greater than 5.5 per cent as showed shrinkage less than 5.5 per cent. At the 1 per cent level of significance, is there reason to think that the radial shrinkage of such wood exceeds 5.5 per cent?

9. Before placing a large order for road flares to be carried by all company cars, a major fleet operator life tests flares of Brand P and Brand Q. The following data show the burning times (in minutes) observed in random samples of 12 flares of each kind:

```
Brand P:  16,  18,  16,  13,  16,  15,  11,  15,  14,  16,  25,  17
Brand Q:  17,  13,  15,  19,  18,  14,  20,  14,  19,  17,  19,  21
```

Test, at a level of significance of 0.01, whether it is reasonable to conclude that there is no difference in the actual mean burning times of the two brands of flares.

10. Random samples of 11 pieces of Cord A and 11 pieces of Cord B are drawn from large populations and tested for breaking strength (in pounds) with the following results:

```
Cord A:  55,  59,  61,  58,  62,  55,  56,  60,  54,  58,  52
Cord B:  56,  54,  53,  51,  49,  63,  55,  54,  54,  54,  60
```

(a) Test, at a level of significance of 0.05, the hypothesis that the two samples came from identical populations against the alternative that the two populations have unequal means.

(b) Test, at a level of significance of 0.05, the hypothesis that the two samples came from identical populations against the alternative that the two populations have unequal dispersions.

11. The following are the Rockwell hardness numbers obtained for 15 aluminum die castings randomly selected from Lot S and for 12 castings randomly selected from Lot T:

    Lot S: 72, 76, 54, 68, 82, 57, 75, 71, 57, 90, 98, 60, 74, 88, 74
    Lot T: 71, 76, 66, 53, 68, 82, 72, 89, 95, 76, 87, 56

(a) Test, at a level of significance of 0.05, the hypothesis that the samples came from identical populations against the alternative that the two populations have unequal means.

(b) Test, at a level of significance of 0.05, the hypothesis that the samples came from identical populations against the alternative that the two populations have unequal dispersions.

12. For the example of page 337, use the U test to verify that we cannot reject, at a level of significance of 0.05, the hypothesis that the mean pattern recognition time is the same for Methods P and Q. Also, verify all the calculations in the example leading to the conclusion that the variabilities in the pattern recognition times are not the same for the two methods.

13. Rank the 18 performance scores shown on page 338 and verify all the calculations leading to the conclusion that the means of the populations from which the three samples came are not all equal.

14. Apply the Kruskal–Wallis test to the data of Exercise 4 on page 320 to determine at the 5 per cent level whether the differences among the average weekly earnings of the salesmen are significant.

15. Apply the Kruskal–Wallis test to the data of Exercise 5 on page 321 to test, at a level of significance of 0.05, the null hypothesis that the average level of knowledge of current events is the same for the four schools.

16. Refer to Exercise 7(b) on page 322 and use the Kruskal–Wallis test to determine at the 5 per cent level of significance whether differences in dosage have a real effect on the times required to fall asleep in the populations from which the samples came.

A WORD OF CAUTION

    Methods of analysis which require no (or virtually no) assumptions concerning the form of population distributions from which samples come are usually less efficient than the corresponding standard methods. Turning again to Chebyshev's Theorem for illustration, we can say that there is a probability of at least 0.75 that the mean of a random sample of size 64 drawn from *some* (infinite) population with $\sigma = 20$ is not "off" its population mean by more than 5. However, assuming that the sample came from a population having the shape of a normal distribution with $\sigma = 20$, we can make the same statement with a probability of 0.95. To put it another way, assertions made with equal confidence require larger samples if they are made without "knowledge" of the form of the underlying distribution than if they are made with such "knowledge." It is generally true that the more one is willing to assume, the more he can infer from a sample; unfortunately, however, the more one assumes, the less believable the inferences are.

13

DECISION MAKING: REGRESSION AND CORRELATION

CURVE FITTING

The foremost objective of many scientific investigations is to make predictions. Whenever possible scientists strive to express, or approximate, relationships between known quantities and quantities to be predicted in terms of mathematical equations. This approach has been very successful in the natural sciences. It is known in chemistry, for instance, that at constant temperature the relationship between the volume (y) of a gas and its pressure (x) is given by the formula

$$y = \frac{k}{x}$$

where k is a numerical constant. Also, it has been discovered in biology that the size of a culture of bacteria (y) can be expressed in terms of the time (x) it has been exposed to certain favorable conditions by means of the formula

$$y = a \cdot b^x$$

where a and b are numerical constants.

Business and economic statisticians have borrowed and continue to borrow liberally from the tools of the natural sciences. The first equation above, for instance, is often used to express the relationship between the demand (y) for a commodity and the price (x) of the commodity. The second equation is also often used to express, among other things, the way in which company or industry production or sales (y) grow with time (x). Specifically, a cotton textile mill has found that its total annual sales can be predicted reasonably well by means of the formula

$$y = 22(1.05)^x$$

where y represents sales in millions of dollars and x is the difference in years between the year for which sales are to be predicted and the year 1950. To predict sales for the year 1975, for example, we substitute $x = 1975 - 1950 = 25$ in the equation, getting

$$y = 22(1.05)^{25} = 74.5 \text{ million dollars}$$

In any situation where we want to use observed data to construct a mathematical equation and use it to predict the value of one variable from a given value of another—a procedure known generally as *curve fitting*—there are essentially three problems to be solved: We must (1) decide what kind of equation is to be used, then (2) find the particular equation itself which is in some sense the best of all those of its kind, and (3) settle certain questions of interest with respect to the goodness of the particular equation, or of predictions made from it.

With respect to the first of these problems of curve fitting, there are many different kinds of curves (and their equations) which may be used for predictive purposes. Designating in the work which follows the variable to be predicted by y and the predictor variable by x, we might, for instance, predict y from x by using a straight line of the form $y = a + bx$, a parabola of the form $y = a + bx + cx^2$, an exponential curve of the form $y = a \cdot b^x$, a power function of the form $y = a \cdot x^b$, or one of various other possible curves. Somewhat involved methods which we shall not discuss exist for putting this decision on a more or less objective basis. However, graph papers are available having ordinary x- and y-scales (arithmetic paper), an arithmetic x-scale and a logarithmic y-scale (semi-log or ratio paper), and logarithmic x- and y-scales (log-log paper). Visual study of a plot (or of plots) on graph paper of the sample data from which the predicting equation is to be obtained usually suggests the most appropriate kind of equation to use.

Of the various kinds of equations we might use to predict values of one variable (y) from associated values of another variable (x), the simplest and most widely used is the linear equation in two unknowns, which is of the form

$$y = a + bx$$

where a and b are numerical constants. Ordinarily, these constants are estimated from sample data, and once they are available, we have merely to substitute a given value of x in the equation and calculate the corresponding predicted value of y. *Linear equations are useful and important not only because there exist many relationships that are actually of this form, but also because they often provide close approximations to relationships which would otherwise be difficult to describe in mathematical terms.*

The term "linear equation" arises from the fact that, when plotted on ordinary graph paper (arithmetic paper), all pairs of values of x and y which satisfy an equation of the form $y = a + bx$ will fall on a straight line. To illustrate, let us suppose that a large national manufacturer wants to predict the first-year sales in thousands of dollars of new salesmen (y) on the basis of scores on a sales aptitude test (x) administered by its personnel department. From sales and scores data collected in a random sample of 10 of its salesmen, the firm has derived (by methods to be explained later) the following predicting equation

$$y = -23.20 + 6.88x$$

whose graph is shown in Figure 13.1. If, for example, a man scores $x = 30$ points on the test, his predicted first-year sales are $-23.20 + 6.88(30) = 183.2$ or \$183,200; similarly, the predicted first-year sales of a man who scores $x = 40$ points on the test are $-23.20 + 6.88(40) = 252.0$ or \$252,000. Any pair of x and y values satisfying the above equation constitutes a point (x, y) which falls on the line of Figure 13.1.

To illustrate the general procedure in curve fitting, suppose that a chemical company wishes to study the effect of various factors on the efficiency of an extraction operation. In the overall study there are several processing conditions (extraction time, moisture content of the raw material, solvent

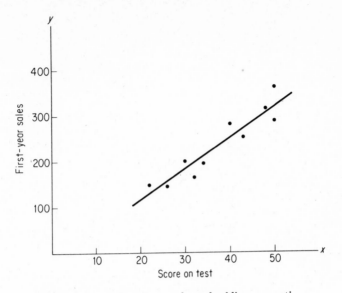

FIGURE 13.1   Scattergram and graph of linear equation

pumping rate, and temperature of extraction) whose effects on efficiency are to be examined, but we choose for study here only the extraction time. The following data show the results of 10 tests; the two numbers given for each test are the extraction time x (in minutes) and a measure of the extraction efficiency y (in per cent, or "points"):

Extraction Time x	Extraction Efficiency y
14	55
34	60
18	44
40	78
48	75
18	50
44	62
30	65
26	55
38	70

Plotting the points corresponding to these 10 pairs of values on arithmetic paper (see Figure 13.2), we observe that, although the points do not fall on a straight line, the overall pattern of the relationship is reasonably well

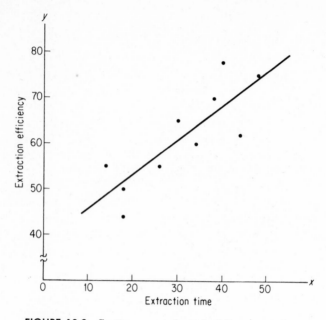

FIGURE 13.2 Scattergram and graph of linear equation

described by the line of the figure. There is no noticeable departure from linearity in the scatter of the points, so we feel justified in deciding that a straight line of the form $y = a + bx$ is a suitable description of the underlying relationship between extraction time and extraction efficiency.

Having settled the first problem of curve fitting by choosing a linear predicting equation to describe this relationship, we now face the second problem: finding the equation of the line (more generally, the curve) which in some sense provides the best possible fit to the pairs of observations. Logically speaking, there is no limit to the number of straight lines which can be drawn on a piece of graph paper. Some of these lines would be such obviously poor fits to the data that we could not consider them seriously, but there are many lines which would seem to provide more or less "good" fits, and the problem is to find that one line which fits the data "best" in some well-defined sense. If all the points actually fell on a straight line, the criterion would be self-evident, but this is an extreme case which we would rarely expect to encounter in practice. In general, we shall have to be satisfied with a line having certain desirable, but not perfect, properties.

The criterion that we shall use to define the line of "best" fit is the criterion of *least squares*. This criterion was originated early in the nineteenth century by the French mathematician, Adrien Legendre, and least-squares methods are now widely used in many branches of statistics.

THE METHOD OF LEAST SQUARES

In fitting a straight line to data plotted as points on a piece of graph paper it is possible to draw a freehand line which, at least to the eye, fits the data well. It is then possible to find from this line values of a and b and, hence, a predicting equation. Unfortunately, though, freehand curve fitting is largely subjective, and for this reason, there is no scientific way of evaluating the "goodness" of predictions made from a freehand line. Least squares methods permit such evaluations, and this is one reason why these methods are so widely used.

Since we shall now be dealing with both actual and predicted values corresponding to a given x value, we shall from this point on write y' for the *predicted* value (the value obtained by substituting x in the predicting equation) so as to distinguish this value from the *actual* (i.e., observed) value y associated with a given value x. Now, as it will be applied in this chapter, the least squares criterion requires that the line which we fit to our data be such that *the sum of the squares of the vertical deviations (distances) from the points to the line is a minimum.* For the extraction example above, the method of least squares requires that the sum of the squares of the distances represented by the solid-line segments of Figure 13.3 be as

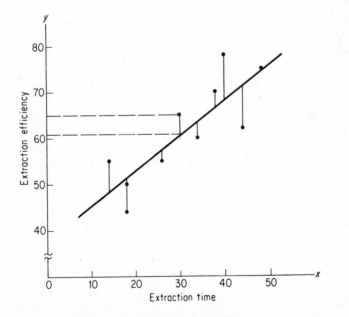

FIGURE 13.3   Line fit to extraction data

small as possible. The logic behind this approach can be explained as follows: For the case where the extraction time was $x = 30$ minutes, the actual efficiency was $y = 65$; reading the value which corresponds to $x = 30$ directly off the line of Figure 13.3, we see that the predicted efficiency is about $y' = 61$. Hence, the error of the prediction in this one case, represented by the vertical distance from the point to the line, is $y - y'$ or about $65 - 61 = 4$. There are 10 such errors corresponding to the 10 tests, and the least squares criterion requires that we minimize the sum of their squares.

To show how a least squares line is actually fitted to a set of data, let us consider n pairs of numbers (x_1, y_1), (x_2, y_2), . . . , (x_n, y_n), which might represent such things as the reading rate and reading comprehension of n financial analysts, the number of labor units per acre and crop yield per acre on n farms, the score on a paper and pencil personality test and vocational success of n college graduates, and product advertising expenditures and product sales of n consumer goods manufacturers.

As before, we use the letter x for the predictor variable, the letter y for the value of the associated variable actually observed for a given value of x, and the letter y' for the value calculated from the equation. Consequently, we write the predicting equation as

$$y' = a + bx$$

and note that for each of the n given values of x we have a given (or observed) value of y and a calculated (or predicted) value y'.

The least squares criterion requires that we minimize the sum of the squares of the differences between the observed values of y and the corresponding predicted values y' (see Figure 13.4). This means that we must

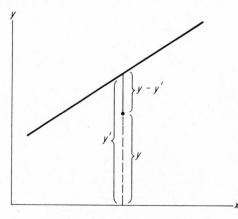

FIGURE 13.4 Least squares line

find the numerical values a and b appearing in the equation $y' = a + bx$ for which

$$\Sigma (y - y')^2$$

is as small as possible. We shall not go through the derivation of the two equations, called the normal equations, which provide the solution to this problem. Instead, we merely state that minimizing (either by the calculus or by completing the square) $\Sigma (y - y')^2$ yields the following two equations in the unknowns a and b:

$$\Sigma y = na + b(\Sigma x)$$
$$\Sigma xy = a(\Sigma x) + b(\Sigma x^2)$$

★

In these equations, whose solution gives the desired least squares values of a and b, n is the number of pairs of observations, Σx and Σy are the sums of the given x's and y's, Σx^2 is the sum of the squares of the x's, and Σxy is the sum of the products obtained by multiplying each of the given x's by the corresponding observed value of y.

We show in the table below the calculation of these sums, the data of the

Extraction Time x	Extraction Efficiency y	x^2	xy
14	55	196	770
34	60	1,156	2,040
18	44	324	792
40	78	1,600	3,120
48	75	2,304	3,600
18	50	324	900
44	62	1,936	2,728
30	65	900	1,950
26	55	676	1,430
38	70	1,444	2,660
310	614	10,860	19,990

first two columns having been copied from page 347. Substituting $n = 10$ and the appropriate column totals into the two normal equations we get

$$614 = 10a + 310b$$
$$19,990 = 310a + 10,860b$$

and we must now solve these two simultaneous linear equations. That is, we must find a pair of values a and b such that 10 times a plus 310 times b equals 614, and also 310 times a plus 10,860 times b equals 19,990. There are several ways in which this can be done; using elementary algebra we could solve the system of equations by the method of elimination or by determinants. In either case we obtain $a = 37.69$ and $b = 0.7648$.

To simplify this work, however, we can solve the two normal equations symbolically for a and b, so that the solution can then be obtained by direct substitution. Performing the necessary algebra, we get

$$a = \frac{(\Sigma\, y)(\Sigma\, x^2) - (\Sigma\, x)(\Sigma\, xy)}{n(\Sigma\, x^2) - (\Sigma\, x)^2}$$

$$b = \frac{n \Sigma\, xy - (\Sigma\, x)(\Sigma\, y)}{n(\Sigma\, x^2) - (\Sigma\, x)^2}$$

★

and for our example we have

$$a = \frac{(614)(10{,}860) - (310)(19{,}990)}{10(10{,}860) - (310)^2} = 37.69$$

$$b = \frac{10(19{,}990) - (310)(614)}{10(10{,}860) - (310)^2} = 0.7648$$

There is still another way of getting these values which is often used because of its convenience: We solve the first normal equation for a, getting $a = \dfrac{\Sigma\, y - b \Sigma\, x}{n}$; then we calculate b using the formula above and substitute it into this formula for a. Since $b = 0.7648$ we get in the example $a = \dfrac{614 - 0.7648(310)}{10} = 37.69$ as before.

We can now write the equation of the least squares line for our numerical example as

$$y' = 37.69 + 0.7648x$$

This is the "predicting" equation, or the equation by which we predict extraction efficiency (y) for given values of extraction time (x), and the graph of this equation is shown in Figure 13.2. The predicted value of y for any given value of x is obtained by multiplying x by 0.7648 and adding

37.69. For $x = 25$, we have $y' = 37.69 + 0.7648(25) = 57$ and for $x = 35$ we have $y' = 37.69 + 0.7648(35) = 64$. In other words, we predict that, when the extraction time is 25 minutes, the extraction efficiency of the process under study will be 57 points, and when the time is 35 minutes, the efficiency will be 64 points. Furthermore, we assert that these values, and other values of y' calculated by substituting given values of x in the predicting equation, are the "best" predictions we can make of the actual extraction efficiencies for the corresponding extraction times. They are the "best," of course, in the least squares sense, and in the next section we shall consider the matter of determining just how good predictions in studies of this sort are.

In the discussion of this section we have considered only the problem of fitting a straight line to paired data. More generally, the method of least squares can also be used to fit other kinds of curves and also to derive predicting equations in more than two unknowns. Of the many equations that can be used to express relationships among more than two variables, the most widely used are linear equations of the form

$$y' = a + bx_1 + cx_2 + dx_3 + \cdots$$

in which y is the variable to be predicted and the x's (of which there may be several) are the known variables on which the predictions are to be based. For example, the equation

$$y' = 3.4892 - 0.0899x_1 + 0.0637x_2 + 0.0187x_3$$

arose in a study of the demand for different meats. Here y stands for the total consumption of federally inspected beef and pork in millions of pounds, x_1 stands for a composite retail price of beef in cents per pound, x_2 represents a composite retail price of pork in cents per pound, and x_3 denotes income as measured by a certain payroll index. The numerical constants of this equation were obtained by the method of least squares. In a situation like this, the least squares method requires the solution of as many normal equations as there are variables, but this is no problem to a digital computer. Least squares equations involving many variables, whose solution by hand would require months of work, can be solved in a matter of seconds on a fast machine. Since problems like this are of great importance, powerful machine programs for their solutions are readily available for various computers.

EXERCISES

1. A study is made of the relationship between the conditioning period (in days) of a laminating glue and the storage life of the glue (in months) at 70°F. The following results are observed in a random sample of 10 specimens:

   Conditioning period, x: 1 2 3 4 4 5 5 6 6 7
   Storage life, y: 4 5 6 5 6 8 7 10 11 10

   (a) Plot the data.
   (b) Find the equation of the line of best fit to these data and plot the line on the graph of part (a).
   (c) Predict the storage life of a glue which has been conditioned for five days.

2. A study is made to determine the relationship between the ages of a large group of machines in a mill and the efficiencies of the machines. The following are the ages (in years) and a composite index of efficiency for a random sample of eight of these machines:

   Age, x: 2 4 11 9 4 6 7 8
   Efficiency, y: 90 65 25 40 80 60 35 50

   (a) Plot the data.
   (b) Fit a least squares line to these data and plot the line on the graph of part (a).
   (c) What is the best estimate of the efficiency of a 10-year-old machine?

3. Raw material used in the production of a synthetic fiber is stored in a place without humidity control. Measurements of the relative humidity in the storage place and the moisture content of a sample of the raw material (both in percentages) on 10 days (randomly selected over a given period of time) showed the following results:

   Humidity, x: 46 30 34 52 38 44 40 45 34 60
   Moisture content, y: 10 7 9 13 8 12 11 11 7 14

   (a) Plot the data.
   (b) Fit a least squares line to these data and plot the line on the graph of part (a).
   (c) Predict the moisture content of a batch of stored raw material when the humidity in the storage place is $x = 50$ per cent.

4. A study is made to determine the relationship between the area burned in forest fires in U.S. counties having a formal fire-control organization (with a county ranger, smoke-chaser assistants during the fire season, fire-crew members, etc., supported by annual fire-control appropriations) and the variable operating costs of the fire protection. (The ranger's salary is a fixed operating cost and is unrelated to the area burned. Money over and above the ranger's salary which determines how much he can travel, the equipment and facilities he can buy, and how many men he can hire—the variable operating funds— directly affect the quality of fire protection.) A random sample of the annual figures for 12 counties shows the following variable operating costs (in cents per acre) and the area burned (as a percentage of the protected area):

Variable Operating Costs x	Protected Area Burned y
2.8	0.09
1.7	0.24
1.9	0.36
2.5	0.12
2.7	0.20
0.8	0.40
1.5	0.34
2.8	0.14
2.2	0.22
1.2	0.42
2.1	0.30
0.7	0.48

(a) Plot the data.

(b) Fit a least squares line to these data and plot the line on the graph of part (a).

(c) Predict the percentage of protected area burned in a county having 2.0 cents per acre variable operating cost.

5. The following shows the advertising expenses (as a percentage of total expense) and net operating profits (as a percentage of sales) in a random sample of 10 small jewelry stores:

Advertising expense, x: 1.2   0.7   1.5   1.8   0.5   3.4   1.0   3.0   2.8   2.5
Profit, y: 2.7   2.4   2.7   3.3   1.1   5.8   2.2   4.2   4.4   3.8

(a) Plot the data.

(b) Find the equation of the line of best fit to these data and plot the line on the graph of part (a).

(c) Predict the net profit of a store whose advertising expense is $x = 2.0$ per cent.

6. In a study devoted to determining the composition of brandy, a random sample of six brandy specimens showed the following tannin content and esters content (both in grams for 100 liters at 100 proof):

$$
\begin{array}{lcccccc}
\textit{Tannin, x:} & 19 & 15 & 9 & 10 & 11 & 19 \\
\textit{Esters, y:} & 31.7 & 32.3 & 8.5 & 14.3 & 14.0 & 17.8
\end{array}
$$

(a) Plot the data.

(b) Find the equation of the line of best fit to these data and plot the line on the graph of part (a).

(c) What is the best estimate of the amount of esters in a brandy with tannin content of $x = 12$?

7. In the brandy sample referred to in Exercise 6, the following are the aldehyde content and esters content (both in grams for 100 liters at 100 proof):

$$
\begin{array}{lcccccc}
\textit{Aldehydes, x:} & 9.8 & 8.9 & 7.3 & 6.3 & 6.6 & 10.3 \\
\textit{Esters, y:} & 31.7 & 32.3 & 8.5 & 14.3 & 14.0 & 17.8
\end{array}
$$

(a) Plot the data.

(b) Fit a least squares line to these data and plot the line on the graph of part (a).

(c) What is the best estimate of the amount of esters in a brandy whose aldehyde content is $x = 8.0$?

8. One phase of a large-scale sociological and economic study is devoted to the relationship between educational level and income. As part of this work, an investigator takes a random sample of eight cities in a 13-state geographical region and determines from census data the percentage of college graduates in the city population and the city median income (in thousands of dollars). The following are the results:

$$
\begin{array}{lcccccccc}
\textit{Per cent college graduate, x:} & 7.2 & 6.7 & 17.0 & 12.5 & 6.3 & 23.9 & 6.0 & 10.2 \\
\textit{Median income, y:} & 4.2 & 4.9 & 7.0 & 6.2 & 3.8 & 7.6 & 4.4 & 5.4
\end{array}
$$

(a) Plot the data.

(b) Find the equation of the least squares line which best fits this data and plot the line on the graph of part (a).

(c) Estimate the median income of a city having 11 per cent college graduates in its population.

9. In the study referred to in Example 8, the following data are gathered on

the median income (in thousands of dollars) for the sample of eight cities and the percentage of families in the cities who own at least one car:

Median income, x:	7.0	4.4	5.4	4.9	7.6	4.2	3.8	6.2
Per cent car owners, y:	68	49	53	51	70	53	48	60

(a) Plot the data.
(b) Fit a least squares line to these data and plot the line on the graph of part (a).
(c) Estimate the percentage of families owning at least one car in a city in which the median income is $6,000.

10. In connection with planning servicing facilities for machines that require attention, a company wants to study the relationship between the number of machines waiting for attention at a given time and the average time required by operators to service the machines. More specifically, the company wants to know whether there is a tendency for operators to work faster (and reduce the service time) when the number of machines waiting for service is large. Accordingly, the company randomly selects eight records showing the number of machines in line at the beginning of a given time period and the number of services completed by the operator during the period. The following are the data:

Machines in line, x:	3	6	5	4	4	6	8	7
Number of completed services, y:	3	2	3	5	3	6	6	4

(a) Plot the data.
(b) Find the least squares line which best fits the data and plot the line on the graph of part (a).
(c) Predict the average number of services an operator will complete during a period when there are five machines in line at the beginning of the period.

THE GOODNESS OF PREDICTIONS

Having obtained the equation of a curve and used it to make predictions, we now come to the third problem of curve fitting: "How good is a predicting equation and how good are the predictions made from it?" Turning again to the extraction problem for definiteness, we might ask such questions as these:

"How good are the values we obtained for the constants a and b in the equation $y' = a + bx$? After all, the numbers 37.69 and 0.7648 are only estimates based on sample data."

"How are we to interpret the predicted efficiency of the process when the extraction time is, say, 25 minutes? It does not seem likely that the process efficiency is exactly the same for all batches of raw materials processed for 25 minutes; in the tests themselves the efficiency was 44 points for one batch processed 18 minutes and 50 points for another batch processed 18 minutes. This certainly suggests that other factors than extraction time affect the extraction efficiency and explains the company's interest in such processing conditions as moisture content of the raw material, solvent pumping rate, and extraction temperature."

"How can we obtain limits (two numbers) and an associated probability which measure the goodness of a prediction in the same way in which a confidence interval measures the goodness of, say, an estimate of the mean of a population?"

When, in the (first) question above, we said that the numbers 37.69 and 0.7648 were "only estimates based on sample data," we implied the existence of two corresponding true values, usually designated α and β, and therefore of a true (linear) equation of the form $\alpha + \beta x$.* Thus we were in effect asking, "How good are the estimates a and b which we obtained for the coefficients—called *regression coefficients*—α and β?" Although the technical detail is fairly involved, we can construct confidence limits for both α and β and evaluate the "goodness" of the estimates a and b of these values. In most problems the value of α has little, if any, practical importance because α is only the value on the line corresponding to $x = 0$. On the other hand, there is often much practical interest in β; in some studies, in fact, the most important objective of the study is to estimate β. This is because β, the slope of the least squares line, is the average (or expected) change in y associated with a change of one unit in x. In the extraction example, β measures the average change in extraction efficiency which is associated with a change of one unit (one minute) in the extraction time, and this is important to know. We show how to construct confidence limits for β in Exercise 13 on page 369.

With respect to the (second) question of how we are to interpret our predictions, we observe that in the extraction problem it is obviously possible for the extraction efficiencies to vary from batch to batch even among batches processed exactly the same amount of time; and in the sales problem it is also obviously possible for first-year sales to vary from man to man even among groups of men who made exactly the same score on the personality test. Furthermore, it is obviously possible for a number of people who read at exactly the same rate to read with different comprehension, and for a number of parts made under the same processing conditions and

---

* The coefficients, α and β, should *not* be confused with the probabilities α of rejecting a true hypothesis and β of accepting a false hypothesis.

having the same hardness to have different tensile strengths. All of this suggests that in statistical prediction problems such as we are discussing, although predictions may sometimes providentially agree with the actual values they are intended to predict, there is no reason to expect them to.

In the sales problem, we predicted first-year sales of $252,000 for a salesman who scored 40 points on the test, and in the extraction problem we predicted an extraction efficiency of 57 points for a batch processed 25 minutes. Both first-year sales and extraction efficiency are random phenomena, however, and therefore are subject to chance variation. If, in fact, many tests were conducted, in all of which the extraction time was 25 minutes, we would observe many different values of extraction efficiency. Thus, if we consider a single test for $x = 25$ minutes, we realize that its extraction efficiency y may be any one of the values in a frequency distribution of efficiencies corresponding to an extraction time of 25 minutes. The *mean* of this frequency distribution depends on the processing time, however, and we can say in this case that, for extraction runs of 25 minutes, *on the average* the extraction efficiency will be 57 points. Similarly, we can say that, for extraction runs of 35 minutes, *on the average* the extraction efficiency will be 64 points; that, for salesmen who score 40 points on the test, *on the average* their first-year sales will be $252,000; and so on. Generally speaking, we look on predicting equations as *curves of means*—or *regression equations* —which give for each x value an estimate of the *mean* of the corresponding frequency distribution of the y's.*

To give one more example, suppose that a study of the ages and used-car-lot asking prices of second-hand cars of a certain make and type yielded the following predicting equation,

$$y' = 1,708 - 136x$$

where x is the age in years and y is the price. Substituting $x = 5$ in this equation, we obtain $y' = \$1,028$, and according to what we have just said, we can predict that *on the average* the price of such cars is $1,028; some are priced less and some more, and $1,028 is the best estimate we can give of their actual average price.

This leads us directly to the (third) question of how we can construct limits which measure the goodness of predictions made from least squares regression equations. In all linear regression studies we use an expression

---

* The term "regression," as it is used here, is due to Sir Francis Galton who first used it in a study of the heights of fathers and sons. Observing that the mean heights of sons increased less rapidly than the mean heights of their fathers, he called this phenomenon "regression" (i.e., a turning back).

of the form $a + bx$ (with a and b determined from sample data) to predict the true mean of y for a given value of x. From the discussion (and in the notation) of this section, this true mean is $\alpha + \beta x$. In Technical Note 6 we shall show how to put confidence limits on this mean and also how to put limits, called *limits of prediction*, on a single value of y for a given value of x. In the first case, we shall find an interval which, with a specified high probability, covers the true mean extraction efficiency for *all conceivable* extraction runs of a given time $x = x_0$. And in the second case we shall find an interval which, with a specified high probability, covers the true efficiency measure of a *single* extraction run of a given time $x = x_0$.

EXERCISES

(The exercises which follow are based on the material in Technical Note 6 on page 374.)

1. Referring to Exercise 3 on page 354,
   (a) Construct 0.95 limits of prediction for the moisture content of a batch of raw material stored in the room when the humidity is 45 per cent.
   (b) Construct 0.95 confidence limits for the mean moisture content of a (conceptually large) population of batches of raw material stored in the room when the humidity is 45 per cent.

2. Referring to Exercise 4 on page 355,
   (a) Construct 0.99 limits of prediction for the percentage of protected area burned in a county with 2.0 cents per acre variable operating cost.
   (b) Estimate with 0.99 confidence the mean percentage of protected area burned in a (conceptually large) population of counties with 2.0 cents per acre variable operating cost.

3. Referring to Exercise 10 on page 357,
   (a) Construct 95 per cent limits of prediction for the number of services an operator will complete when there are five machines in line at the beginning of the period.
   (b) Estimate with 95 per cent confidence the mean number of services an operator will complete during periods when there are five machines in line at the beginning of the period.

CORRELATION

We turn now to the problem of determining how well a least squares regression line fits a given set of paired data. Referring again to the extraction problem, we observe from the original data that there is considerable variation among the y's, the smallest efficiency measure being 44 and the largest 78. Upon examination of these data and Figures 13.2 and 13.3, it seems reasonable to assert that this variation is not due entirely to chance. The fact that the extraction efficiency of the third run was only 44 while that of the fourth run was 78 must, it would seem, be due at least partly to the fact that the third run was only 18 minutes long and the fourth run was 40 minutes long. These observations raise the following question: *"Of the total variation among the y's, how much can be attributed to chance and how much can be attributed to the relationship between the two variables x and y, that is, to the fact that the observed y's correspond to different values of x?"*

The answer to this question requires that we analyze, or separate, the total variation of the observed y's (in our example, the variation in the extraction efficiencies of the 10 runs) into two parts corresponding to the two different sources which give rise to it. A study of Figure 13.5 will help to

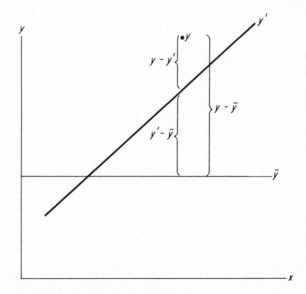

FIGURE 13.5 An illustration that $(y - \bar{y}) = (y' - \bar{y}) + (y - y')$

understand what we mean by this. Referring to this figure, we see that we can write the deviation of any single value y from its mean, $y - \bar{y}$, as the sum of two parts: the deviation of the corresponding value on the regression line from the mean of the y's, $y' - \bar{y}$, and the deviation of the actual y value from the corresponding value on the line, $y - y'$. For any value y we write

$$(y - \bar{y}) = (y' - \bar{y}) + (y - y')$$

and if we now square both sides of this identity and sum both sides over all n values of y, we find that algebraic simplifications lead to

$$\Sigma \, (y - \bar{y})^2 = \Sigma \, (y' - \bar{y})^2 + \Sigma \, (y - y')^2$$

We shall use as our measure of the "total" variation in the y's the quantity on the left above, $\Sigma \, (y - \bar{y})^2$. This measure of the total variation is just $n - 1$ times the *variance* of the y's, and as the equation shows, it has been separated into two additive components. From Figure 13.5 we observe that the first of these components, $\Sigma \, (y' - \bar{y})^2$, consists of the sum of the squares of the deviations of values on the line from the mean $\bar{y}$. This quantity, called the *regression variation*, measures that portion of the total variation in y which would exist if differences in x were the only cause of differences in y, or, in other words, if all the y's lay directly on the line. In any problem where the y's actually do all lie on the line the quantities $\Sigma \, (y - \bar{y})^2$ and $\Sigma \, (y' - \bar{y})^2$ are identical (since for each x value y and y' are equal), and the value of the second component on the right above is zero. This means that the total variation in y is completely explained by the relationship of y to x.

This is hardly ever the case in practice, though, and the fact that all the points do not lie on a regression line is an indication that there are other factors than the one designated x which affect the values of y. It is customary to lump together the totality of these other factors, which are not separately considered in the study, under the general heading of "chance." Chance variation thus depends on the amounts by which the points deviate from the line, and its measure, $\Sigma \, (y - y')^2$, is the sum of the squared discrepancies between the actual y values and their corresponding values y' on the line.

Using the definition formulas given above (but see Exercise 14 on page 369) to calculate these three quantities for the extraction example, we have, for the *total variation*,

$$\Sigma \, (y - \bar{y})^2 = (55 - 61.4)^2 + (60 - 61.4)^2 + \cdots + (70 - 61.4)^2$$
$$= 1,064.40$$

for the *regression variation*,

$$\Sigma \, (y' - \bar{y})^2 = (48.40 - 61.4)^2 + (63.69 - 61.4)^2 + \cdots + (66.75 - 61.4)^2$$
$$= 731.15$$

and for the *chance variation,*

$$\Sigma \, (y - y')^2 = (55 - 48.40)^2 + (60 - 63.69)^2 + \cdots + (70 - 66.75)^2$$
$$= 333.25$$

(The y' values used here are computed by substituting the values 14, 34, ..., and 38 for x in the regression equation $y' = 37.6912 + 0.7648x$; $\bar{y} = \dfrac{614}{10} = 61.4$ comes from the original data.)

From this we see that

$$\Sigma \, (y - \bar{y})^2 = \Sigma \, (y' - \bar{y})^2 + \Sigma \, (y - y')^2$$

or

$$1{,}064.40 = 731.15 + 333.25$$

Hence, we can claim that

$$\frac{\Sigma \, (y - y')^2}{\Sigma \, (y - \bar{y})^2} \cdot 100 = \frac{333.25}{1{,}064.40} \cdot 100 \;\; = 31.30 \text{ per cent}$$

of the total variation in the extraction efficiencies can be attributed to chance (all factors other than the extraction time which affect the efficiency of the process). It follows that the remaining $100 - 31.30 = 68.70$ per cent of the variation in the extraction efficiencies is accounted for by differences in extraction time.

In general, if we divide $\Sigma \, (y - \bar{y})^2 = \Sigma \, (y' - \bar{y})^2 + \Sigma \, (y - y')^2$ through by $\Sigma \, (y - \bar{y})^2$, we get two ratios whose sum is 1, as follows:

$$1 = \frac{\Sigma \, (y' - \bar{y})^2}{\Sigma \, (y - \bar{y})^2} + \frac{\Sigma \, (y - y')^2}{\Sigma \, (y - \bar{y})^2}$$

The first of these ratios on the right is the ratio of the regression variation to the total variation, or the proportion of the total variation in y which is determined (accounted for) by differences in x. This ratio is called the *coefficient of determination* and is designated by r^2. Noting that r^2 is equal to 1 minus the second ratio on the right, we may now define the following measure r, called the *coefficient of correlation,*

$$r = \pm\sqrt{1 - \frac{\Sigma \, (y - y')^2}{\Sigma \, (y - \bar{y})^2}}$$

where the sign attached to r is the sign of b in the predicting equation. Thus, r is *positive* when the least squares line has an *upward slope*, that is, when the relationship between x and y is such that small values of y tend to go with small values of x and large values of y tend to go with large values of x. Correspondingly, r is *negative* when the least squares line has a *downward slope*, that is, when large values of y tend to go with small values of x and small values of y tend to go with large values of x. Corresponding to the sign of r, we say that there is a *positive correlation* or a *negative correlation* between x and y (see Figure 13.6). When $r = 0$, we say that there is *no*

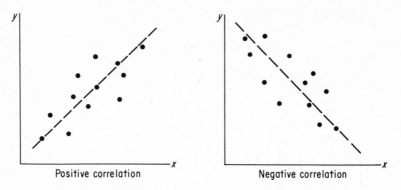

Positive correlation   Negative correlation

FIGURE 13.6 Positive and negative correlation

correlation between x and y. It may be seen from the definition formula for r that a correlation coefficient must lie on the interval from -1 to $+1$. If all the points actually fall on a straight line, the measure of chance variation in the formula, $\Sigma (y - y')^2$, is zero and the resulting value of either -1 or $+1$ for r indicates that the fit of the line to the paired observations is perfect. If, however, the scatter of the points is such that the least squares line is a horizontal line coincident with $\bar{y}$ (i.e., a line with slope $b = 0$ which intersects the y-axis at height $a = \bar{y}$), then $\Sigma (y - y')^2$ and $\Sigma (y - \bar{y})^2$ are identical and $r = 0$. In this case none of the variation of the y's can be attributed to their relationship with x, and the fit is so poor that knowledge of x is of no help in predicting y; the predicted value of y is $\bar{y}$ for any x.

For the extraction problem we can calculate r by taking the square root of r^2, the coefficient of determination (the proportion of the variation in efficiency that is accounted for by differences in time), and get $r = +\sqrt{0.6870} = 0.829$. In practice, however, it is much easier to use the computing formula

$$r = \frac{n(\Sigma\, xy) - (\Sigma\, x)(\Sigma\, y)}{\sqrt{n(\Sigma\, x^2) - (\Sigma\, x)^2}\ \sqrt{n(\Sigma\, y^2) - (\Sigma\, y)^2}} \qquad \bigstar$$

Then $100\, r^2 = \%$

which is equivalent to the definition formula above and also gives r its correct sign. This formula may look imposing, but it is easy to use. With the exception of $\Sigma\, y^2$, all the quantities which must be substituted in it were required in calculating the coefficients a and b. Squaring and summing the y's in the table on page 351, we get $\Sigma\, y^2 = 38{,}764$, and find that

$$r = \frac{10(19{,}990) - (310)(614)}{\sqrt{10(10{,}860) - (310)^2}\,\sqrt{10(38{,}764) - (614)^2}} = 0.829$$

which is identical with the result from the definition formula.

As we have defined the coefficient of correlation r, $100r^2$ (i.e., 100 times the coefficient of determination) gives the percentage of the total variation of the y's which is due to their relationship with x. This is an important measure in the study of the relationship between two variables, and it also permits valid comparisons to be made. If, for instance, in a given study $r = 0.80$, then 64 per cent of the variation in y is accounted for by its relationship with x; if in another study $r = 0.40$, only 16 per cent of the variation in y is accounted for by its relationship with x. Thus, in the sense of "percentage of variation accounted for" we can say that the correlation of 0.80 is *four times as strong* as the correlation of 0.40. In the same way, we say that a correlation of 0.90 is *nine times as strong* as a correlation of 0.30, and so on.

In interpreting a correlation coefficient it is sometimes assumed that a numerically high value of r establishes a cause-and-effect relationship running from x (often called the "independent" variable) to y (often called the "dependent" variable). We want to make it clear that this is not so. There are cases in which changes in one variable are considered to be the "cause" of changes in another variable (called its "effect"), but only logical argu-

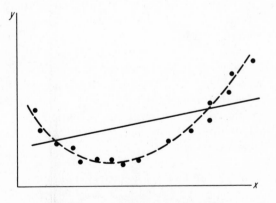

FIGURE 13.7 Nonlinear relationship

ment, not a high value of r, can establish such a relationship. (See "A Word of Caution" at the end of this chapter for some further remarks on this matter.)

We also want to make it clear that r measures only the strength of *linear* relationships. If we calculated r for the data of Figure 13.7, for example, we would get a value near zero even though the points are all very close to the dotted curve. There is an obvious strong *curvilinear* relationship between the two variables, but virtually no linear relationship.

A SIGNIFICANCE TEST FOR r

When r is calculated on the basis of a sample, it is possible to get a strong positive or negative value even though there is actually no linear relationship whatever between the two variables in the population from which the sample came. To illustrate this point, suppose that we take a pair of dice, one red and one green, and roll them five times, letting x and y be the number of points showing on the red and green die, respectively. Presumably, there should be no relationship between x and y for this experiment. It is hard to see why, for example, large values of x should tend to appear with large values of y and small values of x with small values of y. After all, how can one die know what the other is doing?

We performed this experiment, though, with the following results,

$$
\begin{array}{llccccc}
\textit{Red Die, x:} & 5 & 3 & 1 & 5 & 3 \\
\textit{Green Die, y:} & 5 & 2 & 3 & 6 & 1
\end{array}
$$

getting the surprisingly high value of $r = 0.66$, and we wonder whether we are wrong in assuming that there is really no relationship between x and y. Let us, therefore, set up and test the null hypothesis that, in the (infinite) population from which the sample came, there is no relationship between x and y against the alternative that there really is a relationship.

There are several ways we might test a null hypothesis of no relationship between two variables x and y. We shall refer to a special table which has been calculated on the basis of the sampling distribution of r when the x's and y's are sample values coming from populations having roughly the shape of normal distributions. Assuming normality of the populations, and referring to Table VI,

we reject the null hypothesis of no correlation at the level of significance α if the value of r calculated for a set of paired data exceeds $r_{\alpha/2}$ or if it is less than $-r_{\alpha/2}$.

If the value we obtain for r falls between $-r_{\alpha/2}$ and $r_{\alpha/2}$, we say that the correlation coefficient is not significant. Using Table VI we can make our tests at levels of significance of 0.05, 0.02, and 0.01.

Applying the test to the dice example above, we find that the calculated r would have to be numerically greater than $r_{.025} = 0.878$ for $n = 5$ in order for the result to be significant at the 5 per cent level. Since $r = 0.66 < 0.878$ we attribute this result to chance; the relationship observed in this small sample is not strong enough to persuade us that there is a real relationship between x and y.

Applying the test to the extraction example, however, for which $r = 0.829$, we find that $r_{.005} = 0.765$ for $n = 10$. Therefore, we can reject at the 1 per cent level of significance the hypothesis that extraction efficiency and extraction time are not related. It appears almost certain that these two variables are related.

EXERCISES

1. Referring to Exercise 1 on page 354,
   (a) Calculate r for the conditioning period-storage life data. At the 0.05 level of significance, is there a relationship between the two variables?
   (b) What percentage of the total variation in the storage life of the glue in the sample is accounted for by variation in the conditioning times?

2. Calculate r for the data of Exercise 2 on page 354 and determine, at the 5 per cent level of significance, whether there is a relationship between machine age and machine efficiency.

3. Referring to Exercise 3 on page 354,
   (a) Calculate r for the humidity-moisture content data and test, at the 0.01 level of significance, the hypothesis that there is no relationship between these two variables.
   (b) What percentage of the total variation in moisture content of the raw material in the sample is accounted for by variation in the humidity of the storage place?

4. Calculate r for the data of Exercise 4 on page 355 and test, at the 5 per cent level of significance, whether there is a real relationship between the percentage of protected area burned and variable operating costs.

5. Calculate r for the data of Exercise 5 on page 355 and test, at the 5 per cent level of significance, the null hypothesis that there is no relationship between advertising expenses and operating profits in the population from which the stores came.

6. In teaching computer programming, instructional materials are given to a large group of business administration students, and the students are allowed

to proceed at their own pace, handing in required assignments as they are completed. At the end of the course, 10 students are randomly selected and given an achievement test. The following are the numbers of hours the students required to complete the work and their grade on the test:

Hours spent, x:	30	25	50	38	20	70	35	24	60	45
Test score, y:	80	80	45	70	95	20	50	90	25	50

(a) Plot the data.

(b) Fit a least squares line to these data and plot the line on the graph of part (a).

(c) Calculate r and test the relationship for significance at the 0.05 level.

(d) Of the total variation in the sample in test scores, what proportion of it is accounted for by variation in the number of hours spent on the work?

7. Calculate r separately for each of the following sets of data:

(a)

x	y
15	6
5	1

(b)

x	y
10	30
20	5

Is there a way to determine the values of r without making any calculations at all?

8. For the following sample of data

x	y
4	3
2	4
1	2
3	1

verify that there is no correlation between x and y, and that the least squares straight line which fits these data best is a horizontal line coincident with $\bar{y}$.

9. In a study concerned with predicting success in business, data were collected on a number of male M.B.A. graduates of a leading graduate school of business. Success for the men working in large organizations was expressed as a composite of organizational level of authority, degree of participation in deciding overall company policy, and remuneration. Test, at the 5 per cent level of significance, whether there is a real relationship in the population studied between success in business and the following four variables:

(a) Grade-point average in a four-year college: $n = 72$ and $r = 0.14$.

(b) Score on M-F scale of Strong Vocational Interest Blank: $n = 92$ and $r = 0.19$.

(c) Leadership in undergraduate school: $n = 100$ and $r = 0.24$.

(d) Score on college admissions test: $n = 80$ and $r = -0.01$.

10. State in each case whether you would expect to find a positive correlation, a negative correlation, or no correlation between the following:

(a) Men's intelligence and their hat sizes.

(b) The ages of husbands and their wives.

(c) The number of years of education of husbands and their wives.

(d) The sizes of families and their incomes.

(e) Corporate earnings per share and dividends paid per share.

11. An American Cancer Society study of about 800,000 men and women showed that the death rate from strokes was considerably higher, and from coronary heart disease generally higher, among persons who usually sleep 9 to 10 hours a night than among those who usually sleep 7 hours. On the basis of these statistically significant results is it reasonable to argue that longer sleep is the cause of deaths from strokes and coronary heart disease?

12. A study made at a major university showed that a significantly higher proportion of heavy cigarette smokers than of light smokers or nonsmokers failed in college. Based on this evidence, is it reasonable to conclude that heavy smoking is the cause of failure in college?

13. We can write $1 - \alpha$ confidence limits for β (the slope of the least squares regression line) as

$$b \left[1 \pm t_{\alpha/2} \cdot \frac{\sqrt{1 - r^2}}{r\sqrt{n - 2}} \right]$$

where b and r are calculated from the sample observations and there are $n - 2$ degrees of freedom for $t_{\alpha/2}$.

(a) Construct a 95 per cent confidence interval for the slope of the regression line obtained in Exercise 3 on page 354. (The value $r = 0.91$ was calculated for these data in Exercise 3 above.)

(b) Construct a 0.90 confidence interval for the slope of the regression line calculated in part (b) of Exercise 6 above.

14. On page 362 we calculated for the extraction example the total variation, regression variation, and chance (often called "residual") variation using the definition formulas. However, it is easy to show that $\Sigma (y - \bar{y})^2 = \Sigma y^2 - n \cdot \bar{y}^2$ and that $\Sigma (y - y')^2 = \Sigma y^2 - a \cdot \Sigma y - b \cdot \Sigma xy$. These two formulas on the right are much better than the definition formulas for computing the total variation and the chance variation, respectively, and from these two values we can also get the regression variation by subtraction. Use these computing formulas to verify that in the extraction example the total variation is 1,060.40, the regression variation is 731.15, and the chance variation is 333.25.

RANK CORRELATION

When dealing with a large set of paired data, the calculation of a correlation coefficient can be quite tedious. To simplify matters, we sometimes compute a measure of relationship based on the *ranks* of the observations instead of their actual numerical values. We first rank the x's among themselves, giving Rank 1 to the largest (or smallest) value, Rank 2 to the second largest (or smallest), and so on; then we rank the y's similarly among themselves and calculate the *Spearman coefficient of rank correlation* by means of the formula

$$r' = 1 - \frac{6(\Sigma\, d^2)}{n(n^2 - 1)} \qquad \star$$

Here n is the number of pairs of observations, and the d's represent the differences between the ranks of the corresponding x's and y's. In case there are ties, we assign to each of the tied values the mean of the ranks which they jointly occupy. Thus, if the third and fourth largest values of a variable are identical, we assign each a rank of $\frac{3+4}{2} = 3.5$, and if the fifth, sixth, and seventh highest values of a variable are identical, we assign each a rank of $\frac{5+6+7}{3} = 6$.

To illustrate the technique of rank correlation, let us refer again to the problem of measuring the relationship between extraction time (x) and extraction efficiency (y). The following table shows the original data, the rankings (from high to low) of the two variables, columns for d and d^2, and $\Sigma\, d^2$.

x	y	Rank of x	Rank of y	d	d^2
14	55	10	7.5	2.5	6.25
34	60	5	6	−1	1
18	44	8.5	10	−1.5	2.25
40	78	3	1	2	4
48	75	1	2	−1	1
18	50	8.5	9	−0.5	0.25
44	62	2	5	−3	9
30	65	6	4	2	4
26	55	7	7.5	−0.5	0.25
38	70	4	3	1	1
					29

Substituting $n = 10$ and $\Sigma \, d^2 = 29$ into the formula for r' we get

$$r' = 1 - \frac{6(29)}{10(100 - 1)} = 0.824$$

which is very close to the value of 0.829 which we calculated for r.

When there are no ties, r' gives the exact value of r between the ranks; that is, r' is actually equal to the correlation coefficient calculated for the two sets of ranks using the formula on page 364. When ties exist, there may be a small (but usually negligible) difference. When we use the ranks of the x's and y's instead of the actual values, we naturally lose some information, but r' is usually quite close to r. Moreover, rank correlation methods have the advantage that the calculation of r' is often much simpler than the calculation of r, that they can be used to measure relationships in problems where items cannot be measured on a numerical scale but can be ranked (see the exercises following), and that tests of significance based on them are relatively unrestrictive. In testing the null hypothesis of no relationship between two variables x and y, we do not have to make any assumptions about the nature of the populations from which the samples came. We can reject this null hypothesis at a level of significance α if $r' > z_{\alpha/2}/\sqrt{n - 1}$ or if $r' < -z_{\alpha/2}/\sqrt{n - 1}$, where as always, $z_{\alpha/2}$ is the value of the standard normal distribution such that the area to its right is $\alpha/2$.

For the example above, this critical value at the 5 per cent level of significance is

$$\frac{z_{.025}}{\sqrt{n - 1}} = \frac{1.96}{\sqrt{10 - 1}} = 0.65$$

and since $r' = 0.824$ exceeds this value, we can say that there is a significant relationship at the 5 per cent level between extraction efficiency and extraction time.

EXERCISES

1. In a study devoted to the problem of salesmen turnover, an investigator takes a random sample of 11 large companies and ranks them according to their average out-of-pocket hiring costs per man and their separation rate (with Rank 1 assigned to the companies with the highest hiring cost and with the highest separation rate). On the basis of the rankings, which follow, is there evidence,

at the 0.01 level of significance, that companies having the highest hiring cost have the lowest separation rate?

COMPANY

Expense	A	B	C	D	E	F	G	H	I	J	K
Hiring cost	1	2	3	4	5	6	7	8	9	10	11
Separation rate	10	11	9	7	8	6	4	5	1	3	2

2. A random sample of eight railway districts is taken, and the number of derailments occurring in each district for the past year and the year preceding that are recorded. The districts are then ranked on number of derailments (with Rank 1 assigned to the districts with the fewest derailments). Based on these rankings, which follow, is there a relationship at the 0.05 level of significance between the number of derailments in the districts for the two years?

DISTRICT

Year	A	B	C	D	E	F	G	H
Past	1	2	3	4	5	6	7	8
Previous	7	5	2	6	1	8	3	4

3. In the sample referred to above, the number of I.C.C. reportable casualties to employees on duty in maintenance-of-way work (in casualties per million man-hours) in the eight districts were

DISTRICT

Year	A	B	C	D	E	F	G	H
Past	0.8	14.9	8.8	11.3	4.9	14.5	24.0	7.4
Previous	6.0	5.8	4.7	3.8	7.4	4.1	10.0	11.0

(a) Calculate the coefficient of correlation r and test, at the 0.05 level of significance, whether there is a real relationship between casualties in the two years.

(b) Calculate the coefficient of rank correlation r' for these data and test the relationship for significance at the 0.05 level.

4. As part of its product development work, a large food processor asks two panels of judges to rank, in order of their overall preference, samples of minestrone made from 10 different trial mixes. The rankings are shown below:

MIX

	A	B	C	D	E	F	G	H	I	J
Panel 1	1	2	3	4	5	6	7	8	9	10
Panel 2	1	4	5	2	7	3	6	9	10	8

Calculate r' as a measure of the consistency of the two panels and test the relationship for significance at a level of 0.05.

5. The following are the scores on a written examination for driver's licenses (number of correct answers in 10 questions) and the scores on an eye-hand reaction test (number of seconds to complete the test) observed in a random sample of 10 license applicants at a state motor vehicle office:

APPLICANT

Test	A	B	C	D	E	F	G	H	I	J
Written	9	8	8	10	5	10	4	7	6	9
Reaction	3	10	5	9	8	5	3	6	4	8

Does this constitute evidence at the 0.05 level of significance of a real relationship between the scores on these two tests in the population sampled?

6. Referring to Exercise 7 on page 356, calculate the coefficient of rank correlation r' for the aldehyde-esters data, and also calculate the coefficient of correlation r for the *ranks* of these two variables. The two measures should be identical.

7. For the extraction example on page 370, $r' = 0.824$. Verify that the coefficient of correlation calculated from the *ranks* of the two variables is $r = 0.823$. How do you account for this slight discrepancy?

8. Calculate r' for the humidity-moisture content data of Exercise 3 on page 354, and test for a significant relationship between the two variables at the level $\alpha = 0.05$.

9. Calculate r' for the area burned-variable operating cost data of Exercise 4 on page 355, and determine whether or not there is a significant relationship between the two variables at the level $\alpha = 0.05$.

A WORD OF CAUTION

One must always take special care in interpreting the results of correlation studies. We call attention again to the fact that the correlation coefficient r measures only the strength of linear relationships (see Figure 13.7), and that it is possible to find a numerically high degree of correlation in a sample drawn from a population in which there is no correlation whatever. When high r's do result entirely from chance, the correlation is often said to be "spurious." Spuriously high, or inflated, correlations also result from an overlap of two variables that are being correlated (e.g., when scores on two multiple-choice tests consisting partly of the same items are correlated).

We also call attention again to the danger of presuming that a high correlation coefficient establishes a cause-and-effect relationship between two variables x and y. For instance, a high positive correlation has been observed in the study of the relationship between teachers' salaries and liquor consumption, and a high negative correlation has been observed in a study of the annual per capita consumption of chewing tobacco in the United States and the number of automobile thefts reported in a sample of urban areas in the same years. Moreover, in another study, a strong positive correlation was observed between the number of storks seen nesting in English villages (x) and the number of children born in these same villages (y). We leave it to the reader's ingenuity to explain why there might be strong correlations observed in these instances in the absence of any cause-and-effect relationships.

Finally, we observe that there is a significantly high correlation between the advertising expenditures (x) and net operating profits (y) of the jewelry stores of Example 5 on page 355. But which way (if either) should we argue? That high advertising outlays lead to high profits? Or that high profits lead to high advertising outlays?

TECHNICAL NOTE 6 (Confidence Limits and Limits of Prediction)

In order to judge the "goodness" of a prediction made from a least squares regression equation, of either the mean $\alpha + \beta x$ of some *conditional distribution* of y for fixed values of x, or of a y value itself lying in this distribution, it is necessary to have some measure of the variability of the distribution. To construct such a measure, we assume that the y we obtain for any given value of x is a value of a random variable (a random sample of size 1) from a population which can be closely approximated by a normal curve. Furthermore, we assume that *all* the conditional distributions for fixed values of x have the same standard deviation σ. We shall use this standard deviation as the measure of variability of the conditional distributions, estimating it by means of the quantity

$$s_e = \sqrt{\frac{\Sigma (y - y')^2}{n - 2}}$$

called the *standard error of estimate*. In this estimate the quantity $\Sigma (y - y')^2$, which measures that portion of the total variation of the y's which can be attributed to chance, is the sum of the squared deviations of the actual sample y values from their predicted (expected) values, the sum of the squared deviations represented, for example, by the solid-line segments of Figure 13.3. The divisor $n - 2$ is the number of degrees of freedom associated with this estimate of the variability of the conditional distributions. (We have lost two degrees of freedom here because we have calculated two constants, a and b, for the equation of y' from the original n pairs of values.)

Calculation of the standard error of estimate from the definition formula above is tedious work, and s_e is usually calculated from the simpler (but equivalent) formula

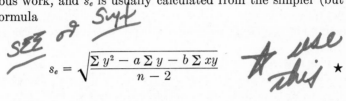

$$s_e = \sqrt{\frac{\Sigma y^2 - a \Sigma y - b \Sigma xy}{n - 2}}$$

Substituting $n = 10$, $a = 37.69$, $b = 0.7648$, $\Sigma y = 614$, $\Sigma y^2 = 38{,}764$, and $\Sigma xy = 19{,}990$ in the equation we find that for the extraction example

$$s_e = \sqrt{\frac{38{,}764 - 37.69(614) - 0.7648(19{,}990)}{10 - 2}} = 6.5$$

We can now determine how good predictions based on a least squares equation are, taking first the case of predicting a single y value corresponding to a given value of x. If we find the equation of a least squares line from a set of paired sample data and use this equation to predict the value y_0' for a given value x_0 of x, we can assert with a probability of $1 - \alpha$ that the value we will actually obtain will lie between the limits $y_0' - A$ and $y_0' + A$, called the *limits of prediction*, where

$$A = t_{\alpha/2} \cdot s_e \sqrt{\frac{n + 1}{n} + \frac{(x_0 - \bar{x})^2}{\Sigma x^2 - n\bar{x}^2}} \qquad \star$$

and $t_{\alpha/2}$ is obtained from Table II with $n - 2$ degrees of freedom. Here x_0 is the value for which we want to make the prediction, y_0' is the value we obtain by substituting x_0 into the equation of the least-squares line, and n, $\bar{x}$, and Σx^2 come from the original data.

Let us now apply this theory to the problem of predicting the extraction efficiency of an extraction run of length $x_0 = 30$ minutes. The predicted efficiency is $y_0' = 37.69 + 0.7648(30) = 61$ points. To find 95 per cent limits of prediction, we substitute $s_e = 6.5$, $n = 10$, $\bar{x} = \dfrac{310}{10} = 31$, $\Sigma x^2 =$

10,860, and $t_{.025} = 2.306$ (for 8 degrees of freedom) into the expression for A, getting

$$A = 2.306(6.5) \sqrt{\frac{11}{10} + \frac{(30 - 31)^2}{10,860 - 10(31)^2}} = 16$$

and the limits of prediction are $61 - 16 = 45$ and $61 + 16 = 77$. In other words, we can assert with a probability of 0.95 that the efficiency of an extraction run 30 minutes long will be contained in the interval from 45 to 77.

This may seem like a fairly wide interval, and in practice it may or may not be adequate for predictive purposes. One reason for the width of the interval is that it is based on only 10 pairs of associated extraction efficiency-extraction time values. Another reason for this width is that $s_e = 6.5$. Evidently, there is considerable variation in extraction efficiencies among extraction runs of exactly the same length. This variation contributes directly to the width of the limits of prediction, and of course, it is an inherent part of the process itself.

In cases where we want to predict the *mean* value of y for a given value x_0, appropriate $1 - \alpha$ confidence limits can be written in the form $y_0' - B$ and $y_0' + B$, where $y_0' = a + bx_0$ and

$$B = t_{\alpha/2} \cdot s_e \sqrt{\frac{1}{n} + \frac{(x_0 - \bar{x})^2}{\sum x^2 - n\bar{x}^2}} \qquad \star$$

(This formula differs from the formula for A only in the first term under the radical.) We leave it as an exercise for the reader to show that we may assert with probability 0.95 that the range of values 56 to 66 covers the true mean extraction efficiency of all conceivable extraction runs of 30 minutes each.

The techniques of prediction outlined above must be used with care. In both instances we have assumed that the true regression curve is a straight line, and also that the conditional distributions are normal curves having equal variances. Moreover, in the case of limits of prediction, a probability $1 - \alpha$ applies to the entire procedure of obtaining the sample data, calculating the limits, and drawing the additional observation whose y value is to be predicted. The probability $1 - \alpha$ applies only to *one* prediction based on a particular set of data.

14

DECISION MAKING: TIME SERIES ANALYSIS

FORECASTING—I

It has been said in many different ways that the future belongs to those who plan for it best. This is certainly true in business and in economics. Of course, business planning is not an end in itself, but organized planning utilizing various statistical techniques (intended to assess past performance and estimate the success or failure of proposed strategies) seems to have everything in its favor. Aside from its intuitive appeal, there are the achievement records of many highly successful companies which treat planning as an organized activity and analyze exhaustively the many factors bearing on planning decisions. Marketing strategy is often planned in great detail for several years ahead, with enough flexibility built in to allow for whatever changes market conditions may require. Financial strategy is also carefully planned, so that operating plans can be carried out and a proper balance maintained between distributed earnings and retained earnings necessary for future growth. Many manufacturing companies attempt to make their long-range planning (e.g., beyond a year) more effective by maintaining projections of 10 years or longer on sales, profits, cash needs, and so on, for all of their major product groups. No intelligent planning of future needs for raw materials and production facilities, for instance, can be done without predictions of such basic variables as product or service demand, production costs, and restrictions on capacity.

Predictions of the sort that involve explaining events which will occur at some future time are called *forecasts*, and the process of arriving at such explanations is called *forecasting*. There are various ways of forecasting the future values of economic variables, including the so-called *intrinsic methods* in which the future values of variables ᵤᵢe predicted from their past values. One important statistical technique included in the intrinsic methods is *time series analysis*. By a *time series* we mean statistical data that are collected, observed, or recorded at regular intervals of time. The term "time series," or simply "series," applies, for example, to data recorded periodically showing the total annual sales of retail stores, the total quarterly value of construction contracts awarded, the total amount of unfilled orders in durable goods industries at the end of each month, and the daily clearings in the Chicago Clearing House. We shall restrict ourselves to the analysis of business and economic data, but neither the term time series nor the methods of analysis which we shall discuss are limited to these kinds of data.

Although in forecasting our concern is with the future, time series analysis begins by looking backward. After all, it would be silly not to put relevant experience from the past to use in planning for an uncertain future. Thus we search for observable regularities and patterns in historical series which are so persistent that they cannot be ignored. If we subsequently base our forecasts on such regularities and patterns, we are simply expressing the feeling that the future follows the past with some degree of consistency, that what has happened in the past will, to a greater or lesser extent, continue to happen or will happen again in the future.

THE COMPONENTS OF A TIME SERIES

Sometimes, when we look at the graph of a time series, we get the impression that it has been scrawled by a small child, and it is hard to believe that any kind of analysis could bring order into the seemingly haphazard movement of the data through time. Nevertheless, if we make some simplifying assumptions it becomes possible to identify, explain, and measure the fluctuations that appear in time series. Specifically, let us assume that there are four basic types of variation in a series which, superimposed and acting in concert, account for the observed changes over a period of time and give the series its erratic appearance. These four *components* are

(1) secular trend,
(2) seasonal variation,
(3) cyclical variation, and
(4) irregular variation,

and we shall assume further that there is a multiplicative relationship between these four components; that is, any particular value in a series is the product of factors which can be attributed to the four components.

This is the traditional approach to time series analysis, but it is only one of many possible models (or schemes) which might be used in studying time series. Although it ignores the hidden interactions and interrelationships in the data and the entire "complex of individually small shifts and nuances," this approach has been and continues to be widely used in practice where, in many instances, it provides entirely satisfactory results. It is possible to construct mathematically sophisticated forecasting models, but there is much of fundamental importance to be learned about the movements of data through time from a study of the traditional methods.

By the secular (or long-term) trend of a time series we mean the smooth or regular underlying movement of a series over a fairly long period of time. Intuitively speaking, the trend of a time series characterizes the gradual and consistent pattern of changes in the series which are thought to result from persistent forces affecting growth or decline (changes in population, income, and wealth; changes in the level of education and technology; etc.) that exert their influence more or less slowly. For example, Figure 14.1 shows the over-all *upward trend* in group life insurance in the United States, and

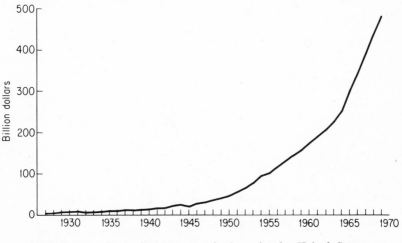

FIGURE 14.1 Group life insurance in force in the United States, 1927–1969

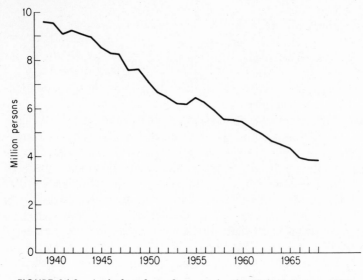

FIGURE 14.2 Agricultural employment in the United States, 1939–1968

Figure 14.2 shows the persistent *downward trend* in agricultural employment in the United States.

The problem in trend analysis is to describe the underlying movement or general sweep of time series in quantitative terms. In many series the patterns of gradual growth or decline can be described reasonably well by means of a straight line, but in others more complicated curves are required.

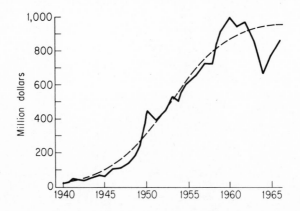

FIGURE 14.3 Growth curve fitted to factory sales of electron tubes and semi-conductors in the United States, 1940–1966

For example, the series on factory sales of electron tubes and semiconductors in the United States shown in Figure 14.3 has the general shape of an elongated letter S, and no straight line can describe its trend even passably well. The curve shown fitted to these data is one of the so-called *growth curves*, and it reflects a type of growth that is frequently observed in time series. In this book we shall study mainly linear trends, although we shall give a method, the *method of moving averages*, which can be used more generally. (See also Exercises 18 and 19 on page 392.)

Strictly speaking, *seasonal variation* is the movements in a time series, like those shown in Figure 14.4, which recur year after year in the same months (or the same quarters) of the year with more or less the same intensity. Thus the month-to-month variation observed in retail sales and the quarter-to-quarter variation observed in consumer installment debt are both examples of seasonal variation in time series. Sometimes the term

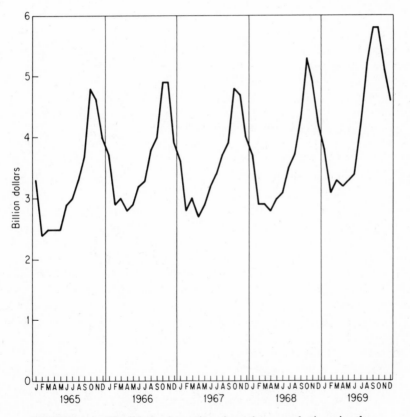

FIGURE 14.4 Monthly cash receipts from farm marketings in the United States, 1965–1969

"seasonal variation" is also applied to other inherently periodic movements, such as those occurring within a day or week or month, whose period is at most one year. In any case, the movement described is a most obvious one.

Few businesses are free from the effects of seasonal variation. The examination of almost any series of economic data recorded on a quarterly, monthly, weekly, daily, or hourly basis shows movements within the series which seem to occur period after period with some definite degree of regularity. About two thirds to three fourths of the total annual business in the jewelry trade is done in the two months before Christmas; except for the holiday season, airline passenger traffic normally drops during the winter months; electric power and natural gas consumption is normally higher during the winter months; city traffic on workdays is always heaviest in the early morning and late afternoon; and so on. To businessmen responsible for realistic planning of such activities as production, purchasing, inventory, personnel, advertising, and sales, an understanding of seasonal patterns is of primary concern. Even in cases where the seasonal variation is not of basic concern, it must often be measured statistically in order to facilitate the study of other types of variation.

After the trend and seasonal variation have been eliminated statistically from a time series, in the model we are discussing here there remain the *cyclical and irregular variations*. Irregular variations in time series are of two types: (1) variations which are caused by such readily identifiable special events as elections, wars and peaces, floods, earthquakes, strikes, and bank failures and (2) random or chance variations whose causes cannot be definitely assigned. Most of the time, irregular variations resulting from the occurrence of special events can be easily recognized and identified with the phenomena which caused them; then the data reflecting their impact can simply be eliminated before measuring the other time series components. As for those essentially random kinds of fluctuations, there is little to be said except that they usually tend to average out in the long run.

It is conceivable that in any given time series which one is studying there are really no systematic movements of any sort, that all the observed fluctuations in the series are in fact irregular ones. Ordinarily, before attempting to measure, say, a trend, we like to test whether or not there actually is a significant movement of this kind in the series. There are several tests of significance which can be applied in these cases. One of them, a test based on runs above and below the median of the data, is described in Technical Note 8 on page 409.

Cyclical variation is sometimes defined as the variation which remains in a time series after the trend, seasonal, and irregular variations have been eliminated. Actually, there is much more to it than that, but in classical time series analysis such a process of elimination is the usual way of measuring (business) cyclical variation, or *business cycles*. Generally speaking, business cycles consist of recurring up and down movements of business or economic activity which differ from seasonal variations in that they extend over longer periods of time and, supposedly, result from an entirely different set of causes. The recurring periods of prosperity, recession, depression, and recovery, which constitute the four phases of a complete business cycle, are considered to be due to other factors than the weather, social customs, and so on, which account for seasonal variations. Because of the great importance of cyclical swings of business to the economic and social life of this country, an enormous amount of effort has been spent in studying the business cycle. Many theories have been proposed to account for it, but no generally accepted explanation of this complicated phenomenon has appeared.

SECULAR TRENDS

The most widely used method of fitting trends to time series is the method of least squares. As we have seen, the problem of fitting a least squares line $y' = a + bx$ is essentially that of determining values of a and b which, for a given set of data, make $\Sigma (y - y')^2$ as small as possible. We can find these two quantities in any problem either by solving the two normal equations on page 351 or by using the special formulas on page 352.

In a time series, however, the x's practically always refer to successive periods (usually years). In this case, the problem of fitting a trend line by the method of least squares can be simplified considerably by performing the following change of scale (coding): Letting x be the variable which measures time and taking the origin (the zero) of the new scale at the middle of the series, that is, at the middle of the x's, we number the years (or other time periods) so that in the new scale $\Sigma x = 0$. If the series has an odd number of years, we assign $x = 0$ to the middle year and number the years ..., $-5, -4, -3, -2, -1, 0, 1, 2, 3, 4, 5, \ldots$. If the series has an even number of years, there is no middle year, and we assign successive years the numbers ..., $-7, -5, -3, -1, 1, 3, 5, 7, \ldots$, with -1 and 1 assigned to the two middle years. When we substitute $\Sigma x = 0$ in the two formulas for a and b, we get

$$a = \frac{\Sigma y}{n} \qquad \text{and} \qquad b = \frac{\Sigma xy}{\Sigma x^2} \qquad\qquad \star$$

and the advantage of this coding is evident.

To illustrate how this simplification works, let us fit a least squares trend

line to the total annual (net) sales of the B. F. Goodrich Company for the
years 1960–1968:

Year	Net Sales (millions of dollars)
1960	764
1961	758
1962	812
1963	829
1964	872
1965	980
1966	1,039
1967	1,006
1968	1,140

Since we have here nine (an odd number of) years, we label them -4, -3, -2, -1, 0, 1, 2, 3, 4, and the sums needed for substitution into the formulas for a and b are obtained in the following table:

Year	x	y	xy	x^2
1960	-4	764	$-3,056$	16
1961	-3	758	$-2,274$	9
1962	-2	812	$-1,624$	4
1963	-1	829	-829	1
1964	0	872	0	0
1965	1	980	980	1
1966	2	1,039	2,078	4
1967	3	1,006	3,018	9
1968	4	1,140	4,560	16
	0	8,200	2,853	60

Substituting $n = 9$, $\Sigma y = 8,200$, $\Sigma xy = 2,853$, and $\Sigma x^2 = 60$ into the new formulas for a and b, we get

$$a = \frac{8,200}{9} = 911.1$$

$$b = \frac{2,853}{60} = 47.6$$

and the equation of the trend line may be written as

$$y' = 911.1 + 47.6x$$

In order to avoid confusion, it is advisable to state precisely the origin of x and the units of both x and y in an explanatory legend added to the trend equation. For our example, we write

$$y' = 911.1 + 47.6x$$

(origin, 1964; x units, 1 year; y, annual
sales in millions of dollars)

This makes it clear that 911.1 is the trend value for 1964 and that the annual trend increment (the year-to-year growth) in the Goodrich annual sales is estimated at \$47.6 million for the given period of time.

Using the equation we have obtained, we can now determine the trend value for any year by substituting the corresponding value of x. For instance, for 1960 we substitute $x = -4$ and get a trend value of $y' = 911.1 + 47.6(-4) = 720.7$, and for 1968 we substitute $x = 4$ and get a trend value of $y' = 911.1 + 47.6(4) = 1,101.5$. Plotting these two trend values and joining them by a straight line, we obtain the least squares trend line shown drawn through the original sales data in Figure 14.5.

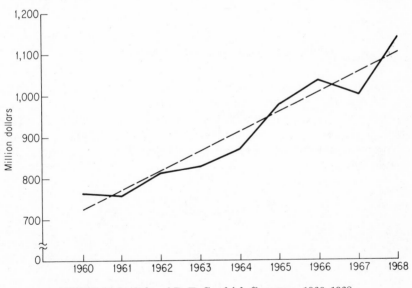

FIGURE 14.5 Sales of B. F. Goodrich Company, 1960–1968

It is sometimes desirable, or necessary, to modify a trend equation in some way. For instance, we might want to modify an equation like the one above so that it could be used with monthly data. This will require changing the units of both x and y, and it will be convenient to shift the origin of the equation to the middle of some month. To illustrate how this is done, let us first change the y units in the above equation from annual sales to *average monthly sales*. Since the average monthly sales figures for the nine years are one twelfth the corresponding total annual sales figures (i.e., $y/12$), we must write a new equation with a and b replaced by $a/12$ and $b/12$. The modified equation and its legend are

$$y' = 75.93 + 3.97x$$

(origin, 1964; x units, 1 year; y, average
monthly sales in millions of dollars)

Now let us modify the trend equation further by changing the x's so that they refer to successive months instead of successive years. Since b measures the *trend increment*, the increase or decrease of trend values corresponding to one unit of x, we shall have to divide b by 12, changing it from an *annual trend increment* to a *monthly trend increment*. Leaving the constant a unchanged, we thus get

$$y' = 75.93 + 0.33x$$

(origin, 1964; x units, 1 month; y, average
monthly sales in millions of dollars)

Finally, let us change the origin of x from the middle of 1964, where it is now, to, say, the middle of January 1964. (This kind of modification will be helpful later in connection with some problems of forecasting.) Since the middle of January 1964 is $5\frac{1}{2}$ months earlier than the middle of the year 1964, we shall have to subtract 5.5 monthly trend increments from the 1964 trend value of 75.93. The new value of a is thus

$$75.93 - 5.5(0.33) = 74.11$$

and we finally get

$$y' = 74.11 + 0.33x$$

(origin, January 1964; x units, 1 month; y,
average monthly sales in millions of dollars)

Earlier we described a secular trend as being indicative of the "general sweep" of the development of a time series. If it is uncertain whether the trend is linear or whether it might be better described by some other kind of curve, if we are not sure whether we are actually dealing with a trend or part of a cycle, and if we are not really interested in obtaining a mathematical equation, we can describe the overall "behavior" of a time series quite well by means of an artificial series called a *moving average*. A moving average is obtained by replacing each value in a series by the mean of itself and some of the values directly preceding and directly following it. For instance, in a three-year moving average calculated for annual data, each annual figure is replaced by the mean of itself and the annual figures corresponding to the two adjacent years; in a five-year moving average each annual figure is replaced by the mean of itself, those of the two preceding years, and those of the two following years. If the averaging is done over an even number of periods, say, 4 years or 12 months, the moving average will initially fall between successive years or months. In such cases, the values are customarily brought "back in line" (or "centered") by taking a subsequent two-year (or two-month) moving average. We shall use this procedure later on in measuring seasonal variation.

The basic problem in constructing a moving average is choosing an appropriate period for the average. This choice depends largely on the nature of the data and the purpose for which the average is constructed. Ordinarily, the purpose of fitting a moving average is to eliminate, insofar as possible, some sort of unwanted or distracting fluctuations in the data. In describing the trend of annual data by a moving average, for example, the main problem is to eliminate those up and down departures of the data from the basic trend which result from business cyclical influences. If all business cycles were exactly alike both in duration and amplitude, their influences could be easily removed from (averaged out of) a series because any absolutely uniform periodic movements are completely eliminated by a moving average whose period is equal to (or a multiple of) the period of the movement. This means also that, if the seasonal movements in a series of monthly data were exactly uniform, the seasonal (and also most of the irregular) variations could be removed from the series by fitting to it a 12-month moving average. However, uniformly periodic cyclical, seasonal, and irregular movements do not appear in economic time series, so the effect of fitting moving averages to series is to *smooth out*, but not eliminate completely, certain fluctuations from the series.

To illustrate the construction of a moving average, let us calculate a five-year moving average to smooth the time series of the production of creosote oil in the United States for the years 1939 through 1968. The original data are shown in the second column of the table on page 389. The third column of the table shows the *five-year moving totals*, which for any given year consists of the sum of that year's figure and those of the two preceding and two succeeding years. The five-year moving averages, shown in the last column, are obtained by dividing each of the corresponding moving totals by five. Both the original data and the five-year moving averages are shown in Figure 14.6, and it is evident that the moving average has substantially reduced the fluctuations in the series and given it a much smoother appearance. It should be noted that in using any sort of moving average we lack some figures at each end of the series, but this is usually no problem unless a series is very short or unless values are needed for each year for further calculations.

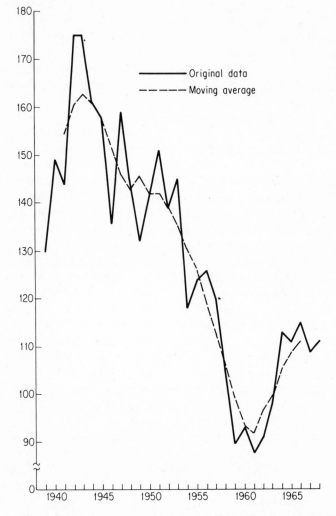

FIGURE 14.6   Creosote oil production in the United States, 1939–1968

Year	Production of Creosote Oil	Five-Year Moving Totals	Five-Year Moving Averages
1939	130		
1940	149		
1941	144	773	154.6
1942	175	804	160.8
1943	175	813	162.6
1944	161	805	161.0
1945	158	789	157.8
1946	136	759	151.8
1947	159	730	146.0
1948	145	714	142.8
1949	132	729	145.8
1950	142	709	141.8
1951	151	709	141.8
1952	139	695	139.0
1953	145	677	135.4
1954	118	652	130.4
1955	124	633	126.6
1956	126	593	118.6
1957	120	565	113.0
1958	105	534	106.8
1959	90	496	99.2
1960	93	467	93.4
1961	88	460	92.0
1962	91	483	96.6
1963	98	501	100.2
1964	113	528	105.6
1965	111	546	109.2
1966	115	559	111.8
1967	109		
1968	111		

EXERCISES

1. The number of major private construction projects begun in the central area of a city during the years 1966–1971 were, respectively, 3, 2, 4, 6, 6, and 8.

   (a) Plot the series on arithmetic paper.

   (b) Fit a least squares straight line to the data and plot the line on the chart showing the original data.

   (c) Modify the least squares equation by shifting the origin of x to the year 1966.

2. The number of manufacturing companies establishing new plants in a certain city during the years 1965–1971 were, respectively, 10, 8, 7, 5, 5, 3, and 2.

(a) Plot the series on arithmetic paper.

(b) Fit a least squares straight line to the data and plot the line on the chart showing the original data.

(c) Modify the least squares equation by shifting the origin of x to the year 1966.

3. The consolidated net sales of the Libbey-Owens-Ford Company for the years 1965–1969 were, respectively, 3.7, 3.9, 3.8, 4.2, and 4.5 (in hundreds of millions of dollars).

(a) Plot the series on arithmetic paper.

(b) Fit a least squares straight line to the series and plot the line on the chart showing the original data.

(c) Modify the trend equation for use with monthly data using January 1965 as the origin of the new equation.

4. The revenues of the Niagara Mohawk Power Corporation from the sales of electricity to other electric systems for the years 1965–1969 were, respectively, 21.0, 18.9, 14.9, 9.3, and 8.0 (in millions of dollars).

(a) Plot the series on arithmetic paper.

(b) Fit a least squares straight line to the series and plot it on the chart with the original data.

(c) Modify the trend equation for use with monthly data using January 1968 as the origin of the new equation.

5. The net investments of the International Business Machines Corportation in plant, rental machines and other property at the end of the years 1962–1970 were, respectively, 1.5, 1.6, 1.7, 2.3, 3.1, 3.5, 3.4, 3.9, and 4.7 (in billions of dollars).

(a) Plot the series on arithmetic paper.

(b) Fit a least squares straight line to the series and plot it on the chart showing the original data.

(c) Modify the trend equation of part (b) by shifting the origin to the (middle of the) year 1969.

(d) Modify the trend equation of part (b) by shifting the origin to the end of the year 1964.

6. Gross income from sales, service and rentals of the International Business Machines Corporation for the years 1961–1970 were, respectively, 2.2, 2.6, 2.9, 3.2, 3.6, 4.2, 5.3, 6.9, 7.2, and 7.5 (in billions of dollars).

(a) Plot the series on arithmetic paper.

(b) Fit a least squares straight line to the series and plot it on the chart showing the original data.

(c) Modify the trend equation of part (b) for use with monthly data using January 1963 as the origin of the new equation.

7. Total annual sales of the Caterpillar Tractor Company for the years 1960–1969 were, respectively, 716, 734, 827, 966, 1,217, 1,405, 1,525, 1,473, 1,707, and 2,002 (in millions of dollars).

   (a) Plot the series on arithmetic paper.

   (b) Fit a least squares straight line to the series and plot it on the chart showing the original data.

   (c) Modify the trend equation of part (b) for use with monthly data using January 1960 as the origin of the new equation.

8. On pages 395–398 the monthly sales of shoe stores are given for the years 1964–1968. Fit a least squares straight line to the five annual totals and then write an equation with origin January 1964 which can be used with the monthly data.

9. The new investments in U.S. government bonds made by U.S. life insurance companies for the years 1950–1969 were, respectively, 2.0, 7.3, 4.3, 3.6, 4.6, 5.7, 4.2, 3.4, 4.8, 4.5, 4.0, 5.5, 5.4, 4.7, 3.9, 3.4, 2.9, 3.1, 4.5, and 3.6 (in billions of dollars).

   (a) Calculate a three-year moving average for this series and plot it and the original series on arithmetic paper.

   (b) Calculate a five-year moving average for this series and compare its graph with that of the three-year moving average.

10. Construct a three-year moving average for the creosote oil production series shown on page 389. Plot the original series, the three-year moving average, and the five-year moving average shown on page 389 and compare the graphs of the two averages.

11. The following figures are the U.S. total imports of green (raw) coffee (in millions of bags) for the years 1939–1968: 15.2, 15.5, 17.0, 13.0, 16.6, 19.7, 20.5, 20.6, 18.9, 20.9, 22.1, 18.4, 20.3, 20.3, 21.0, 17.1, 19.6, 21.3, 20.9, 20.2, 23.2, 22.1, 22.3, 24.5, 23.8, 22.8, 21.3, 22.1, 21.3, and 25.4.

   (a) Construct a three-year moving average and plot it on a chart showing the original data.

   (b) Construct a five-year moving average, plot it on the chart obtained in part (a), and compare the graphs of the two averages.

12. The following figures show the U.S. production of rice (in millions of bags) for the years 1939–1968: 24, 24, 23, 29, 29, 31, 31, 32, 35, 38, 41, 39, 46, 48, 53, 64, 56, 49, 43, 45, 54, 55, 54, 66, 70, 73, 76, 85, 89, and 105.

   (a) Construct a five-year moving average and plot it on a chart showing the original data.

   (b) Construct a seven-year moving average, plot it on the chart obtained in part (a), and compare the graphs of the two averages.

(The exercises which follow are based on the material in Technical Note 8.)

13. The number of net development oil wells completed by the Union Oil Company for the years 1960–1969 were, respectively, 200, 161, 174, 213, 208, 264, 179,

194, 200, and 180. Test at a level of 0.05 whether there is a significant trend in this series.

14. The number of net exploratory oil wells completed by the Union Oil Company for the years 1960–1969 were, respectively, 20, 13, 12, 11, 17, 17, 10, 8, 10, and 5. Test at a level of significance of 0.05 whether there is a real trend in this series.

15. The total revenues of the Union Oil Company for the years 1960–1969 were, respectively, 1,190, 1,232, 1,284, 1,363, 1,450, 1,508, 1,656, 1,703, 1,852, and 2,002 (in millions of dollars). Test at the 0.01 level of significance whether there is a real trend in this series.

16. The total number of restaurants opening for business and also quitting business within the calendar years 1950–1971 in a large city were, respectively, 107, 105, 123, 140, 145, 120, 114, 151, 142, 160, 141, 124, 143, 127, 132, 135, 141, 146, 150, 123, 104, and 110. At the 5 per cent level of significance, is there a real trend in this series?

17. Test, at a level of significance of 0.05, whether there is a real trend in
   (a) the new investments series of Exercise 9 this set;
   (b) the coffee imports series of Exercise 11 this set;
   (c) the creosote oil production series shown on page 389.

(The exercises which follow are based on the material in Technical Note 7.)

18. A technological research and development firm founded in 1956 has received the following contract awards (in thousands of dollars) for the years 1956–1970: 95, 106, 93, 102, 119, 135, 210, 200, 198, 197, 245, 240, 275, 325, and 340.
   (a) Plot this series on arithmetic paper.
   (b) Verify that a parabola fit to this series by the method of least squares is $y' = 179.53 + 17.79x + 0.6678x^2$ (origin, 1963; x units, 1 year; y, total annual contract awards in thousands of dollars). In this equation $a = 179.53$ is the trend value for 1963, $b = 17.79$ is the slope of the curve at $x = 0$, and $2c = 1.3356$ is the rate at which the slope changes at this particular point.
   (c) Set up a table with columns for the year, x, $a + bx$, cx^2, and y'. Calculate $a + bx$ and cx^2 separately for each x, then add them to get the annual trend values y'. Plot the trend values on the chart showing the original data and draw a smooth curve through them to indicate the parabolic trend.

19. The contract award series of Exercise 18 straightens out quite well when plotted on ratio paper.
   (a) Taking the logarithms of the values of this series from Table XI, verify that a straight line fit to these logarithms is $\log y' = 2.2435 + 0.0424x$ (origin, 1963; x units, 1 year; y, total annual contract awards in thousands of dollars).
   (b) Find the logarithms of the trend values for the years 1956–1970 by substituting the corresponding values of x in the equation given in part (a),

then find the trend values themselves by looking up in Table XI the numbers corresponding to these logarithms. Plot the trend values on the chart drawn in part (c) of Exercise 18 and draw a smooth curve through them to indicate the exponential trend. Compare the parabolic and exponential trends visually.

Looking up in Table XI the numbers whose logarithms are 2.2435 and 0.0424, the values of log a and log b, we can write the equation given in part (a) as $y' = 175(1.10)^x$ (where the legend is the same). In this form 175 is the trend value for 1963, and 1.10 is 1 plus the average annual rate of growth in contract awards over the given period. Hence, the company has experienced an average annual growth of 0.10, or 10 per cent, in contracts awarded over this 15-year period.

SEASONAL VARIATION

Let us turn now to the problem of measuring those movements in a time series which recur more or less regularly in the same months of successive years. We call the measure of this *seasonal variation* an *index of seasonal variation*, or a *seasonal index*. For monthly data a seasonal index consists of 12 numbers, one for each month, each of which expresses that particular month's activity as a percentage of the average month's activity. For instance, if the June seasonal index of sales of a wholesaler is 92, this means that June sales are typically 92 per cent of sales in the average month. We use the word "typically" here because the actual percentage for a given month varies more or less widely from year to year, and 92 per cent is an average of these percentages.

There are many ways in which seasonal variation can be measured or a seasonal index can be constructed. These range from rather crude measures based on very simple calculations to highly refined measures based on involved computer techniques. We shall illustrate the construction of a seasonal index by using the so-called *ratio-to-moving-average*, or *percentage-of-moving-average*, method. Until certain refinements were made possible by the use of high-speed computers, this method was probably the most widely used and the most generally satisfactory one available.

In constructing a seasonal index all our efforts are aimed at eliminating trend, cyclical, and irregular variations from the series. The way this is done in the basic ratio-to-moving-average method is relatively simple. We begin by calculating a 12-month moving average of the data in order to remove the seasonal movements from the series. Since an n-period moving average will completely eliminate any absolutely uniform n-period recurring movement, a 12-month moving average would eliminate all the seasonal movements from the series, provided these movements recurred with complete regularity year after year. Of course, in actual practice seasonal patterns vary somewhat from year to year, so the moving average cannot be expected to eliminate all of the seasonal variation. It will eliminate most of it, however, as well as most of the irregular variation, so the 12-month moving average is an estimate of the trend and cyclical components of the series. In the classical model we are discussing, each value in the original series is assumed to be the product of factors attributed to the four basic components (secular trend, seasonal variation, cyclical variation, and irregular variation). Therefore, dividing each value by the corresponding value of the 12-month moving average gives an estimate of the seasonal and irregular components in the series. In other words, dividing $T \cdot S \cdot C \cdot I$ by $T \cdot C$ leaves us with $S \cdot I$, the product of the factors attributed to seasonal and irregular variations. All that is left to do then is to eliminate, insofar as possible, the irregular fluctuations.

When one knows a good deal about the series under study, it may be possible to identify and eliminate directly the monthly $S \cdot I$ values which reflect the impact of extraordinary events (e.g., a crippling nine-day tule fog, or a strike in a supplier's plant). Irregular variations of this sort, as well as those which are due to chance (nonassignable causes) can also be effectively eliminated by averaging, in some way, the $S \cdot I$ figures for the different Januaries, for the different Februaries, and so on. We can, for instance, reduce the effect of the irregular forces by using the median of the values given for each month, or perhaps by using the *modified (arithmetic)* mean which is the mean of the values remaining after the smallest and largest values have been cast out. Moving averages can also be used for

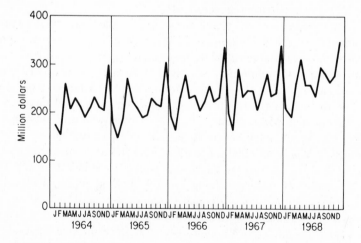

FIGURE 14.7 Monthly sales of shoe stores in the United States, 1964–1968

smoothing out the irregular variations remaining at this stage of the calculation. In any case, by some sort of averaging process we finally arrive at an estimate of the way seasonal factors alone influence the values of a series. This estimate, consisting of the 12 monthly values, is the seasonal index.

The series we have chosen to illustrate the calculation of a seasonal index by the ratio-to-moving-average method is the sales of shoe stores in the United States. The original data, plotted in Figure 14.7 and given in Column 1 of the following table, are the monthly sales in millions of dollars for the years 1964–1968. The pronounced seasonal pattern in the sales and an upward trend are evident from the graph.

SALES OF SHOE STORES
(*millions of dollars*)

Year/Month	Sales (1)	12-Month Moving Total (2)	2-Month Moving Total (3)	Centered 12-Month Moving Average (4)	Percentages of 12-Month Moving Average (5)
1964 January	174				
February	154				
March	262				
April	208				
May	229				
June	214				
		2,591			
July	192		5,188	216.2	88.8
		2,597			
August	209		5,190	216.3	96.6
		2,593			
September	231		5,111	213.0	108.5
		2,518			
October	211		5,099	212.5	99.3
		2,581			
November	207		5,155	214.8	96.4
		2,574			
December	300		5,143	214.3	140.0
		2,569			
1965 January	180		5,137	214.0	84.1
		2,568			

SALES OF SHOE STORES (*Continued*)

Year/Month	Sales (1)	12-Month Moving Total (2)	2-Month Moving Total (3)	Centered 12-Month Moving Average (4)	Percentages of 12-Month Moving Average (5)
February	150		5,122	213.4	70.3
		2,554			
March	187		5,107	212.8	87.9
		2,553			
April	271		5,112	213.0	127.2
		2,559			
May	222		5,124	213.5	104.0
		2,565			
June	209		5,136	214.0	97.7
		2,571			
July	191		5,155	214.8	88.9
		2,584			
August	195		5,186	216.1	90.2
		2,602			
September	230		5,250	218.8	105.1
		2,648			
October	217		5,304	221.0	98.2
		2,656			
November	213		5,321	221.7	96.1
		2,665			
December	306		5,356	223.2	137.1
		2,691			
1966 January	193		5,395	224.8	85.9
		2,704			
February	168		5,435	226.5	74.2
		2,731			
March	233		5,487	228.6	101.9
		2,756			
April	279		5,518	229.9	121.4
		2,762			
May	231		5,542	230.9	100.0
		2,780			
June	235		5,591	233.0	100.9
		2,811			
July	204		5,628	234.5	87.0
		2,817			
August	222		5,634	234.8	94.5
		2,817			
September	255		5,692	237.2	107.5
		2,875			

SALES OF SHOE STORES (*Continued*)

Year/Month		Sales (1)	12-Month Moving Total (2)	2-Month Moving Total (3)	Centered 12-Month Moving Average (4)	Percentages of 12-Month Moving Average (5)
	October	223		5,707	237.8	93.8
			2,832			
	November	231		5,681	236.7	97.6
			2,849			
	December	337		5,710	237.9	141.7
			2,861			
1967	January	199		5,727	238.6	83.4
			2,866			
	February	168		5,750	239.6	70.1
			2,884			
	March	291		5,794	241.4	120.5
			2,910			
	April	236		5,830	242.9	97.2
			2,920			
	May	248		5,851	243.8	101.7
			2,931			
	June	247		5,865	244.4	101.1
			2,934			
	July	209		5,879	245.0	85.3
			2,945			
	August	240		5,915	246.5	97.4
			2,970			
	September	281		5,912	246.3	114.1
			2,942			
	October	233		5,960	248.3	93.8
			3,018			
	November	242		6,047	252.0	96.0
			3,029			
	December	340		6,070	252.9	134.4
			3,041			
1968	January	210		6,109	254.5	82.5
			3,068			
	February	193		6,191	258.0	74.8
			3,123			
	March	263		6,249	260.4	101.0
			3,126			
	April	312		6,284	261.8	119.2
			3,158			
	May	259		6,351	264.6	97.9
			3,193			

SALES OF SHOE STORES (*Continued*)

Year/Month	Sales (1)	12-Month Moving Total (2)	2-Month Moving Total (3)	Centered 12-Month Moving Average (4)	Percentages of 12-Month Moving Average (5)
June	259		6,395	266.5	97.2
		3,202			
July	236				
August	295				
September	284				
October	265				
November	277				
December	349				

The first step in the calculations is to obtain the 12-month moving totals, which are shown in Column 2. The first entry in this column, 2,591, is the sum of the 12 monthly sales figures for 1964, and it is recorded at the middle of the period, between June and July 1964. The second entry in this column, 2,597, is obtained by subtracting the January 1964 figure from 2,591 and adding the January 1965 figure; in other words, 2,597 is the sum of the 12 monthly sales from February 1964 through January 1965, and it is recorded at the middle of this period. The third and succeeding entries in the column are found by continuing this process of subtracting and adding monthly values.

In order to obtain a 12-month moving average which is centered on the original data, we next calculate two-month moving totals of the entries of Column 2. These are shown in Column 3, with the first entry being the sum of the first two values in Column 2, the second entry being the sum of the second and third values in Column 2, and so on. These entries in Column 3 are recorded between those of Column 2, and hence, they are in line with (or centered on) the original data.

Since each entry of Column 2 is the sum of 12 monthly figures and each entry of Column 3 is the sum of two entries of Column 2, or altogether the sum of 24 monthly figures, we finally obtain the centered 12-month moving average shown in Column 4 by dividing each entry of Column 3 by 24. These moving average values are the trend-cycle estimates, and we use them now to eliminate the $T \cdot C$ components from the original series. This is done by dividing the original $T \cdot S \cdot C \cdot I$ data month by month by the corresponding $T \cdot C$ estimates (that is, the corresponding values of the moving average) and multiplying these ratios by 100. In this way, we arrive at the percentages of moving average shown in Column 5.

All that remains to be done is to eliminate the irregular variations as best we can, and to this end we rearrange the entries of Column 5 as in the first five columns of the following table:

Month	1964	1965	1966	1967	1968	Median	Seasonal index
January		84.1	85.9	83.4	82.5	83.8	83.8
February		70.3	74.2	70.1	74.8	72.2	72.2
March		87.9	101.9	120.5	101.0	101.4	101.4
April		127.2	121.4	97.2	119.2	120.3	120.3
May		104.0	100.0	101.7	97.9	100.4	100.4
June		97.7	100.9	101.1	97.2	99.3	99.3
July	88.8	88.9	87.0	85.3		87.9	87.9
August	96.6	90.2	94.5	97.4		95.6	95.6
September	108.5	105.1	107.5	114.1		108.0	108.0
October	99.3	98.2	93.8	93.8		96.0	96.0
November	96.4	96.1	97.6	96.0		96.2	96.2
December	140.0	137.1	141.7	134.4		138.6	138.6
						1,199.7	1,200.

Of the various ways in which we can average the figures given for each month, we choose the median here. (In this case where we have four values for each month, the median is, in fact, equivalent to the modified mean.) The 12 medians are shown in the second column from the right. Now, since the seasonal index for each month is supposed to be a percentage of the average month, the sum of the 12 values should equal 1,200. Actually, the medians total 1,199.7, and so we adjust for this by multiplying each of the medians by $\dfrac{1,200}{1,199.7} = 1.00025$. In some cases this adjustment is substantial (see Exercise 1 on page 404), but here it has no effect since we are carrying the index out to only one decimal place. Consequently, the final values of the seasonal index, shown in the last column, are the same as the 12 medians.

The interpretation of this index is straightforward. For instance, January sales of shoe stores are typically 83.8 per cent of those of the average month; sales are usually relatively low in February, and December sales are typically 38.6 per cent above sales of the average month.

In using a seasonal index for any purpose, we must always be mindful of its limitations. An index is based on historical (past) data, and we cannot reasonably expect seasonal patterns to remain completely constant over long periods of time. The method we have illustrated here applies to the description of constant seasonal patterns or seasonal patterns which do not change very much. If there are pronounced changes in the seasonal pattern with the passage of time, the sort of index we have discussed will not be suitable.

Seasonal indexes are extremely important in various practical applications. We shall briefly explain two of these, the first in *deseasonalizing* data and the second in *forecasting*. Leaving the use of seasonal indexes in forecasting to the section which follows, let us now describe the process of removing seasonal influences from a given set of data, or deseasonalizing a series. Of course, when seasonal fluctuations have actually occurred, nobody knows how things would have been if a series had been uninfluenced by seasonal factors. So, when we speak of what things would have been like without seasonal fluctuations, we are speaking rather loosely. Nevertheless, the notion of a series of data free from seasonal influences is a useful way of understanding the concept.

The process of removing seasonal variation, or deseasonalizing data, consists merely of dividing each value in a series by the corresponding value of the seasonal index and multiplying the result by 100 (or by dividing by the corresponding value of the seasonal index written as a proportion). The logic of this process is quite simple: If April shoe sales are 120 per cent of those in the average month, taking $\frac{100}{120} \cdot 100 = 83$ per cent of the April sales would tell us what these sales should have been if there had been no seasonal variation.

We shall illustrate this process by deseasonalizing the 1968 sales of shoe stores, using the seasonal index obtained above. In the table on page 401, the sales figures and the seasonal index are copied from pages 397 and 399, with the values of the seasonal index given as proportions and rounded to two decimals. The values in the right-hand column are the deseasonalized sales, and they are computed by dividing each month's actual sales by the corresponding value of the seasonal index.

Inspecting this table we discover several interesting facts. For instance, there was an increase of $70 million in sales from February to March. This is nothing to rejoice over, however, since the deseasonalized data show that this increase is actually less than might have been expected in accordance with typical seasonal patterns. Also, there was a drop of $19 million in sales from September to October. This is no cause for alarm, though, since the deseasonalized data show that this drop is actually less than might be expected in accordance with typical seasonal patterns. The need for taking

1968	Sales	Seasonal Index	Deseasonalized Sales
January	210	0.84	250
February	193	0.72	268
March	263	1.01	260
April	312	1.20	260
May	259	1.00	259
June	259	0.99	262
July	236	0.88	268
August	295	0.96	307
September	284	1.08	263
October	265	0.96	276
November	277	0.96	289
December	349	1.39	251

seasonal variation into account in the analysis of business and economic time series should be apparent from these observations.

The principal source of seasonally adjusted (deseasonalized) series is the Federal government. In constructing these series, seasonal indexes are derived by highly sophisticated computer methods which are essentially a refinement and an extension of the basic ratio-to-moving-average method. The computational burden in this work is tremendous, and without the great arithmetic ability of digital computers, the necessary calculations could never be made for a large number of series within any reasonable limits of time and cost. The Bureau of the Census, however, has developed computer methods to the point where it can now virtually mass produce seasonally adjusted series of a high order of quality for almost all important series of data. Other computer approaches to the seasonal adjustment problem, including further refinements in the ratio-to-moving-average technique and, what is an entirely different concept, seasonal adjustments based on regression analysis, are being studied, and these may result in still further improvement in seasonally adjusted series.

FORECASTING—II

Now that we are familiar with time series and know how to measure some of their components, let us discuss briefly the tremendously complicated problem of business forecasting. The rationale of basing forecasts on time series is that, having observed some regularity in the movement of data through time, we are hopeful that "what has happened in the past will, to a greater or lesser extent, continue to happen or will happen again in the future." Thus, the obvious way to forecast the trend of a time series is to extrapolate from the trend equation describing the historical data. By "extrapolate" we mean extend the trend into the future so as to estimate a value that lies beyond the range of the values used to derive the trend equation.

To illustrate this, suppose that we are given the following trend equation which describes the long-term growth of a discount department store in a large city:

$$y' = 7,200 + 360x$$

(origin, 1965; x units, 1 year; y, total
annual sales in thousands of dollars)

This equation has been calculated by the method of least squares from the 1957–1970 sales data, taken from the store's records. The problem is to estimate both total and monthly sales for the year 1972.

In 1965 the trend value is \$7,200,000 and the growth forces are estimated to be producing an increase in sales of \$360,000 per year. Thus, we shall extrapolate by substituting $x = 7$ into the equation and say that, based on the long-term growth forces alone, 1972 total annual sales are expected to be $y' = 7,200 + 360(7) = 9,720$, or \$9,720,000.

If the sales of the store were uninfluenced by seasonal fluctuations, we could estimate monthly sales in precisely the same way. Modifying the trend equation for use with monthly data by dividing a by 12 and b by 144 and then shifting the origin to January 1965, we get

$$y' = 586.25 + 2.5x$$

(origin, January 1965; x units, 1 month; y, average
monthly sales in thousands of dollars)

Substituting $x = 84, 85, 86, \ldots$, and 95 into this last equation and solving for y' we get the 12 estimated sales values (shown in the first column of the table on page 403) for the months January, February, $\ldots$, and December 1972. However, there is a very pronounced seasonal variation in the store's sales. In fact, a seasonal index calculated from recent historical data by the ratio-to-moving average method shows that typical January sales are 82 per cent, February sales are only 74 per cent, $\ldots$, and December sales are 138 per cent of what they would be if there were no seasonal variation.

Assuming that the seasonal pattern is not changing and is adequately defined by the seasonal index, we shall multiply the trend values, month

Month	Trend Values	Seasonal Index	Predicted Monthly Sales for 1972
January	796.25	0.82	652,925
February	798.75	0.74	591,075
March	801.25	0.98	785,225
April	803.75	1.06	851,975
May	806.25	1.11	894,937
June	808.75	0.95	768,313
July	811.25	0.84	681,450
August	813.75	0.90	732,375
September	816.25	1.02	832,575
October	818.75	1.06	867,875
November	821.25	1.14	936,225
December	823.75	1.38	1,136,775

by month, by the corresponding seasonal index values (shown in the second column of the table); that is, we multiply the January trend figure by 0.82, the February figure by 0.74, and so on, getting the 1972 predicted monthly sales (shown in the right-hand column of the table). If there were no seasonal variation, we would expect December sales to be about 3.5 per cent higher than January sales because of the upward trend in sales; taking into account the seasonal influence, though, December sales are projected at 74 per cent higher than January sales.

What we have done in arriving at these monthly sales predictions is precisely the opposite of deseasonalizing data. We have introduced the seasonal patterns into the data (rather than removed them from it) by multiplying the trend values by the corresponding values of the seasonal index, written as proportions. These products of a measure of the trend and of the seasonal, or $T \cdot S$, are the values we would expect if trend and seasonal forces were the only factors influencing the values of a series, and they are often called the *normal* values.

In the predicted sales values above, we have taken into account the effect of trend and seasonal patterns on the store's sales, but we have not yet considered the possible effects of cyclical and irregular influences. The latter, we have said, are essentially unpredictable; in the store's case, a large freight shipment lost in transit for five weeks, for instance, or a fire which closes a competing store for six months cannot be foreseen. The effects of such events may tend to average out in the long run, but they can substantially affect sales in particular months and cause even the most careful forecasts to go astray. In connection with the major problems of forecasting both short-run and long-run business cycles and their effect at the level of the economy as a whole and at industry and firm levels, we shall only remark that much help is available to everyone from both private and public sources. One very important aid is the Bureau of the Census monthly report *Business Conditions Digest*. Among the wealth of economic data contained in this report are various *economic indicators*. These are series of data which tend to turn up or down before overall economic activity does (the *leading indicators*), series which tend to move at about the same time as overall activity (the *roughly coincident indicators*), and series which tend to move somewhat behind overall activity (the *lagging indicators*) and hence to confirm or refute the earlier directional signals.

EXERCISES

1. In constructing a seasonal index for the sales of a branch department store by the ratio-to-moving-average method, a statistician has arrived at the following medians (in the same way we arrived at the values in the column entitled *Median* on page 399): 78.4, 75.5, 93.5, 96.1, 103.9, 96.4, 83.8, 87.9, 100.6, 120.1, 111.7, and 134.3. Complete the calculation of the seasonal index.

2. The following data show the monthly demand for kerosene in the United States (in millions of barrels) for the years 1964–1969:

Year	Jan	Feb	Mar	Apr	May	Jun	Jul	Aug	Sep	Oct	Nov	Dec
1964	14.7	11.5	8.3	5.9	4.4	3.2	4.6	5.2	6.4	7.9	7.5	13.3
1965	13.0	12.0	11.0	6.3	4.3	4.6	4.9	5.9	6.0	7.7	9.4	12.7
1966	14.1	12.1	8.7	6.1	5.9	4.9	4.6	5.9	7.5	7.9	10.7	13.0
1967	13.6	12.4	9.6	5.7	6.2	4.3	5.5	6.1	7.1	7.7	10.5	11.4
1968	16.3	12.2	9.7	5.6	5.9	4.8	4.3	6.2	6.6	7.8	10.5	13.5
1969	15.5	11.9	10.2	5.8	5.5	4.5	5.6	5.2	7.3	7.1	9.3	12.6

Compute a seasonal index for this series by the ratio-to-moving-average method, using the median to average the percentages of moving average obtained for the individual months.

3. Use the seasonal index obtained in Exercise 2 to deseasonalize the 1964–1969 monthly kerosene demand shown in the exercise. Plot the deseasonalized data together with the original data on one chart.

4. Referring to Exercise 2, fit a trend equation by the method of least squares to the total annual kerosene demand for the years 1964–1969. Use this equation (suitably modified) and the seasonal index computed in Exercise 2 to forecast the monthly demand for kerosene in the year 1975.

5. The following data show the numbers of privately owned new housing units (in thousands of houses) started each month in the United States for the years 1964–1969:

Year	Jan	Feb	Mar	Apr	May	Jun	Jul	Aug	Sep	Oct	Nov	Dec
1964	95	102	128	142	153	158	142	137	119	144	112	96
1965	82	81	120	149	153	152	139	128	125	133	110	101
1966	79	76	118	141	130	121	99	102	89	77	73	60
1967	59	61	92	114	132	125	125	127	122	135	118	80
1968	80	85	127	162	141	138	140	137	134	141	127	96
1969	102	90	132	159	156	147	125	125	129	122	94	84

Compute a seasonal index for these data, using the modified mean (see page 394) to average the percentages of moving average obtained for the different months.

6. Use the seasonal index obtained in Exercise 5 to deseasonalize the 1969 data on new housing starts.

7. Referring to Exercise 5, fit a trend equation by the method of least squares to the total annual new housing starts for the years 1964–1969. Use this equation (suitably modified) and the seasonal index computed in Exercise 5 to forecast the monthly number of new housing starts in the year 1974.

8. Often, deseasonalized monthly data are multiplied by 12 and then referred to as *annual rates*. The use of seasonally adjusted annual rates is particularly helpful in facilitating the analysis of month-to-month changes in series of data which are best understood on an annual basis. Accordingly, we often see reported such statements as, "Americans had more money income in June on an annual rate basis than in any other month in history." Based on the seasonally adjusted figure for June 1969 obtained in Exercise 6, at what annual rate were new housing starts then running?

9. The following data show the total monthly cash receipts from farm marketings in the United States (in billions of dollars) for the years 1965–1969:

Year	Jan	Feb	Mar	Apr	May	Jun	Jul	Aug	Sep	Oct	Nov	Dec
1965	3.3	2.4	2.5	2.5	2.5	2.9	3.0	3.3	3.7	4.8	4.6	4.0
1966	3.7	2.9	3.0	2.8	2.9	3.2	3.3	3.8	4.0	4.9	4.9	3.4
1967	3.6	2.8	3.0	2.7	2.9	3.2	3.4	3.7	3.9	4.8	4.7	4.0
1968	3.7	2.9	2.9	2.8	3.0	3.1	3.5	3.7	4.3	5.3	4.9	4.2
1969	3.8	3.1	3.3	3.2	3.3	3.4	4.2	5.2	5.8	5.8	5.1	4.6

(a) Compute a seasonal index by the ratio-to-moving-average method, using the median to average the percentages of moving averages obtained for the individual months.

(b) Plot the original data together with the values of the centered 12-month moving average on one diagram.

(c) Use the seasonal index obtained in part (a) to deseasonalize the 1969 data.

(d) At what annual rate (see Exercise 8) were cash receipts running in March 1969?

(e) Fit a trend equation by the method of least squares to the total annual cash receipts from farm marketings for the years 1965–1969. Use this equation (suitably modified) and the seasonal index computed in part (a) to forecast the monthly cash receipts in the year 1975.

10. A building materials supply company had sales of $30,000 and $36,000 in April and May of a certain year. The company's seasonal index stands at 105 and 140 for these months. The president of the company said the May sales were a disaster but the sales manager said that he was quite pleased with the rise above the April sales. How should the president set the sales manager straight in his thinking? The sales manager also predicted on the basis of the May sales that total sales for the year would be $432,000, but the president predicted sales of only about $309,000. Criticize the sales manager's estimate and explain how the president may have arrived at his figure.

11. The equation describing the trend in sales of a large paint manufacturer is $y' = 14 + 2x$ (origin, 1967; x units, 1 year; y, total annual sales in millions of dollars). The company's seasonal index of sales is 78, 75, 100, 126, 138, 121, 101, 104, 99, 103, 80, and 75 for the months of January–December.

(a) Ignoring the existence of a trend during the year 1972, draw up a tentative monthly sales forecast for the company for 1972.

(b) Taking into account the trend during the year 1972, draw up a monthly sales forecast for the company for 1972.

12. Suppose that the equation which describes the trend in sales of a branch department store in a large metropolitan shopping center is $y' = 1,440 + 72x$ (origin, 1967; x units, 1 year; y, total annual sales in thousands of dollars). The seasonal index of sales is 80, 75, 95, 98, 107, 96, 82, 89, 104, 123, 110, and 141 for the months of January–December. Draw up a monthly sales forecast for the store for 1972.

13. In a study of its sales, a manufacturer of vending machines obtained the following least squares trend equation: $y' = 3,200 + 250x$ (origin, 1965; x units, 1 year; y, total number of machines sold annually). The company has physical facilities to produce 5,450 machines per year, and it believes that, at least for the next decade, its sales trend will continue as before.

(a) What is the average annual increase in number of machines sold?

(b) By what year will the company's expected sales have equaled its present physical capacity?

(c) How much in excess of the company's present capacity is the estimated sales figure for 1977?

A WORD OF CAUTION

The real trouble in forecasting and planning for the future is that there are altogether too many variables which need to be taken into account.

Some of these are at least quantifiable and essentially predictable, but some are not. In analyzing rates of return on various capital investments, for instance, such inputs to the problem as the net installed cost of a machine and earnings from new equipment can often be determined or estimated reasonably well. But sound capital budgeting forces companies also to take into account such things as government policies which might affect the availability and price of money, changes in leasing opportunities, and possible revisions of depreciation and depletion laws as well as future tax decisions on these rates and allowances. Waiting is not much help, either, since there is no possibility of waiting until all, or sometimes even any, of the basic questions are settled finally. The projection of past experience to the uncertain future is speculative and hazardous, but at some time decisions must be made on the basis of the available incomplete knowledge; otherwise, nothing gets done.

Except for a few irreversible decisions, however, no one is irrevocably committed to a forecast, to survive or perish with it once it has been made. In the illustration of the preceding section, for instance, the store may have to make adjustments to take account of improved area transportation facilities, the opening of a new competing store nearby, an increase in the sales tax, and other things which could not be foreseen at the time the forecast was made. Actually, forecasts are tentative things—special kinds of hypotheses, so to speak—which can be modified or revised in response to changing conditions. When forecasts are revised in the light of new information, all those concerned must take whatever steps are necessary to translate revised production, sales, or other goals into action. Intelligent forecasting and planning demand one's continual attention to changing conditions.

Generally speaking, it seems clear that realistic forecasts, which contribute greatly both to individual success and to the stability of the economy, are the results of applying sound business experience and judgment to relevant and timely statistical analyses.

TECHNICAL NOTE 7 (Parabolic and Exponential Trends)

When the data seem to depart more or less widely from linearity in regression or time series analysis, we must consider fitting some other curve than a straight line. One of the most useful of these other curves is the *second-degree polynomial* (or *parabola*) whose equation is $y' = a + bx + cx^2$. In fitting this curve by the method of least squares, we are required to determine a, b, and c so that $\Sigma (y - y')^2$ is a minimum. In general, this requires solving the following set of three normal equations

$$\Sigma\, y = na + b(\Sigma\, x) + c(\Sigma\, x^2)$$
$$\Sigma\, xy = a(\Sigma\, x) + b(\Sigma\, x^2) + c(\Sigma\, x^3) \qquad \star$$
$$\Sigma\, x^2y = a(\Sigma\, x^2) + b(\Sigma\, x^3) + c(\Sigma\, x^4)$$

but when dealing with equally spaced series of data, the work may be simplified by using the same convention we have illustrated in fitting linear trend lines to time series. Putting the zero of a new scale at the middle of the series of data will make $\Sigma\, x = 0$ and $\Sigma\, x^3 = 0$, and the normal equations reduce to

$$\Sigma\, y = na + c(\Sigma\, x^2)$$
$$\Sigma\, xy = b(\Sigma\, x^2) \qquad \star$$
$$\Sigma\, x^2y = a(\Sigma\, x^2) + c(\Sigma\, x^4)$$

We can then find b directly from the second equation, $b = \dfrac{\Sigma\, xy}{\Sigma\, x^2}$, and we can find a and c by solving the first and third equations simultaneously. Polynomial equations of higher degree in x than two, such as $y' = a + bx + cx^2 + dx^3$, can also be fitted by the method of least squares. Where x is equally spaced, special methods can be used to simplify the calculations.

Often a set of data which does not seem linear when plotted on ordinary graph paper (arithmetic paper) appears to "straighten out" when plotted on paper with a logarithmic vertical scale (semi-log or ratio paper). On arithmetic paper, equal intervals on the vertical scale represent equal amounts of change, and the y' values calculated from the equation $y' = a + bx$ plot as a straight line on arithmetic paper. On ratio paper, however, equal intervals on the vertical scale represent equal rates of change, and the y' values calculated from the equation $y' = ab^x$ plot as a straight line on ratio paper. This latter curve is called *exponential* since the x appears in the equation as the exponent of b, and the trends of time series which appear linear when plotted on ratio paper are called *exponential trends*.

Taking the logarithm of both sides of the equation $y' = ab^x$ we have

$$\log y' = \log a + x \cdot \log b$$

which is a linear equation in x and $\log y'$. (Writing A, B, and Y for $\log a$, $\log b$, and $\log y'$, the equation becomes $Y = A + Bx$ which is the usual equation of a straight line.) In order to fit an exponential trend by the method of least squares (that is, to fit a straight line to the logarithms of the y values), we can find numerical values for $\log a$ and $\log b$ from the formulas

$$\log a = \frac{\Sigma \log y}{n} \quad \text{and} \quad \log b = \frac{\Sigma\,(x \cdot \log y)}{\Sigma x^2} \qquad \bigstar$$

provided by the usual change of scale $\Sigma\,x = 0$, and then find a and b. The work proceeds exactly the same as when fitting a straight line to the y values themselves with the exception that $\log y$ is used instead of y.

Data which seem to be nonlinear when plotted on either arithmetic or ratio paper may sometimes straighten out reasonably well when plotted on paper with logarithmic scales for both x and y (log-log paper). In this case we may fit a *power function* of the form $y' = ax^b$ which, when written in its logarithmic form $\log y' = \log a + b \cdot \log x$, is seen to be a linear equation in $\log x$ and $\log y'$. This curve is more widely used in regression and correlation work than in time series analysis, however. In marine resources studies, for example, the mean weights of albacore have been estimated from length samples using regression equations of log weight on log length.

TECHNICAL NOTE 8 (Runs Above and Below the Median)

There are several tests that can be used to determine whether an apparent trend in a series of data is significant or whether it can be attributed to chance. The one we shall discuss in this note is based on the idea that if there is an *upward trend* most of the small values come first and the large values come later, while if there is a *downward trend* most of the large values come first and the small values come later. To make the distinction between large values and small values precise, we divide the data into values falling *above the median* (represented by the letter a) and values falling *below the median* (represented by the letter b). We then base our decision on the resulting sequences of a's and b's; specifically, we base it on the *total number of runs* of a's and b's, where a run is defined as a sequence of identical letters (or other symbols) which is followed and preceded by different letters or no letters at all.

To illustrate this procedure, let us decide whether there is an upward trend in the purchases of group life insurance certificates in the United States for the years 1950–1969. The data are shown on the following page; their median is

$$\frac{3{,}498 + 3{,}534}{2} = 3{,}516$$

and we have indicated for each year's figure whether it is above or below the median. (If one of the figures had actually equaled the median, we would have omitted it.)

| | Number of policies | |
Year	(000 omitted)	
1950	2,631	b⎫
1951	1,629	b⎪
1952	2,039	b⎬
1953	2,195	b⎭
1954	3,611	a}
1955	2,217	b⎫
1956	2,935	b⎪
1957	3,035	b⎬
1958	2,999	b⎪
1959	3,090	b⎭
1960	3,734	a⎫
1961	3,971	a⎭
1962	3,498	b}
1963	3,534	a⎫
1964	4,225	a⎪
1965	7,007	a⎪
1966	4,055	a⎬
1967	4,353	a⎪
1968	4,875	a⎪
1969	5,068	a⎭

Inspecting the table, we find that there is first a run of four b's, followed by a run of one a, a run of five b's, a run of two a's, a run of one b, and finally a run of seven a's. The fact that there are relatively few runs, with most of the b's coming first, suggests that there is a trend in the purchases of group life certificates, but we may put the decision on a rigorous basis by using the following criterion to test the null hypothesis that the arrangement of a's and b's is random against the alternative that there is a trend:

Reject the null hypothesis at a level of significance α if u, the total number of runs above or below the median, is less than or equal to u_α, where u_α is to be obtained from Table VII on page 481.

In our example $u = 6$ and the number of a's = the number of b's = 10. From Table VII we find that the corresponding $u_{.05} = 6$, so we conclude that there is a significant trend at the level 0.05; the trend is not significant at the 0.01 level, however, since $u_{.01} = 5$.

15

PLANNING
BUSINESS RESEARCH

SOURCES OF BUSINESS DATA

Regardless of whether we merely describe things numerically or whether we generalize beyond our data, locating, assembling, or collecting data can raise many problems. This is a serious matter because access to a good supply of high-quality data is fundamental to all of statistics. Data required to solve practical everyday business problems can come from many sources, sometimes classified broadly as *internal* and *external*. Internal data are generated from the activities within a firm; they may be taken from a firm's order book or inventory, payroll, personnel, service, inspection, or accounting records; they may be collected in experiments and tests of product quality characteristics, or they may be gathered by agents, by telephone, or by questionnaires from customers as well as suppliers. External data are obtained from sources outside the firm; they may come from records of state, local, and national governments and their agencies and regulatory bodies, from trade associations, private institutions, other firms, and so on. Some problems require the combination of internal and external data, say, when a company compares its own operating performance with that of its competitors or the industry as a whole.

External data are sometimes classified as *primary*, meaning that the organization gathering the data also publishes or releases them, or as *secondary*, meaning that the data are published by an organization other than the one which gathered them. There are many important and highly respected sources of primary as well as secondary data. Among the nongovernmental sources we find private statistical services, trade associations, trade publications, university research bureaus, commercial and financial periodicals, and specialized reporting agencies. From these sources come data on employment, farm prices and marketings, construction contracts, store sales, bank debits, and the like, often broken down on a regional, county, or city basis. In addition, there are reports on prices, production, sales, employment, and the like in different industries, and all this information is supplied to fill the needs of individuals and groups for reliable statistics.

By far the biggest collector and publisher of business data, and generally the most important single source of external data, is the Federal government. Within the great mass of statistical material flowing from the government it is possible to find information relating to virtually every aspect of the life of the nation. For instance, the Department of Commerce publishes each year the *Statistical Abstract of the United States*, an immense storehouse of data referring to many things and gathered from many sources. Through the Bureau of the Census, the Commerce Department periodically takes and publishes the results of censuses of population, manufactures, distribution, housing, and agriculture, and also publishes monthly trade reports giving data on inventories, sales, and so on, in various wholesale and retail lines of business for the entire country and for selected cities. The Office of International Trade, the Office of Industry and Trade, and the Bureau of Foreign and Domestic Commerce, all of the Commerce Department, collect and publish data on the trade of the United States and other countries. Through the Office of Business Economics, the Department issues one of the most important of all statistical publications, the monthly *Survey of Current Business* (which is supplemented by weekly data on some of the major series of data). The regular monthly issues contain indicators of general business conditions and series relating to wholesale and retail commodity prices and trade, construction activity, population and employment, payrolls, wages and hours, finance, foreign trade, transportation, and so on. Without much doubt, this publication constitutes one of the most valuable collections of current business data.

Another Commerce Department publication of major importance is the *Business Conditions Digest*, published monthly by the Bureau of the Census, which contains almost 500 economic indicators in a form which permits various types of analyses of current and prospective business conditions.

Other important statistics are collected and published by the Department of Labor, whose Bureau of Labor Statistics issues the *Monthly Labor Review*, the primary source for the indexes of consumer prices and wholesale prices, and data on construction contracts and costs, employment, payrolls, wages and hours, and work stoppages. In addition to the publications of many other Departments, agencies, and commissions of the government, the Board of Governors of the Federal Reserve System publishes monthly the *Federal Reserve Bulletin* and the *Federal Reserve Chart Book on Financial and Business Statistics* and releases other data periodically. Besides containing a wealth of financial information, the *Bulletin* is the primary source of the Index of Industrial Production.

Most of the data collected by the government are needed by the government itself in discharging its many responsibilities; data relative to some phenomena are not needed specifically by the government, but are collected and published in response to the needs of such large groups of individuals as to justify their collection at public expense.

EXERCISES

1. Determine whether the *Survey of Current Business* is a primary or a secondary source for the following data:
   (a) gross national product or expenditure;
   (b) newspaper advertising linage;
   (c) employees on payrolls of nonagricultural establishments;
   (d) U.S. merchandise trade;
   (e) major U.S. government transactions;
   (f) U.S. international transactions;
   (g) total life insurance premiums collected.

2. Answer each of the following questions for which numerical data and other information are summarized (from various sources) in *Business Statistics:*
   (a) What is the definition of gross national product or expenditure?
   (b) Do the figures concerning newspaper advertising linage refer to total advertising linage for *all* newspapers in the United States?
   (c) Do world production figures for gold include production in the USSR?
   (d) Does the *total labor force* include the armed forces of the United States?
   (e) What constitutes *long term unemployment?*

3. Refer to the *Monthly Labor Review* to obtain
   (a) the annual values of the Consumer Price Index for the past five years;
   (b) the annual values of the Wholesale Price Index for the same period.

4. Refer to the *Statistical Abstract of the United States* to determine the appropriate source for the number of municipal and county 9- and 18-hole golf courses.

5. According to the *Federal Reserve Bulletin*, what are the coupon rate and current amount of issue of the following U.S. Treasury Bonds: (a) February 15, 1980, (b) February 15, 1990, and (c) November 15, 1998?

6. Refer to the *Handbook of Labor Statistics 1970* to determine where the Bureau of the Census obtains its data for *Consumer Income*. Are the data completely reliable?

7. As of the most recent date or period for which you can locate data, determine

   (a) the total amount of currency in circulation in the United States;

   (b) the total number of publicly financed buildings authorized for construction by all permit-issuing places in the United States;

   (c) the total book value (unadjusted) of the inventories of nondurable goods industries in the United States;

   (d) the total reported net profit after taxes of all manufacturing industries in the United States.

SAMPLING

It is important to recognize that much of the published information available from government and private sources (indeed, much of the world's knowledge) is actually based on samples. Even in problems involving relatively small numbers of items or individuals, it is rarely feasible, practical, or economical to collect and analyze all relevant observations or measurements. As we have suggested earlier, there are usually time, cost, or other limitations which force one to take samples and generalize from them to the whole populations from which they came. Although the word "sample" is used somewhat loosely in everyday language, let us observe again that in statistics it has a very special meaning: A sample is a set of data that can reasonably serve to make generalizations, or inferences, about the population from which it came.

Broadly speaking, there are two different types of samples: *probability samples* and *judgment samples*. By a probability sample from a finite population we mean a sample chosen in such a way that each element of the population has a known, though not necessarily equal, probability of being included in the sample. Simple random samples of size n drawn from populations of size N are probability samples, since each element of the population has a known probability n/N of being included in the sample. There are still other ways to take probability samples from populations and all such samples possess one great advantage: Only probability samples enable us to calculate sampling errors and, hence, judge the "goodness" of estimates or decisions to which statistical analyses lead.

In contrast to probability samples, we shall refer to samples as judgment samples if, in addition to (or instead of) chance, personal judgment plays a significant role in their selection. There are many situations where, for practical reasons, investigators use judgment samples to gain needed information. One important use of such sampling is in testing markets for new products. Because of the tremendous cost of market testing on a national scale, many products are first tested in one or a few cities. Such test cities are usually not selected at random; instead they are carefully chosen because in someone's considered judgment they are "typical" or "average" American cities. Subsequently, when generalizations are made to national markets, samples of public reaction to advertising, packaging, and the like constitute judgment samples.

Regardless of how the conclusions or actions based on judgment samples ultimately turn out, judgment samples have the undesirable feature that standard statistical theory cannot be applied to evaluate the accuracy and reliability of estimates or to calculate the probabilities of making various kinds of erroneous decisions. Whenever elements of judgment enter in the selection of a sample, the evaluation of the "goodness" of estimates or decisions based on the sample is again largely a matter of personal judgment.

SAMPLE DESIGNS

There are a great many ways to select a sample from a population, and there is an extensive literature devoted to the subject of designing sampling procedures. In statistics, a sample design is a definite plan, completely determined before any data are collected, for obtaining a sample from a given population. Thus, a plan to take a simple random sample from among the 268 members of a trade association by using a table of random numbers in a prescribed way constitutes a sample design. Of the many ways in which a sample can be taken from a given population, some are quite simple while others are relatively involved. In what follows, we shall discuss briefly some of the most important kinds of sample designs; detailed treatments of this subject are referred to in the Bibliography at the end of this book.

SYSTEMATIC SAMPLING

There are many situations where the most practical way of sampling is to select, say, every 10th voucher in a file, every 20th name on a list, every 50th piece coming off an assembly line, and so forth. Sampling of this sort is referred to as *systematic sampling*, and an element of randomness is usually introduced into this kind of sample by using random numbers or some gambling device to pick the unit with which to start. Although a systematic sample may not be a random sample in accordance with our definition, it is often reasonable to treat systematic samples as if they were random samples. Whether or not this is justified depends entirely on the structure (order) of the list, or arrangement, from which the sample is obtained. In many instances, systematic sampling actually provides an improvement over simple random sampling inasmuch as the sample is "spread more evenly" over the entire population.

The real danger in systematic sampling lies in the possible presence of *hidden periodicities*. For instance, if we inspect every 40th piece made by a particular machine, our results would be biased if, because of a regularly recurring failure, it so happened that every 20th piece had blemishes. Also, a systematic sample might yield biased results if we interviewed the residents of every 10th house along a certain route and it so happened that every 10th house selected was a corner house on a double lot.

STRATIFIED SAMPLING

As we saw in Chapters 8, 9, and 10, the goodness of a generalization or the closeness of an estimate depends mostly on the standard error of the statistic being used, which in turn depends on both the size of the sample and the variability of the population. In theory, at least, we can increase the precision of a generalization by increasing the sample size and in practice, if the cost of sampling is largely overhead and the items to be sampled are readily at hand, it may be about as easy to take a sample of size 200 as it is to take a sample of size 50. On the other hand, if the cost of sampling is more or less proportional to the size of the sample (e.g., in *destructive testing* where the cost of sampling is largely the cost of the items tested), it may be too costly to increase the size of the sample. Also, as we saw in Chapter 8, a sample which is four times as large as another yields estimates which are only two times as reliable as those based on the other.

One relatively simple scheme for reducing the size of the standard error of a statistic is *stratification*. This is a procedure which consists of stratifying (or dividing) the population into a number of nonoverlapping subpopulations, or *strata*, and then drawing random samples independently from the different strata. If the items selected from each of the strata constitute a simple random sample, the entire procedure (first stratification and then simple random sampling) is called *stratified (simple) random sampling*. Note that samples obtained in this way *are* probability samples.

Although the concept of stratifying is relatively simple, several substantial problems immediately arise: What should be the basis of stratification? How many strata should be formed? What sample sizes should be allocated to the different strata? How should the samples within the strata be taken? Stratification does not guarantee good results, but if successful and properly executed, a stratified sample will generally lead to a higher degree of precision, or reliability, than a simple random sample of the same size drawn from the whole population.

To illustrate the general idea behind stratified sampling, let us consider the following oversimplified, though concrete, example. Suppose we want to estimate the mean weight of four persons on the basis of a sample of size 2; the weights of the four persons are, respectively, 110, 130, 180, and 200 pounds, so that μ, the mean weight we are trying to estimate, is equal to 155 pounds. If we take an ordinary random sample of size 2 from this population and weigh these two persons, it can easily be seen that $\bar{x}$ can vary from 120 to 190. In fact, the six possible samples of size 2 that can be taken from this population have means of 120, 145, 155, 155, 165, and 190, so that $\sigma_{\bar{x}} = 21.0$ (see Exercise 2 on page 420).

Now suppose that we make use of the fact that among the four persons there are two men and two women and *stratify* our sample by sex, then randomly select one of the two women and one of the two men. Assuming that the two smaller weights are those of the two women, we now find that $\bar{x}$ varies on the much smaller interval from 145 to 165. In fact, the means of the four possible samples are now 145, 155, 155, and 165, so that $\sigma_{\bar{x}} = 7.1$ (see Exercise 3 on page 420). *This illustrates how, by stratifying the sample, we were able to reduce $\sigma_{\bar{x}}$ from 21.0 to 7.1.*

Essentially, the goal of stratification is to form strata in such a way that there is some relationship between being in a particular stratum and the answers sought in the statistical study, and that within the separate strata there is as much homogeneity (uniformity) as possible. Note that in our example there was such a connection between sex and weight and that there was much less variability in weight within the two groups than there was within the entire population. A more detailed and more theoretical treatment of some of the problems of stratified sampling is taken up in the Technical Note at the end of this chapter.

Stratification is not restricted to a single variable of classification, or characteristic, and populations are often stratified according to several characteristics. In a system-wide survey designed to determine the attitude of its students, say, toward a new tuition plan, a state college system with 17 colleges might stratify the students with respect to class standing, sex, major, and the college which they attend. Thus, part of the sample would be allocated to sophomore women majoring in English at College A, part to senior men majoring in engineering at College N, and so on. Up to a point, stratification like this, called *cross stratification*, will often increase the precision (reliability) of estimates, and it is widely used, particularly in opinion sampling and market surveys.

QUOTA SAMPLING

In stratified sampling, the cost of obtaining random samples for the individual strata is often so expensive that interviewers are simply given *quotas* to be filled from the different strata, with few (if any) restrictions on *how* they are to be filled. For instance, in determining voters' attitude toward increased tax refunds to elderly persons, an interviewer working a certain area might be told to interview 5 male self-employed homeowners under 30 years of age, 10 female wage earners in the 50–60 year bracket who live in apartments, 4 retired males over 60 who live in trailers, and so on, *with the actual selection of the individuals being left to the interviewer's discretion.* This is called *quota sampling*, and it is a convenient, relatively inexpensive, and often necessary procedure, but as it is often executed, the resulting samples are not probability samples. In the absence of firm restrictions on their choice, interviewers naturally tend to select individuals who are most readily available—persons who work in the same building, shop in the same store, or perhaps reside in the same general area. Quota samples are thus essentially judgment samples, and although it may be possible to guess at sampling errors by using experience or corollary information, quota samples generally do not lend themselves to any sort of formal statistical evaluation.

CLUSTER SAMPLING

To illustrate another important kind of sampling, suppose that a large foundation wants to study the changing patterns of family expenditures in the Los Angeles area. In attempting to complete schedules for 2,000 families, the foundation finds that simple random sampling is practically impossible, since suitable lists are not available and the cost of contacting families scattered over a wide area (with possibly two or three callbacks for the not-at-homes) is very high. One way in which a sample can be taken in this case is to divide the total area of interest into a number of smaller, nonoverlapping areas, say, city blocks. A number of these blocks are then randomly selected, with the ultimate sample consisting of all (or samples of) the families residing in these blocks. Generally speaking, in this kind of sampling, called *cluster sampling*, the total population is divided into a number of relatively small subdivisions, which are themselves *clusters* of still smaller units, and then some of these subdivisions, or clusters, are randomly selected for inclusion in the overall sample. If the clusters are geographic subdivisions, as in our example, this kind of sampling is also called *area sampling*.

To give two further illustrations of cluster sampling, suppose that the management of a large chain store organization wants to interview a sample of its employees to determine their attitude toward a proposed pension plan. If random methods were used to select, say, five stores from the list and if all employees of these stores were interviewed, the resulting sample would be a cluster sample. Also, if the Dean of Students of a university wanted to know how fraternity men at the school feel about a certain new regulation, he would obtain a cluster sample if he interviewed the members of several randomly selected fraternities.

Although estimates based on cluster samples are generally not as reliable as estimates based on simple random samples of the same size (see Exercise 4 on page 420), they are usually more reliable *per unit cost*. Referring again to the survey of family expenditures in the Los Angeles area, it is easy to see that it may well be possible to obtain a cluster sample several times the size of a simple random sample at the same expense. It is much cheaper to visit and interview families living close together in clusters than families selected at random over a wide area.

In practice, several of the methods we have discussed may well be used in the same survey. For instance, if government statisticians wanted to study the attitude of American farmers toward marketing cooperatives, they might first stratify the country by states or some other geographic subdivision. To obtain a sample from each stratum, they might then use cluster sampling, subdividing each stratum into a number of smaller geographic subdivisions, and finally they might use simple random sampling or systematic sampling to select a sample of farmers within each cluster.

EXERCISES

1. The following amounts of money (rounded to the nearest dollar) were spent by the first 12 customers who patronized a college book store on the first day of the semester:

$51, $20, $43, $38, $68, $27, $67, $55, $76, $61, $90, $52

(a) List the three possible systematic samples of size 4 that can be taken from this list by starting with one of the first three amounts and then taking each third amount on the list.

(b) Calculate the means of the three samples obtained in part (a), and assuming that the starting point was randomly selected from among the first three amounts, show that the mean of this sampling distribution of $\bar{x}$ equals the population mean μ, namely, the mean of the 12 amounts.

2. Verify for the six sample means on page 417 (which are assigned equal probabilities of $1/6$) that $\sigma_{\bar{x}} = 21.0$.

3. Verify for the four sample means on page 417 (which are assigned equal probabilities of $1/4$) that $\sigma_{\bar{x}} = 7.1$.

4. To generalize the example given in the text, suppose that in a group of six athletes there are three tennis players whose weights are 140 pounds, 150 pounds, and 160 pounds, and three football players whose weights are 210 pounds, 220 pounds, and 230 pounds.

(a) List all possible random samples of size 2 which may be taken from this population, calculate the means of these samples, and show that $\sigma_{\bar{x}} = 22.7$.

(b) List all possible stratified random samples of size 2 which may be obtained by selecting one tennis player and one football player, calculate the means of these samples, and show that $\sigma_{\bar{x}} = 5.8$.

(c) Suppose that the six athletes are divided into clusters according to their sports, each cluster is assigned a probability of $1/2$, and a random sample of size 2 is taken from one of the randomly chosen clusters. List all possible samples, calculate their means, and show that $\sigma_{\bar{x}} = 35.2$.

(d) Compare and discuss the results obtained for $\sigma_{\bar{x}}$ in parts (a), (b), and (c).

5. On the basis of the amount of their total deposits, 60 savings banks in a state are classified into 30 small, 20 medium-sized, and 10 large banks.

(a) In how many ways can we choose a stratified 10 per cent sample of the 60 savings banks if one third of the sample is allocated to each of the three strata?

(b) In how many ways can we choose a stratified 10 per cent sample of the 60 savings banks if the proportion of the sample allocated to each stratum is proportional to the size of the stratum?

(*The exercises which follow are based on the material in Technical Note 9 on page 428.*)

6. In a large factory with a number of similar machines, records are kept of the number of minutes required by an operator to get the machine restarted following machine failures of various kinds. Suppose that 1,400 records are on file, with 700 classed as ordinary stoppages, 400 as moderate stoppages, and 300 as severe stoppages. In order to estimate the mean restart time for the whole population, a 2 per cent stratified random sample (with proportional allocation) is taken from the file with the following results (times are in minutes):

Ordinary stoppages: 4, 5, 3, 6, 4, 3, 7, 2, 5, 4, 3, 5, 3, 2
Moderate stoppages: 9, 10, 8, 12, 9, 10, 12, 10
Severe stoppages: 15, 18, 21, 17, 16, 21

(a) Find the mean of each of these three samples and then determine their weighted mean, using as weights the respective sizes of the three strata.

(b) Verify that the identical result would have been obtained in part (a) if the weights used had been the sizes of the respective samples.

(c) Verify that the results of parts (a) and (b) are identical with the ordinary arithmetic mean of the 28 observations. (That is, verify that in a problem like this, proportional allocation is self-weighting.)

7. A population is divided into two strata so that $N_1 = 5,000$, $N_2 = 15,000$, $\sigma_1 = 30$, and $\sigma_2 = 40$. How should a sample of size 200 be allocated to the strata if we use (a) proportional allocation and (b) optimum allocation?

8. A population is divided into four strata so that $N_1 = 4,000$, $N_2 = 2,000$, $N_3 = 3,000$, $N_4 = 1,000$, $\sigma_1 = 1$, $\sigma_2 = 3$, $\sigma_3 = 2$, and $\sigma_4 = 4$. How should a sample of size 300 be allocated to the four strata if we use (a) proportional allocation and (b) optimum allocation?

DESIGN OF EXPERIMENTS

Even those who have never been actively engaged in research should be able to visualize the problems involved in planning an experiment so that it can actually serve the purpose for which it is designed. It happens all too often that an experiment purported to test one thing tests another or that an experiment that is improperly designed cannot serve any useful purpose at all. Suppose, for instance, that in order to compare the cleansing action of two detergents, someone has soiled 10 swatches of white cloth equally with India ink and oil and then washed 5 swatches in an agitator-type machine using a cup of Detergent Q and the other 5 swatches in the same machine using a cup of Detergent R. Following this, whiteness readings were made on the swatches with the following results:

Detergent Q: 76, 85, 82, 80, 77
Detergent R: 72, 58, 74, 66, 70

Treating these data as independent random samples from two (conceptually infinite) populations of such readings, we want to test, say, at the level of significance 0.01, whether the difference between 80 and 68, the two sample means, is significant. Proceeding as on page 282, we formulate the hypothesis $\mu_1 = \mu_2$ and the alternative hypothesis $\mu_1 \neq \mu_2$, where μ_1 and μ_2 are the "true" average whiteness readings for the two populations. Calculating t according to the formula on page 282, we obtain $t = 3.67$, and since this exceeds $t_{.005} = 3.355$ for eight degrees of freedom, we reject the null hypothesis. In other words, we conclude that there is a real difference in the actual average whiteness of the two populations.

Interpreting this result, we might arrive at the perfectly natural conclusion that Detergent Q is superior in cleansing action to Detergent R. However, a moment's reflection will make us realize that we may not really have a basis for such a conclusion. For all we know, the water temperature could have been different when testing the two detergents, one detergent could have been used in soft water and the other in hard water, the washing times could have differed substantially, and even the machine used to determine the whiteness readings could have gone out of adjustment after the readings were taken for the first detergent. Thus, what may have seemed an obvious conclusion at first turns out to be highly questionable. It is entirely possible, of course, that the difference between the two means is due to quality differences in the two detergents, but we have just listed several other factors which could be responsible. In fact, we could go on indefinitely listing possible causes, any of which—either singly or in combination with others—might have accounted for the results. The significance test which we have performed convinces us that the difference is too large to be attributed to chance, but it does not tell us *why* this difference has occurred.

Questions of this sort are treated in a branch of statistics called the *design and analysis of experiments*. If we want to be able to show that one factor (among many possible ones) can definitely be considered as the *cause* of an observed phenomenon which has been shown to be significant, we must make sure somehow that none of the other factors could possibly be held responsible. In other words, if we hope to show that one particular factor caused certain results, we must make sure that other relevant factors are *controlled* in such a way that they could not reasonably be held accountable. One way of handling this kind of problem is to perform a rigorously controlled experiment in which all variables except the one with which we are concerned are held fixed. In our example, we might thus use the same washing machine (carefully inspected after each use), the same washing time, water of exactly the same temperature and hardness, and we might inspect all of the testing equipment after each use. Under these rigid conditions we know that a significant difference between the (whiteness) means is not due to differences in washing machines, washing times, water temperature or hardness, or the testing equipment. On the positive side, we know that one detergent performs better than the other *if it is used in this narrowly restricted way*. This does not tell us whether the same difference would exist if the tests were performed in a different kind of washer, if the washing time were longer or shorter, if the water had a different temperature or hardness, and so on. Generally speaking, this kind of "overcontrolled" experiment does not provide us with the kind of information we want.

Another way of handling this kind of problem is to design the experiment in such a way that we cannot only compare the merits of the two detergents under more general conditions but can also test whether other important variables might affect their performance. To illustrate how this might be done, suppose we decide that we want to investigate the effects of four factors on the cleanliness of cottons washed in an agitator-type machine: the detergent used, the washing time, the water temperature, and the water hardness. (Such other factors as water level are assumed to be rigorously controlled in the experiment.) Letting Q and R stand for a cup of the respective detergent, S and L for washing times of 10 minutes and 20 minutes, W and H for warm and hot water, and E and F for soft and hard water, there are altogether $2 \cdot 2 \cdot 2 \cdot 2 = 16$ ways in which these four factors can be combined. Using identical samples of cloth and testing equipment that is rigorously checked, we might conduct the following series of test washings in which each of the 16 possible combinations is included once:

Test washing	Detergent	Time	Temperature	Hardness
1	R	L	W	E
2	R	L	H	F
3	Q	L	W	E
4	R	S	W	F
5	Q	S	H	F
6	Q	S	W	E
7	R	S	W	E
8	Q	L	H	F
9	R	S	H	F
10	Q	S	H	E
11	Q	L	W	F
12	Q	S	W	F
13	R	L	H	E
14	R	S	H	E
15	Q	L	H	E
16	R	L	W	F

This means that the first test washing is performed with Detergent R, a washing time of 20 minutes, and warm soft water; the second test washing is performed with Detergent R, a washing time of 20 minutes, and hot hard water; and so on. A scheme like the above is said to be *completely balanced* since each detergent is used once with each possible combination of washing time, temperature, and hardness.

The seeming lack of order in the arrangement of the 16 tests is by no means accidental. When we first wrote down the possible combinations, we filled the *Detergent* column by writing eight Q's followed by eight R's; then we filled the next column by alternately writing down four L's and four S's, the *Temperature* column by alternately writing down two W's and two H's, and the last column by alternating E's and F's. If we actually performed the tests in this order, we would run the first eight tests with Detergent Q, the other eight with Detergent R, and extraneous factors might conceivably upset the results. For instance, there might be a progressive deterioration in machine efficiency which could not be detected by inspection. Similarly, we might get in trouble if we deliberately conducted the first eight test washings using the shorter washing time, the hot water, or the soft water. We protect ourselves against biases which might inadvertently invalidate the results by *randomizing* the order of the tests; after writing down the 16 possible combinations of the four factors, we selected the order shown in the preceding table with the use of random numbers.

Another important consideration in the design of an experiment is that of *replication* or *repetition*. Any time we want to decide whether an observed difference between sample means is significant or whether a sample mean differs significantly from some assumed value, we must have some idea, some estimate, of chance fluctuations. In experiments of the sort we have just described this kind of variation is called the *experimental error*, and it is usually estimated by repeating all (or part) of the entire experiment a number of times. In our example, we might conduct the 16 test washings in the given order, then rerandomize the order and conduct 16 more tests. The 32 tests thus made, including two each of the 16 possible combinations, would yield an estimate of the experimental error, and they would permit a rather detailed analysis of the effects of the four variables on the whiteness of cottons washed with the given kind of equipment.

The purpose of this example has been to introduce some of the basic ideas of experimental design. Generally speaking, the analysis of an experiment depends partly on the design itself, and partly on assumptions concerning the populations from which the data are obtained. Specification of these assumptions (for example, that the populations from which the data are obtained have normal distributions) thus constitutes another important aspect of the proper use of statistics in experimentation. The analysis of a four-factor experiment like the one described above is fairly complicated, but it is similar, in principle, to the analysis of variance techniques described in Chapter 11.

EXERCISES

1. Suppose that a tire manufacturer wants to experiment with three different tread designs, A, B, and C, which he wants to try on two different kinds of road surfaces, I and II, and at two different speeds, S and F. List the 12 tests he has to perform so that each tread design is used once on each combination of road surfaces and speeds.

2. A manufacturer of pharmaceuticals wants to market a new cold remedy which is actually a combination of five drugs, and he wants to experiment first with two dosages for each drug. If A_L and A_H denote the low and high dosage of Drug A, B_L and B_H the low and high dosage of Drug B, C_L and C_H the low and high dosage of Drug C, D_L and D_H the low and high dosage of Drug D, and E_L and E_H the low and high dosage of Drug E, list the 32 preparations he has to test if each dosage of each drug is to be used once in combination with each dosage of each of the other drugs.

3. Suppose that a clothing manufacturer wants to compare three different kinds of sewing machines, and that he wants to know, in particular, how they perform with three different kinds of needles and three different kinds of threads. To try each combination of needles and threads on each machine would require $3 \cdot 3 \cdot 3 = 27$ different tests, but if it is required only that each needle and each thread is tried once on each machine, this can be done with only nine different tests. Verify that this is, in fact, accomplished in the following design, where the three machines are denoted K, L, and M, the three needles are denoted a, b, and c, and the three threads are denoted I, II, and III:

Test	Machine	Needle	Thread
1	K	a	I
2	K	b	II
3	K	c	III
4	L	a	III
5	L	b	I
6	L	c	II
7	M	a	II
8	M	b	III
9	M	c	I

4. With reference to Exercise 3, suppose that the clothing manufacturer has three sewing-machine workers on his payroll: Mrs. P, Mrs. Q, and Mrs. R. Indicate how he might assign three of the nine tests to each of these operators, so that each operator works once with each machine, once with each kind of needle, and once with each kind of thread. (*Hint:* Use trial and error, although there exists

a systematic way of planning this kind of experiment as a so-called *Graeco–Latin square.*)

5. To test their ability to make decisions under pressure, nine executives are to be interviewed by each of four psychologists. As it takes a psychologist a full day to interview three of the executives, the schedule for the interviews is arranged as follows, where the nine executives are denoted A, B, C, D, E, F, G, H, and I:

Day	Psychologist	Executives		
April 2	1	A	B	C
April 3	1	D	E	F
April 4	1	G	H	I
April 5	2	B	F	G
April 6	2	A	E	I
April 9	2	C	D	H
April 10	3	C	F	I
April 11	3	B	E	H
April 12	3	A	D	G
April 13	4	A	F	H
April 16	4	B	D	I
April 17	4	C	E	G

Verify that

(a) each of the nine executives is interviewed once and only once by each of the four psychologists;

(b) each of the nine executives is interviewed once and only once together with each of the other executives on the same day. (This may be important because each executive is thus tested together with each other executive once under *identical conditions*, i.e., by the same psychologist on the same day.)

6. A company has seven vice-presidents, who are assigned to its various administrative committees as shown in the following table:

Committee	Vice-Presidents
Investments	Clark, Evans, Flynn, Gordon
Advertising	Adams, Duncan, Flynn, Gordon
Sales	Adams, Brown, Evans, Gordon
Personnel	Adams, Brown, Clark, Flynn
Purchasing	Brown, Clark, Duncan, Gordon
Insurance	Adams, Clark, Duncan, Evans
Research	Brown, Duncan, Evans, Flynn

Verify that

(a) each of the seven vice-presidents serves on four committees;

(b) each of the seven vice-presidents serves together with each of the other vice-presidents on two committees. (This kind of *balance* may not seem too important, but it might prevent some sort of clique from taking over the operation of the company.)

7. With reference to the committees of Exercise 6, suppose that Clark, Adams, and Brown are (in that order) appointed chairmen of the first three committees. How will the chairmen of the other four committees have to be chosen so that each of the seven vice-presidents is chairman of one committee? Are there more ways than one?

A WORD OF CAUTION

In view of the fantastic amount of "statistical" information that is disseminated to the public for one reason or another, we cannot over-emphasize the point that such information must always be treated with extreme caution. To avoid serious mistakes in the use of published data, it is essential to check the precise definition of all terms (e.g., "employ-ment," "sales," "shipments"), and this usually requires *looking behind* the words themselves. A careful search is often necessary to discover not only what the data are supposed to represent, but also what units are being used, how these units are defined, and whether the definitions are consist-ent throughout so that comparisons can be made. The availability of just such information as this is one of the most valuable features of data published by the Federal government. For all series published by the government, it is possible to find somewhere a complete description of what data (rigorously defined) are contained, how, when, and where they were gathered, how they were processed, and so on. Unfortunately, such in-formation is often unavailable for data supplied by other sources, in which case it is always advisable to proceed with extra care.

Another area in which one must proceed cautiously is the area of public opinion sampling. There are in the United States a number of highly reputable polls—the Gallup Poll, the Roper Poll, and the California Poll, for example—based on carefully designed and executed statistical surveys. The past few years, however, have seen a phenomenal growth of polls of all sorts, and there are now literally hundreds of polls whose existence is hard to justify. In countless radio station and newspaper polls people are invited to phone or write in and register votes for their favorite presidential candidate, and interviewers stationed at busy downtown locations ask people for their preferences. In "popcorn" polls, theater-goers "vote" for their choice by buying popcorn in bags displaying their candidate's picture; in "ice cream" polls, a purchase of chocolate ice cream in a super-market is recorded as a vote for Candidate A while a purchase of vanilla ice cream is recorded as a vote for Candidate B. Actually, this might all come under the heading of good fun, if it were not for the tremendous and growing influence of public opinion sampling on political issues.

TECHNICAL NOTE 9 (Stratified Sampling)

Two of the basic problems in stratified sampling are (1) choosing the individual strata which show promise of making the sampling design effective and (2) allocating the sample to the individual strata. In connection with the first problem, we attempt to construct strata which are such that the items within the separate strata are as nearly alike as possible. It is this homogeneity within strata which leads to small errors of estimate; the variability in the population taken as a whole is of no consequence because (if stratification is successful) the error variance reflects only the relatively small within-stratum variances.

In connection with the second problem, suppose that we wish to take a sample of size n from a finite population of size N and that this population is subdivided into k strata whose sizes are $N_1, N_2, \ldots,$ and N_k, respectively. Suppose further that the mean of the whole population is μ and that the means of the individual strata are $\mu_1, \mu_2, \ldots,$ and μ_k. If we now take a sample of size n_1 from the first stratum, a sample of size n_2 from the second stratum, $\ldots$, and a sample of size n_k from the kth stratum, and the means of these k samples are, respectively, $\bar{x}_1, \bar{x}_2, \ldots,$ and $\bar{x}_k$, an *unbiased* estimate of the mean of the whole population is given by the formula

$$\frac{N_1\bar{x}_1 + N_2\bar{x}_2 + \cdots + N_k\bar{x}_k}{N}$$

This is a weighted mean of the individual $\bar{x}$'s, with the weights equal to the sizes of the respective strata.

In choosing the sample sizes or, in other words, in allocating the sample to the different strata, we note that altogether we are choosing n items from N so that the fraction of the population sampled is n/N. One obvious way to allocate the sample to the different strata then is to take this same fraction from each stratum. This method of allocation is called *proportional allocation* because it requires that we make the number of items taken from each stratum proportional to the number of items in that stratum. Symbolically, in proportional allocation we make

$$\frac{n_1}{N_1} = \frac{n_2}{N_2} = \cdots = \frac{n_k}{N_k}$$

or make these ratios as nearly equal as possible. For instance, if a random sample of size $n = 50$ is to be taken from a population of size $N = 1{,}000$ and the population is divided into three strata whose sizes are $N_1 = 600$, $N_2 = 300$, and $N_3 = 100$, the fraction sampled is $50/1{,}000$ (or 5 per cent), and we would take samples of size $n_1 = 30$, $n_2 = 15$, and $n_3 = 5$ from the strata.

The method outlined above requires that we know the sizes of the different strata but, in many cases, we do not know these numbers; sometimes, in fact, the main reason for sampling a population is to determine how many units of different kinds it contains. An obvious solution to this problem is to estimate the population mean by using as weights the various sample sizes (however the sample is allocated to the different strata) instead of the unknown strata sizes, calculating

$$\frac{n_1 \bar{x}_1 + n_2 \bar{x}_2 + \cdots + n_k \bar{x}_k}{n}$$

This estimate of μ is *identical* with the estimate given by the first formula of this note, *provided we have used proportional allocation*. If we have used any other kind of allocation, though, this will not be the case. Moreover, in proportional allocation both of these estimates are equal to the simple arithmetic mean of the sample, and for this reason proportional allocation is said to be *self-weighting*.

In proportional allocation the importance of different strata sizes is taken into account by the fact that the larger strata contribute relatively more items to the sample. However, strata differ not only in size but also in internal variability, and it might seem reasonable to consider taking somewhat smaller samples from the less variable strata and somewhat larger samples from the more variable strata. We can take into account both the size and the internal variability of strata by taking from each stratum a sample whose size is proportional to the product of the stratum size and the stratum standard deviation. If we designate the standard deviations of the strata $\sigma_1, \sigma_2, \ldots,$ and σ_k, this kind of allocation requires that we make

$$\frac{n_1}{N_1 \sigma_1} = \frac{n_2}{N_2 \sigma_2} = \cdots = \frac{n_k}{N_k \sigma_k}$$

or make these ratios as nearly equal as possible. In this way, the larger and more variable strata will contribute relatively more items to the total sample. This particular method of allocation is called *optimum allocation*, since for a fixed sample size, the sample chosen in this way will have the smallest possible standard error for the estimate of the population mean. To take a sample which meets the above requirement, we find the sample sizes for the different strata by using the formula

$$n_i = \frac{n \cdot N_i \sigma_i}{N_1 \sigma_1 + N_2 \sigma_2 + \cdots + N_k \sigma_k}$$

for $i = 1, 2, 3, \ldots,$ and k. To illustrate the use of this formula, let us refer again to the example on page 429 where we had $N_1 = 600$, $N_2 = 300$, and $N_3 = 100$ and obtained samples of size $n_1 = 30$, $n_2 = 15$, and $n_3 = 5$ through proportional allocation. Now, supposing that $\sigma_1 = 5$, $\sigma_2 = 12$, and $\sigma_3 = 25$ we obtain

$$n_1 = \frac{50 \cdot 600 \cdot 5}{600 \cdot 5 + 300 \cdot 12 + 100 \cdot 25}$$

which, to the nearest integer, is 16, and similarly, we find $n_2 = 20$ and $n_3 = 14$ (both rounded to the nearest integer). Thus, optimum allocation requires only about half as many items from the large first stratum, because of the small value σ_1, but almost three times as many items from the small third stratum, because of the relatively large value σ_3.

One problem in using optimum allocation is that we must know the standard deviations of the different strata. Another problem arises if we want to estimate several population characteristics from the same set of sample data, since what is optimum for one characteristic may not be optimum for another. In situations like these, it may be better to use proportional allocation. Indeed, the gain in reliability due to optimum allocation is often not large enough to offset the obvious practical advantages of (self-weighting) proportional allocation.

16

INTRODUCTION
TO OPERATIONS
RESEARCH

OPERATIONS RESEARCH

In the past few years, the scale of business activity has expanded tremendously, its pace has quickened, corporations have grown in size, in variety of product, in extent of the market, and operations have been decentralized on a large scale. As a result of this, and various environmental factors, there is emerging from an already complex business world an even more complex one. Nowadays, even with timely and high-quality information, planning and control of operations is becoming increasingly difficult, and few companies can hope to survive for long without both sound planning and enough flexibility to change plans rapidly in response to changing conditions.

Over the years, a great deal of research has been conducted on various aspects of business and industry. Besides practical investigations of problems concerning profitability (manufacturing processes, costs and productivity, product design and development, consumer and market characteristics, etc.), much research has been conducted in the area of administration itself. However, none of this work has contributed much to what is management's primary responsibility: the process of making and executing the "right" decisions necessary to achieve enterprise goals. The recognition of this fact accounts partly for present efforts directed toward strengthening the decision-making process by integrating the scientific study of operations through pulling together in a sort of omnibus discipline, called *operations research*, various specialized techniques (e.g., time and motion study, production planning, and quality control), and other powerful analytical methods borrowed from the physical and social sciences. Intuition, judgment, experience, and good intentions have always been, and still are, necessary personal qualities in successful business managers. But it is obvious that these qualities alone are not sufficient to meet today's needs; it has been clearly demonstrated by now that, despite often staggering technical difficulties, carefully designed scientific investigations lead to decisions that are generally superior to those intuited by humans working under the burdens of too little time and/or too little experience.

The new discipline, operations research, is characterized by its concern with operating problems outside the usual fields of science, a broadened view of organizations as entities, and often a team approach to problem solving. Unlike most "pure" research this new type of research is deliberately practical, and it is aimed specifically at providing a quantitative basis for decision making. Among the important techniques usually classified under the general heading of operational research techniques are several we have already studied in this book—statistical sampling, "rational" decision making in the face of uncertainty, game theory, regression and correlation analysis, and time series analysis and forecasting, for instance. In this short final chapter, we conclude our introduction to elementary methods of operations research by discussing briefly the following four topics: linear programming, inventory, waiting lines, and Monte Carlo methods.

LINEAR PROGRAMMING

Largely under the influence of military needs, much work has recently been done on the general problem of allocating limited resources so as best to meet desired objectives. Characteristically, the primary interest in allocation problems is in planning, or programming, activities (e.g., in planning a variable amount of work in such a way that the maximum return can be achieved from a fixed, or limited, amount of resources or in planning a required amount of work so that it can be done most efficiently given certain limitations on the resources). The programming methods we shall discuss apply to situations that can be described by *linear models;* that is, they apply to problems that can be stated using linear expressions, linear equations, and linear inequalities. In general, we are interested in maximizing a linear expression of the form

$$a_1x_1 + a_2x_2 + \cdots + a_kx_k$$

where the a's are known, and the x's are unknowns subject to restrictions expressed by means of linear equations or linear inequalities. This explains why these methods are called *linear programming*.

The use of linear models in the study of economics goes back at least to the work of Francois Quesnay who, in 1758, published his *Tableau Economique*. In 1889, Leon Walras showed for the first time in his general equilibrium theory that (under perfect competition) full employment of resources was possible in a society where each individual sought to maximize the return from his own resources. Both in this and in Cassel's later formulation of the Walrasian system, it is assumed that production functions (which specify what combinations of amounts and kinds of inputs and outputs are possible) are linear and homogeneous; that is, they are linear expressions of the form indicated above. In 1928, the economist Wassily Leontief suggested describing an economy expanding at a constant rate in terms of a linear model. In further studies, he applied general equilibrium theory to certain problems of interindustry economics, providing a new technique called "input-output analysis" to solve the systems of linear equations in terms of which the problems were formulated. This success in applying general equilibrium theory to the analysis of an economic reality greatly stimulated interest among economists in linear economic models.

A large class of linear programming problems consists of so-called "mixing" problems, which require that given resources be combined or "mixed" so as to produce specified outputs most efficiently. For many years such problems were analyzed by conventional marginal analysis, a technique of early nineteenth century origin. However, for various reasons this technique is not well suited to actual problems arising in today's complex technological environment, and a wide variety of mixing problems are much better solved by the powerful new analytical methods referred to generally as "linear programming."

To illustrate, let us consider a problem in the operations of a company making two models, Standard and Deluxe, of a large commercial-sized coffeemaker. Components of these coffeemakers are processed by two different machines, I and II. To complete one unit of the Standard model requires 1 hour on Machine I and 1 hour on Machine II; to complete one unit of the Deluxe model requires 2 hours on Machine I and 5 hours on Machine II. This information is shown in the following table:

Model	Processing Time (hours)	
	Machine I	Machine II
Standard	1	1
Deluxe	2	5

During the scheduling period there are 20 hours of Machine I time and 35 hours of Machine II time available. The company can sell all units it can produce at a profit of $10 for each Standard unit and $30 for each Deluxe unit. Under these conditions, how many units of each model should the company schedule for production in order to maximize its profit from their sale?

In order to translate this problem into the language of mathematics, let us designate the number of units of Standard and Deluxe models to be produced by x and y. (When there are more than two unknowns, it is preferable to write them $x_1, x_2, x_3, \ldots$, but for two-variable problems the reader is probably more familiar with the x and y notation.) Since x and y must both be positive integers or zero, the first two restrictions we impose on these variables are

$$x \geq 0 \quad \text{and} \quad y \geq 0$$

Also, referring to the above table, we find that $x + 2y$ hours of Machine I time (of which there are 20 hours available) are required to process x Standard models and y Deluxe models. Hence, we have a third restriction of the form

$$x + 2y \leq 20$$

Since, similarly, $x + 5y$ hours of Machine II time (of which there are 35 hours available) are required to process x Standard models and y Deluxe models, we have the final restriction

$$x + 5y \leq 35$$

The inequality signs in these last two restrictions indicate that 20 and 35 hours are the *upper limits* on machine time availability. This means that it would be all right to use less, but not more, than these given amounts of time.

There are various values of x and y which will satisfy the four requirements, and our problem is to find the ones which will maximize the company's profit. Since the profit for x Standard models and y Deluxe models is given by the linear expression

$$10x + 30y$$

we will have to find that pair of values of x and y which maximizes $10x + 30y$ subject to the restrictions imposed by the four inequalities.

Let us begin by indicating geometrically the region into which the point (x, y), corresponding to a solution of the problem, must fall. The first two inequalities state that x and y cannot be negative or, in other words, that a solution (x, y) must lie *on or above* the x-axis and *on or to the right of* the y-axis. Since $x + 2y \leq 20$ implies that (x, y) must lie *on or to the left of* the line whose equation is given by $x + 2y = 20$, and since $x + 5y \leq 35$ implies that (x, y) must lie *on or to the left of* the line whose equation is given by $x + 5y = 35$, we find that the region of feasible, or possible, solutions is given by the shaded region of Figure 16.1. *Thus, any*

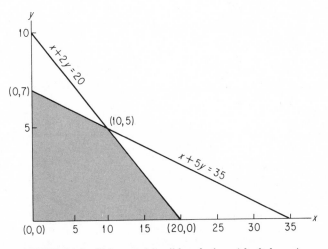

FIGURE 16.1   Polygon of feasible solutions (shaded area)

point (x, y) *lying in or on the edge of the shaded region of Figure 16.1 corresponds to values of* x *and* y *which satisfy the requirements of the problem.*

As can be seen from this diagram, there are many pairs of integers x and y which yield a point (x, y) inside the shaded region: There are $x = 0$ and $y = 0$, $x = 1$ and $y = 1$, $x = 2$ and $y = 3$, $x = 5$ and $y = 1$, and many more. Fortunately, there is no need to consider them all, as there exists the following theorem:

If there is a unique pair of values of x and y which maximizes (or minimizes) a linear expression of the form ax + by, then (x, y) must be a vertex (corner) of the polygon of feasible solutions; if there is more than one solution, at least two will have to correspond to vertices of this polygon.

(The theorem extends immediately to the case where there are more than two unknowns.)

As a result of this theorem, we have only to check the vertices of the shaded region of Figure 16.1. Three of these are easily found to be $(0, 0)$, $(0, 7)$, and $(20, 0)$, while the fourth, $(10, 5)$, is obtained by simultaneously solving the two linear equations $x + 2y = 20$ and $x + 5y = 35$. Substituting into $10x + 30y$ which gives the company's profit, we obtain the results shown in the following table:

Vertex	$10x + 30y$	Profit
$(0, 0)$	$10 \cdot 0 + 30 \cdot 0$	$ 0
$(0, 7)$	$10 \cdot 0 + 30 \cdot 7$	210
$(20, 0)$	$10 \cdot 20 + 30 \cdot 0$	200
$(10, 5)$	$10 \cdot 10 + 30 \cdot 5$	250

Inspection of this table shows that $x = 10$ and $y = 5$ is the optimum (and also the unique) solution; that is, the company will realize a maximum profit of $250 by manufacturing 10 Standard models and 5 Deluxe models. (We were fortunate that the solution was in integers, since with this geometrical method we would otherwise have had to investigate points in the vicinity of the one yielding the "solution"; of course, it is not a requirement of all linear programming problems that the solution must be in terms of integers.)

Another important class of linear programming problems arises in situations where resources must be allocated to meet given demands—for instance, where goods from several sources must be delivered to a number of destinations at minimum cost. To illustrate such *transportation* problems, suppose that a company operates three retail outlets S_1, S_2, and S_3, in the San Francisco Bay region, supplying these stores from two different warehouses, W_1 and W_2. Suppose, furthermore, that the company has exactly 20 television sets in warehouse stock, 8 in W_1 and 12 in W_2, and that it must deliver 10 sets to store S_1, 4 sets to store S_2, and 6 sets to store S_3. It wants to *minimize* the cost of these deliveries, given that the cost of moving a set from W_1 to S_1, S_2, and S_3 is, respectively, $12, $15, and $9, while the cost of moving a set from W_2 to S_1, S_2, and S_3 is, respectively, $8, $16, and $11.

This problem may appear to be a six-variable problem which cannot be represented by a two-dimensional geometric model, but we can readily turn it into a two-variable problem in the following way. If we schedule x sets for shipment from W_1 to S_1 and y sets for shipment from W_1 to S_2, we can then determine the rest of the shipping schedule in terms of these two variables from our knowledge of the supply at the warehouses and the demand at the stores. Shown below is a 2×3 table with the six shipments (and the unit shipping costs) shown in the cells:

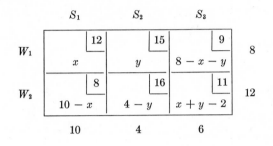

It can be seen from this table that the cost of shipping the sets is

$$12x + 15y + 9(8 - x - y) + 8(10 - x) + 16(4 - y) + 11(x + y - 2)$$
$$= 6x + y + 194$$

and we want to choose x and y in such a way that this cost is minimized. As for the restrictions in the problem, no shipment can be negative, so we formulate the following system of six inequalities: $x \geq 0$, $y \geq 0$, $8 - x - y \geq 0$, $10 - x \geq 0$, $4 - y \geq 0$, and $x + y - 2 \geq 0$. It will facilitate drawing a graph like Figure 16.1 and it will help understand the sense of the restrictions if we state them in the following way:

$$x \geq 0, \quad y \geq 0, \quad y \leq 4,$$
$$x + y \leq 8, \quad x + y \geq 2$$

(Here the original restriction $10 - x \geq 0$ has been dropped since it is unnecessary to write $x \leq 10$ in view of the restriction $x + y \leq 8$.)

We shall leave it to the reader (in Exercise 1 on page 438) to verify that, subject to the above restrictions, the transportation cost is \$196 for $x = 0$ and $y = 2$ and this is the lowest possible cost.

Both of the examples of this section could be solved by means of simple two-dimensional geometric figures. When we move to large and realistically complex problems, however, geometrical solutions become impossible and algebraic methods are required. Among these, the *Simplex Method*, devised by G. B. Dantzig in 1947, has found almost universal acceptance. This method is a highly efficient trial-and-error process which begins by finding a feasible solution and determining whether it maximizes (or minimizes) the linear expression with which we are concerned. If not, which is usually the case, the Simplex Method indicates the proper direction to take in searching for a maximizing (or minimizing) solution, and if such a solution exists it will eventually be reached by this method. Dantzig and others have developed a *Revised Simplex Method* which reduces substantially the computational burden in solving linear programming problems on digital computers and makes more efficient use of these facilities than the original method.

Although there is no apparent connection between linear programming and the material discussed earlier in this book, it is a curious fact that the solution of a zero-sum two-person game (see Chapter 6) is actually equivalent to the solution of a linear program. As the reader will recall, each player in such a game tries to minimize his losses (or expected losses), and expressing these losses linearly in terms of the probabilities with which he selects the various strategies, he arrives at a linear program. We shall not go into this in any detail, but let us point out that games which are more complicated than the ones treated in Chapter 6 are practically always solved by means of the Simplex Method after they have been reformulated as linear programs. With this method we obtain optimum strategies for both players and also the value of the game.

EXERCISES

1. Referring to the example on page 437, draw the polygon of feasible solutions and verify that the transportation cost $6x + y + 194$ has a minimum of $196.

2. A manufacturer of a line of patent medicines is preparing a production run on Medicines A and B. There are sufficient ingredients on hand to make 20,000 bottles of A and 40,000 bottles of B, but there are only 45,000 bottles into which either of the medicines can be put. Furthermore, it takes 3 hours to prepare the ingredients to fill 1,000 bottles of Medicine A, it takes 1 hour to prepare the ingredients to fill 1,000 bottles of Medicine B, and there are 66 hours available for this operation. The profit is 8 cents per bottle for A and 7 cents per bottle for B. How should the manufacturer schedule production of the two medicines so as to maximize his profit from their sale? What is the maximum profit?

3. A processor of animal feeds makes one item for which food supplements I, II, and III are required in quantities of at least 16, 20, and 36 units, respectively.

The supplements are not available in pure form, but the processor can buy Compound A (each ounce of which contains 2, 1, and 1 units, respectively, of the three supplements) and Compound B (each ounce of which contains 1, 2, and 6 units, respectively, of the three supplements).

(a) If Compounds A and B cost 1 cent and 3 cents per ounce, respectively, what quantities each of the two compounds should the processor use so as to meet the minimum standards on the three supplements at the lowest possible cost?

(b) What quantities of the two compounds should he buy in order to meet these minimum dietary standards at minimum cost if Compounds A and B cost 1/2 cent and 3 cents per ounce, respectively?

4. The owner of a large kennel wants to be sure that a certain breed of growing dog gets in his food at least 2,000 milligrans of calcium and at least 1,000 milligrams of phosphorus per day. The owner buys prepared dry foods A, B, and C in large lots at a cost of 0.5, 0.3, and 0.4 cents per ounce. The calcium and phosphorus contents (in milligrams) of one ounce each of the three foods is shown below:

Mineral	A	B	C
Calcium	100	50	80
Phosphorus	40	40	10

If the owner wishes to meet these daily nutritional requirements at the lowest possible cost, what quantities each of the three foods should he give to each dog? What is the minimum daily cost per animal?

5. A company makes two novelty cloth toys, A and B, which must be cut and finished. To make, one dozen of Toy A requires 2 hours in the cutting room and 1 hour in the finishing room; it requires 4 hours and 5 hours, respectively, in the two rooms to make one dozen of Toy B. During the schedule period there are 80 hours available in the cutting room and 70 hours available in the finishing room. The profit to the company is $12 a dozen on Toy A and $25 a dozen on Toy B. How many units (dozens) of each toy should the company schedule for production during this period in order to maximize its profit from their sale? What is the maximum profit?

6. Subject to the restrictions $x \leq 6$, $x + 2y \leq 10$, $y \leq 4$, $x \geq 0$, and $y \geq 0$,

(a) maximize $P = 2x + 7y$;

(b) maximize $P = 3x + 6y$.

7. A large national restaurant chain wants to buy 13,000 No. 10 cans of chili con carne with 6,000, 3,000, and 4,000 cans, respectively, to be delivered to Warehouses A, B, and C. Suppliers 1 and 2 submit bids offering 5,000 and 8,000 cans, respectively. The buyer's procurement evaluation section calculates the delivered unit cost to the company by adding the supplier's unit price (f.o.b. origin), the unit shipping charge, and the unit warehouse handling cost. The

delivered unit costs thus arrived at are the following: for Supplier 1, $2.60, $2.30, and $2.45 at Warehouses A, B, and C, respectively; for Supplier 2, $2.50, $2.40, and $2.55 at Warehouses A, B, and C, respectively. What procurement policy will fill the requirements at the lowest possible cost?

INVENTORY PROBLEMS

For many years, businessmen have been as much concerned with problems of managing inventory as with any other problem of running a business. Inventory is listed in the balance sheet as a current asset, and it is thus considered a part of working capital (the excess of current assets over current liabilities). Certainly, the $170-odd billion of raw materials, work in progress, and finished goods inventories held nowadays by American manufacturers, wholesalers, and retailers represent wealth of a sort. However, inventories are not an end in themselves; they are intended to be sold, not held, (hopefully) at a profit large enough to encourage enterprisers to continue to operate. On the other hand, inventories are often business "assets" in name only, being a severe drain on financial resources and one of the main causes of business failures.

Until recently, researchers have provided little help in solving inventory problems; about all any enterpriser could do was follow some vague, intuitive course leading (again, hopefully) to an inventory of such size that inventory costs were "not unreasonably high" and "not too many" sales were lost because of stockouts. In the last few years, however, the inventory problem has been studied extensively, and many recent developments such as centralized vendor control with decentralized procurement, blanket orders, direct releases, and cubic air space storage at point of use (rather than central storage) have greatly increased the efficiency of total materials management.

Beyond these developments, various quantitative techniques have helped greatly to improve materials handling ability. Restrictions imposed on early mathematical inventory models (which were largely concerned with cost minimization and ignored demand) have been relaxed to some extent, and more generally useful models have been derived. The statistical analysis of daily field sales reports, transmitted to a central computer over a teletype network, has enabled large manufacturers to reduce field, raw materials, and total inventories and to increase operating efficiency by relating manufacturing operations more closely to consumer demand. Aside from its direct contribution to individual companies, improved inventory control has had a stabilizing effect on the total economy; the increasing use of the new quantitative methods has been a major factor in preventing large inventory buildups, thus lessening the risk of an "inventory depression."

Model-building techniques have nowadays advanced to the point where fairly sophisticated models of many practical inventory systems can be devised, leading to answers to the important questions of how much to order and when to order. The one we shall consider first is a simple *deterministic* item inventory model in which, say, over a period of one year a known number of units is to be supplied to "customers" at a constant rate, the time required to replenish stock (the "lead time") is negligible, and no shortage in the item will ever be allowed. After an amount is ordered and received, the stock is permitted to decline to zero, at which point another order (for items immediately available) is placed. During the year, this process is repeated a number of times—items are ordered, received, used up, ordered, received, used up, and so on, as illustrated in Figure 16.2.

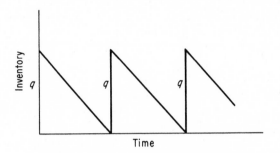

FIGURE 16.2 A deterministic demand

We referred to this model as *deterministic* since there are no random elements, that is, nothing is left to chance. The only question there is in this model is *how much to order each time;* when to order does not present a problem in this kind of situation.

In order to derive a formula for the total variable cost C of maintaining inventory on the item for a year (or some other period), let us introduce the following notation: R is the total demand for the year (or other period), K is the order cost (the cost of placing each order regardless of size), I is the inventory carrying cost (the cost of carrying one unit in inventory for the given period of time), and q is the fixed quantity ordered at any one time.* Since the total demand for the year is R units and each order is of

---

* In connection with I and K, we are concerned only with costs which may change with modifications in inventory policy. The principal costs which may enter into the carrying cost I are interest, insurance, taxes, depreciation, obsolescence, and storage costs; the order cost K is the sum of the various relevant costs of ordering materials or processing an order—forms, envelopes, stamps, personnel, and so on.

size q, the number of orders that will have to be placed is R/q, and the corresponding cost is $K(R/q)$. Furthermore, the average inventory size is $q/2$, as can be seen from Figure 16.2, so that the total inventory carrying cost for the year is $I(q/2)$ and the total variable cost of maintaining inventory on the item for a year is

$$ C = \frac{1}{2} Iq + \frac{KR}{q} \qquad \star $$

Having derived this formula expressing the total variable cost of maintaining inventory for a year (or some other period) in terms of I, K, R, and q, we can now determine that value of q, say q^*, which will minimize this cost. Using elementary calculus we find that

$$ q^* = \sqrt{\frac{2RK}{I}} \qquad \star $$

and substituting this expression for q^* in the formula for C, we also find that the minimum cost itself is

$$ C^* = \sqrt{2RKI} \qquad \star $$

To illustrate the use of these formulas, suppose that a large electrical contractor estimates that it will require 750 dozen switches of a certain type in the coming year. The fixed cost of placing an order is \$6, and the inventory carrying cost is 40 cents per dozen per year. On the basis of this information, we want to find the optimum order size and the minimum cost of inventorying the item for the year. Assuming that the deterministic model above applies, we substitute $R = .750$, $K = 6$, and $I = 0.40$ into the formula for q^*, obtaining

$$ q^* = \sqrt{\frac{2(750)(6)}{0.40}} = \sqrt{22{,}500} = 150 \text{ dozen} $$

Substituting these same constants into the formula for C^*, we have

$$ C^* = \sqrt{2(750)(6)(0.40)} = \sqrt{3{,}600} = \$60 $$

Hence, the switches should be ordered in 150-dozen lots and the (minimum) total inventory cost on this item for the year is \$60, of which \$30 results from placing five orders at a cost of \$6 each, and the other \$30 results from carrying an average inventory of 75 dozen switches.

Although deterministic inventory models are used in the control of many items in practice, there are a great many situations in which both the demand for an item and the lead time (the time required to replenish stock) depend in some way on chance and consequently fluctuate from time to time. In these cases both demand and lead time are random variables whose behavior is characterized by probability distributions. To illustrate a situation in which the *demand* for an item (but not the lead time) is random, suppose that a baker bakes for stock each morning a certain cake which costs him $2 to bake and which he sells for $4, giving him a profit of $2 on each cake sold. The baker guarantees cakes fresh every day, and all unused cakes are in fact disposed of at a loss of $2 each (equal to his production cost). Finally, the baker never rebakes this item on any given day, so it is impossible for him to increase his supply during the day no matter how good early sales might be.

If the daily demand were known, there would be no problem; the baker would maximize his profits on this item by baking each day the exact number of cakes which would be demanded. Unfortunately, the demand for any given day is not known; however, the baker has kept extensive historical records on the demand for all of his bakery items. For this cake the records show that the probabilities that 0, 1, 2, 3, 4, 5, and 6 or more cakes will be demanded on a given day are, respectively, 0.05, 0.05, 0.15, 0.25, 0.30, 0.15, and 0.05. A review of all relevant factors suggests no change in the underlying demand for the cake in the foreseeable future, so this distribution is considered the appropriate one for this item. With this information at hand, we can now determine how many cakes the baker should stock each day so as to maximize his *expected profits* from their sale, and we shall consider this to be an entirely satisfactory solution to this stocking problem.

We show directly below the *demand distribution* in tabular form and since

Demand Number of Cakes	Probability
0	0.05
1	0.05
2	0.15
3	0.25
4	0.30
5	0.15
6 or more	0.05

in calculating the expected daily profits we shall also need the probabilities of a demand for "1 or more cakes," "2 or more cakes," . . . , and "6 or more cakes," we have cumulated the above probabilities and show the results in the following table:

Demand Number of Cakes	Probability
1 or more	0.95
2 or more	0.90
3 or more	0.75
4 or more	0.50
5 or more	0.20
6 or more	0.05

For no cakes stocked, the profit is obviously $0, whatever the demand. For 1 cake stocked, the demand for "1 or more" (or "at least 1") has a probability of 0.95, so that the expected profit is

$$4(0.95) - 2 = \$1.80$$

For 2 cakes stocked, the demand for exactly 1 cake has a probability of 0.05 and the demand for "2 or more" has a probability of 0.90, so that the expected profit is

$$4(0.05) + 8(0.90) - 4 = \$3.40$$

For 3 cakes stocked, the demand for exactly 1 cake has a probability of 0.05, the demand for exactly 2 cakes has a probability of 0.15, the demand for "3 or more" cakes has a probability of 0.75, so that the expected profit is

$$4(0.05) + 8(0.15) + 12(0.75) - 6 = \$4.40$$

Continuing in this way (see also Exercise 6 on page 446), we obtain the expected profits for 4, 5, and 6 cakes. All these results are shown in the following table:

Number of Units Stocked	Expected Profit
0	$0.00
1	1.80
2	3.40
3	4.40
4	4.40
5	3.20
6	1.40

It is not possible to carry these calculations further than this since the probabilities of a demand for exactly 6, 7, and so on, cakes are not known. It is not necessary to go further, though, since the maximum return has already been passed and the expected profits will continue to decrease.

Inspecting the preceding table, we find that the baker can maximize his expected profit on this item by stocking either 3 or 4 cakes each day. We note, however, that if he stocks only 3 cakes there will be a large unsatisfied demand (the probability of a demand for "4 or more" cakes is 0.50). Since this unsatisfied demand may ultimately depress his market, the best policy for the baker is to stock 4 cakes each day.

The purpose of this example has been to show how chance elements can be taken into account in inventory problems. When, as is very often the case, the lead time is also subject to chance variations the problem becomes somewhat more complicated.

EXERCISES

1. A large insurance company requires 1,000 boxes of a certain form each year. The cost of carrying one box in inventory for the year is 50 cents, and the fixed cost of placing an order with a single vendor is $10. What is the optimum order quantity on this item? How many times should lots of this size be bought during the year? What is the minimum total cost of maintaining inventory on this item for a year? Of this cost, how much is the carrying cost and how much is the order cost?

2. Referring to Exercise 1 above, if orders are placed with multiple vendors, the fixed order cost is $14. If it is necessary to procure the item from multiple vendors, what are the optimum order quantity and the minimum total yearly inventory cost on the item?

3. In Exercise 1 above, $R = 1,000$, $K = 10$, and $I = 0.50$. Letting $q = 100, 120, 140, 160, 180, 200, 220, 240, 260, 280$, and 300, respectively, calculate the values of $\frac{1}{2}Iq$ $(= q/4)$ and plot them on a piece of ordinary graph paper with q measured along the horizontal scale. Calculate also the values of KR/q $(= 10,000/q)$ for the same values of q and plot them on the same diagram. Graphically adding the values obtained for $\frac{1}{2}Iq$ and KR/q for the various values of q, plot the corresponding total inventory costs and connect them with a smooth curve. Examination of this curve shows the relative insensitivity of this model. Within a range of ± 25 per cent of the optimum quantity, the curve is relatively flat; thus, the order quantity could be adjusted to a quantity discount lot or to the nearest standard or package lot (dozen, gross, etc.) without too much penalty in cost.

4. A manufacturer uses the trend equation $y' = 1,000 + 120x$ (origin 1966; x units, 1 year; y, total annual usage) to get the estimated 1971 requirements for

a certain mechanical part. The unit cost of the part itself is $5. The interest part of the unit carrying cost is the weighted average of the interest paid on the following capital used in the business: $8 million in capital stock at 10 per cent, $1 million in bonds at 4 per cent, and $1 million in notes at 6 per cent. The carrying cost includes also a 9 per cent charge per unit to cover insurance, taxes, storage, and the like. The fixed cost of placing an order is $12.

(a) What is the optimum order size for this item? What is the minimum total inventory cost of the item for the year?

(b) If the standard lot size is 100 and the order size is rounded to the nearest standard lot size, how many orders will be placed during the year for how many units each? What will be the total inventory cost? Compare this cost to the minimum cost.

5. Suppose that in Exercise 4 above the company has borrowed money at 6 per cent specifically to finance its inventory and so decides to include only 6 per cent as the interest part of the carrying cost. Suppose further that in the company's judgment the estimated requirements based on the trend equation should be increased by 12.5 per cent to take care of an expected higher level of activity on this part in 1971 caused by aging equipment.

(a) Taking these two things into account, recalculate the optimum order quantity and the minimum total inventory cost.

(b) Repeat part (b) of Exercise 4.

6. Referring to the example on page 443, show that the baker's expected profits for stock levels of 4, 5, and 6 cakes are $4.40, $3.20, and $1.40, respectively.

7. A street vendor of corsages in San Francisco's theater district picks up his evening's supply from a wholesale florist who closes at 6 P.M.; consequently, the vendor cannot increase his supply during the sales period no matter how good his early sales might be. The flowers are perishable and the vendor has no storage facilities, so that unsold items are a complete loss. The cost to the vendor is $1 per corsage and the selling price is $4.00. On the basis of past experience and future prospects, it seems reasonable to assume that the probabilities of a daily demand for 0, 1, 2, 3, 4, 5, and 6 or more corsages are 0.05, 0.15, 0.30, 0.25, 0.15, 0.05, and 0.05, respectively. How many corsages should the vendor stock each day in order to maximize his expected profit from their sale?

8. A vendor can buy an item for 70 cents and sell it for $1.50. The probabilities that 0, 1, 2, 3, 4, and 5 or more items will be demanded each day are, respectively, 0.05, 0.15, 0.30, 0.25, 0.15, and 0.10. Show that the daily expected profits resulting from stocking 0, 1, 2, 3, 4, and 5 items are, respectively, $0, $0.725, $1.225, $1.275, $0.95, and $0.40. How many items should the vendor stock each day?

9. Suppose that as many as 2 "boosters," each with a lifetime of only 1 day, may have to be introduced into a certain chemical process each day. All unused boosters must be physically destroyed each day at a cost of $1 for labor, transportation, and the like. It is estimated that the shortage cost (the loss resulting from not having a booster available when it is needed and having to

get one in a real hurry) is $5. The probabilities of a demand for 0, 1, or 2 boosters are, respectively, 0.50, 0.40, and 0.10. Find the stock level which will minimize the expected daily disposal and shortage cost.

10. Repeat Exercise 9 with the modification that the loss resulting from not having a booster available when needed is $10.

11. Referring to Exercise 9, show that it would be preferable not to stock any of the boosters if and only if the loss from not having a booster available when needed is less than $1.

12. Suppose that as many as 5 "boosters," each with a lifetime of only 1 day, may have to be introduced into a certain chemical process each day. The cost to handle a booster in stock and carry it for a day is 10 cents. All unused boosters must be physically destroyed each day at a cost of $1 each. It is estimated that the loss resulting from not having a booster available when it is needed (the shortage cost) is $8 per booster. The respective probabilities of a demand for 0, 1, 2, 3, 4, and 5 boosters are 0.10, 0.15, 0.25, 0.30, 0.15, and 0.05. Show that the expected daily totals of the carrying, disposal, and shortage costs are $19.20, $12.20, $6.55, $3.15, $2.45, and $3.10 for stock levels of 0, 1, 2, 3, 4, and 5 boosters, respectively, so that in order to minimize these expected total costs the company should stock 4 boosters each day.

WAITING LINES

When a customer arrives at a cafeteria for service, he often finds the server busy with an earlier arrival and so he must wait; while he is waiting, others may arrive and a fairly long waiting line, or queue, can develop. Such waiting lines are not restricted to persons waiting for service at a food counter, a market checkout stand, an airline ticket office, a bank window, and so on, but they arise quite generally. Ships and trucks waiting to be unloaded at receiving docks, aircraft to be landed at airports, court cases to be heard, relief applications to be processed, machines to be repaired, to mention but a few, are all examples of waiting lines. If both the arrival times of "customers" and the service times at the "counter" were completely regular (e.g., a customer every 5 minutes and a service completed in exactly 4 minutes), proper scheduling could prevent waiting lines altogether. If, however, there is some chance element in arrival times, service times, or both (e.g., *on the average* a customer every 5 minutes and a service completed *on the average* in 4 minutes), waiting lines will sooner or later appear.

In all situations where units arrive at some facility which serves and eventually releases them, there are three basic elements in terms of which the servicing system can be described: (1) the *distribution of arrivals* (the input process), (2) the *service mechanism* including the distribution of service times, and (3) the *queue discipline*. Schematically, the overall system may be pictured as in Figure 16.3.

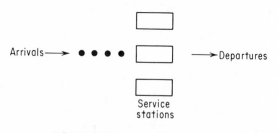

Arrivals ⟶ ● ● ● ● ⟶ Departures

Service stations

FIGURE 16.3 A servicing system

Generally speaking, arrival times are irregular and more or less widely scattered over any fixed interval of time, and the way in which a queue forms may be described in either one of two ways. First, we may give an *arrival distribution* which specifies the probability that x customers will arrive during a given period of time. Among the various theoretical distributions which might describe such a process, the *Poisson distribution* is by far the most widely used; roughly speaking, it corresponds to the assumption that an arrival during one *small* interval of time is as likely as an arrival during any other *small* interval of equal length, and the assumption that successive arrivals are independent. Using a formula similar to that on page 171, we have

$$f(x) = \frac{(\lambda t)^x \cdot e^{-\lambda t}}{x!} \qquad \bigstar$$

for the probability that there will be x "Poisson arrivals" during a time interval of length t. Here λ (*lambda*) is the arrival rate, namely, the average number of arrivals per unit time. Thus, if there are on the average four arrivals per hour and $t = 1$ hour, the probabilities of $0, 1, 2, 3, 4, 5, \ldots$, arrivals are, respectively, $0.02, 0.07, 0.15, 0.20, 0.20, 0.16, \ldots$. Probabilities like these are best obtained from special tables of Poisson probabilities; but they can be calculated directly with the use of the above formula, with $e^{-\lambda t}$ (in this case $e^{-4} = 0.018$) obtained from Table IX at the end of this book.

Instead of describing the way in which a queue forms by means of a probability distribution for the number of arrivals over a fixed period of time, an alternative (and often preferable) way is to give a distribution for the times (or delays) between arrivals for service. When dealing with "Poisson arrivals," we can obtain such a distribution, called an *interarrival distribution*, by means of the following argument: To find the probability that the time between two successive arrivals lies between two arbitrary numbers, say, a and b, we have only to multiply the probability of 0 arrivals up to time a by the probability of at least 1 arrival between a and b. Using the above formula for the Poisson distribution, we find that the probability of 0 arrivals up to time a is $e^{-\lambda a}$, and that the probability of at least 1 arrival on the interval from a to b (1 *minus* the probability of 0 arrivals) is $1 - e^{-\lambda(b-a)}$. Multiplying these two probabilities we get

$$e^{-\lambda a}[1 - e^{-\lambda(b-a)}] = e^{-\lambda a} - e^{-\lambda b} \qquad \star$$

for the probability that the time between two successive arrivals assumes a value between a and b. We refer to such probabilities as *exponential probabilities*, and to the corresponding interarrival distribution as an *exponential distribution*. The graph of such an exponential distribution is shown in Figure 16.4, where the probability of obtaining a value between a and b is,

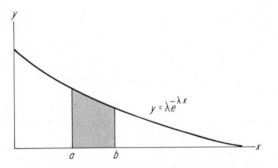

FIGURE 16.4   An exponential distribution

as always, given by the area under the curve between a and b. Referring again to the example where there were on the average 4 arrivals per hour, we find, for instance, that the probability of a delay of less than 6 minutes (0.10 of an hour) between successive arrivals is

$$e^{-0} - e^{-4(0.10)} = 1 - 0.67 = 0.33$$

after substituting $a = 0$ and $b = 0.10$. Similarly, substituting $a = 0.10$ and $b = 0.20$, we find that the probability of a delay between 6 and 12 minutes between successive arrivals is

$$e^{-4(0.10)} - e^{-4(0.20)} = 0.67 - 0.45 = 0.22$$

(The necessary values of e^{-x} were obtained from Table IX at the end of the book.)

So far as the service mechanism is concerned, there are essentially two problems: the first concerns the number of service lines that are available (the number of attendants, servers, unloading docks, bank windows, etc.) and the other concerns the length of time the service requires. Even though the service is nominally the same in a given situation, the actual times required to perform the service are usually not constant. Partly as a matter of mathematical convenience and partly because of its suitability to describe practical service time situations, the exponential distribution we have just introduced is very often used also as the *service time distribution*. The parameter μ is now used instead of λ, and it represents the average number of complete services that can be performed per unit time. In what follows, we shall restrict ourselves to situations where there is only one service line and the service distribution is exponential.

The way in which units in the queue are served when servicing facilities become available is called the "queue discipline." In many instances, service is on a "first come, first served" basis (as when airplanes are "stacked up" over an airfield in a holding pattern waiting to land), sometimes it is on a random basis (as when broken parts are tossed into a bin as they arrive and the next part repaired is selected more or less by chance), or on a "last come, first served" basis (as in bankruptcy proceedings of various sorts). Often, as in the case of messages waiting to be transmitted by telegraph, or work done by a computer, priorities are assigned to customers and some juggling takes place in the line as new arrivals appear. The exact type of discipline prescribed to govern a queue in a given situation is a matter of policy, and it is often chosen so as to meet certain goals.

Once the arrival distribution, the number of service lines, the service time distribution, and the queue discipline have been specified, various probabilities and various averages can be calculated. In what follows, we shall consider a problem arising in a tool-repair center, where one attendant handles the service needs of a large number of machinists. It will be assumed that the arrivals have a Poisson distribution, the service-time distribution is exponential, queue discipline is "first come, first served," and all customers, from wherever they come, must form a single line for service. In the formulas which follow, λ will be the arrival rate (the average number of arrivals per unit time), μ will be the service rate (the average number of services per unit time), and their ratio $h = \dfrac{\lambda}{\mu}$ (called the *utilization factor*) is the proportion of the time the service facilities are in use *or* the probability that a new arrival will have to wait. This assumes that $h < 1$, since otherwise the queue will grow indefinitely long. Using these assumptions, it can be shown that

$$\text{Average number of customers in the system at a given time} = \frac{h}{1-h}$$

$$\text{Average time a customer spends in the system} = \frac{1}{\mu - \lambda}$$

$$\text{Average length of the waiting line} = \frac{h^2}{1-h}$$

$$\text{Average time a customer must wait for service} = \frac{h}{\mu - \lambda}$$

★

Assuming that the arrival rate of machinists at the tool repair center is $\lambda = 2$ per hour and that the service rate is $\mu = 3$ per hour, we find that the facilities are in use $h = \dfrac{2}{3}$ of the time, the average number of machinists waiting for service or being served at any given time is $\dfrac{h}{1-h} = \dfrac{2/3}{1 - 2/3} = 2$, the average time a machinist spends waiting and being served is $\dfrac{1}{\mu - \lambda} = \dfrac{1}{3-2} = 1$ hour, the average number of men waiting is $\dfrac{h^2}{1-h} = \dfrac{4/9}{1 - 2/3} = \dfrac{4}{3}$, and the average time spent waiting for service is $\dfrac{h}{\mu - \lambda} = \dfrac{2/3}{3-2} = \dfrac{2}{3}$ hours.

If we assume, furthermore, that the hourly wage paid the attendant is $1.50 and that the hourly cost of a machinist away from his work is $4.00, we find that the average cost of operating the system for an 8-hour day is

$$16 \cdot 1 \cdot 4 + 8 \cdot 1\tfrac{1}{2} = \$76$$

Here 16, the average number of arrivals per day, is multiplied by 1, the average time a machinist spends in the system, and then by 4, the hourly cost of a machinist away from his work; to this we added $8 \cdot 1\tfrac{1}{2}$, namely, the amount paid to the attendant.

Calculations like these are important because they enable us to decide, for example, whether it might be wise to replace the attendant with one whose hourly rate is $3.50, but who can service on the average $\mu = 4$ customers per hour (see Exercise 4 below). Using a somewhat more complicated theory, we might also consider the possibility of hiring *two or more* attendants at $1.50 an hour, with each being able to handle on the average 3 customers per hour. It is interesting to note that for two such attendants the cost of operating the system for an 8-hour day is reduced from $76 to $48; this is accounted for by the fact that with one attendant a machinist spends on the average $\frac{2}{3}$ hours waiting and only $\frac{1}{3}$ hours being served (see Exercise 2 below).

EXERCISES

1. Suppose that at a small stores supply room of a large company, service is provided by a single attendant on a "first come, first served" basis. Arrivals on the day shift have the Poisson distribution with a mean delay between arrivals of 8 minutes (and hence a mean arrival rate of 1/8 per minute). If service times have an exponential distribution with a mean service time of 5 minutes (and hence a mean service rate of 1/5 per minute), find
   (a) the average number of customers in the system (either in wait or in service) at a given time;
   (b) the average time a customer spends in the system;
   (c) the average length of the waiting line;
   (d) the average time a customer must wait for service.

2. The difference between the average time a customer spends in a service system and the average time a customer must wait for service is the average service time itself. Verify symbolically that, in fact,

$$\frac{1}{\mu - \lambda} - \frac{h}{\mu - \lambda} = \frac{1}{\mu}$$

3. The difference between the average number of customers in the system at a given time and the average length of the waiting line is the fraction of the time the service facility is in use. Verify this symbolically by showing that

$$\frac{h}{1 - h} - \frac{h^2}{1 - h} = h$$

4. In the example on page 451 we found that the average daily cost of operating the system described there was $76. Determine whether or not it would be more economical to replace the attendant with another man who can complete a service on the average in 15 minutes and whose hourly rate is $3.50.

5. For an 8-hour day, the mean arrival rate at a service facility is 4 per hour and the mean service rate is 5 per hour. The customers are machines and the cost of being in the service system and not in production is $10 per hour. If it is possible

to increase the mean service rate to 6 per hour by making certain changes in personnel and equipment costing $15 a day, would this change be worthwhile?

6. Under the assumptions of the example on page 451, it can be shown that the probability that exactly n units are in the system at a given time is given by

$$P_n = (1 - h)h^n$$

for $n = 0, 1, 2, \ldots$. With reference to the example on that page, find

(a) the probability that no machinists are in the system at a given time;

(b) the probability that exactly one machinist is in the system at a given time;

(c) the probability that exactly two machinists are in the system at a given time;

(d) the probability that three or more machinists are in the system at a given time.

7. Referring to Exercise 1, calculate the respective probabilities that at any given time $0, 1, 2, 3, 4, 5, 6,$ or 7 customers are in the system described in that exercise. Use these figures to calculate the approximate mean of this probability distribution and compare the result with that obtained in part (a) of Exercise 1. (Use the formula of Exercise 6.)

MONTE CARLO METHODS

In Chapter 8, particularly in Technical Note 5, we indicated how tables of random digits (or other gambling devices) can be used to simulate sampling experiments. In recent years, similar techniques have been applied to a great variety of problems in the physical, social, and biological sciences under the general heading of *Monte Carlo methods*. They have been applied to the solution of problems leading to mathematical equations which actually cannot be solved by direct means, to the solution of problems whose solution would otherwise be too costly or require too much time, and to the solution of problems where experimental conditions simply cannot be reproduced. (Among the latter we find, for example, studies of the spread of cholera epidemics which, of course, cannot be experimentally induced.) Very often, the use of Monte Carlo methods eliminates the cost of building and operating expensive physical equipment; it is thus used in the study of collisions of photons with electrons, the scattering of neutrons, and other complicated phenomena.

A classical application of Monte Carlo methods to a problem of pure mathematics is the determination of π, the ratio of the circumference to the diameter of a circle, by probabilistic means. Early in the eighteenth century, the French naturalist, George de Buffon, proved that if a very fine needle of length L is thrown at random on a board ruled with equidistant parallel lines, the probability that the needle will intersect one of the lines is $2L/\pi a$, where a is the distance between the lines. The remarkable thing about this result is that it involves the constant $\pi = 3.1415926\ldots$, which in elementary geometry is approximated by the circumferences of regular polygons inscribed in a circle of radius $1/2$. Buffon's result implies that if such a needle is actually tossed a great many times, the proportion of the time it crosses one of the lines provides an estimate of $2L/\pi a$ and hence an estimate of π since L and a are known. Early experiments of this kind conducted in the middle of the nineteenth century yielded estimates of 3.1519 based on 5,000 trials and 3.155 based on 3,204 trials. Much more extensive experiments have been performed since, among them an exhibit at the 1939–1940 New York World's Fair. (An alternate way of estimating π with a Monte Carlo method is given in Exercise 1 on page 461.)

Simulation by means of statistical sampling, that is, Monte Carlo techniques, has found wide applications in business research. Such methods are used to solve inventory problems, questions arising in connection with waiting lines, advertising, competition, the allocation of resources, scheduling of operations, and situations involving overall planning and organization. To conclude our brief introduction to operations research, we shall illustrate the use of Monte Carlo techniques with a simple waiting-line problem arising in a company's planning of a production operation. The company wants to determine whether a certain automatic wash table, designed to perform a final wash of a product at the *constant* rate of one per minute, is adequate. Arrivals at the wash table are characterized by a Poisson distribution with a mean arrival rate of one every two minutes; that is, the arrival distribution is a Poisson distribution with $\lambda = \dfrac{1}{2}$. There is a single service line, service is on a "first come, first served" basis, and the service time is *exactly* one minute.

The only aspect of this problem which has to be simulated using Monte Carlo methods is the arrival of the product at the wash table, and we shall do this by determining the times between successive arrivals with the use of a table of random digits. The method we shall use here is somewhat different from the one discussed on page 230, since the *exponential* interarrival distribution (see page 449), like the normal curve, is a continuous distribution. First we construct the cumulative interarrival distribution by calculating the probability that the time between arrivals will be less than t minutes for various values of t. Substituting $a = 0$ and $b = t$ into the formula on page 449, we find that the probability of a delay between arrivals of less than t is

$$e^{-\lambda \cdot 0} - e^{-\lambda t} = 1 - e^{-\lambda t}$$

and for our particular example this equals

$$1 - e^{-(1/2)t}$$

Substituting $t = 0, 1, 2, 3, \ldots$, and 10 into this last expression and using Table IX, we obtain the values shown in the following table:

Time Between Arrivals, t	Probability That Time Between Arrivals Is Less Than t
0	0.000
1	0.393
2	0.632
3	0.777
4	0.865
5	0.918
6	0.950
7	0.970
8	0.982
9	0.989
10	0.993

We did not go beyond $t = 10$ in this table, since the probability of getting an interarrival time greater than 10 minutes is, for all practical purposes, negligible.

Now we plot this cumulative probability distribution as we have done in Figure 16.5, and we are ready to sample interarrival times with the use

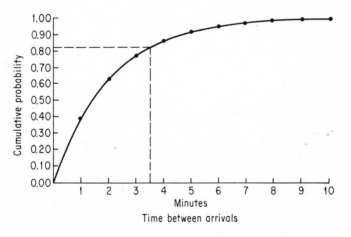

FIGURE 16.5   Cumulative interarrival distribution

Random Number	Time Between Arrivals	Time of Arrival	Serviced		Waiting Time in Queue
			From	Until	
		0.00	0.00	1.00	0.00
29	0.75				
		0.75	1.00	2.00	0.25
45	1.25				
		2.00	2.00	3.00	0.00
80	3.25				
		5.25	5.25	6.25	0.00
19	0.50				
		5.75	6.25	7.25	0.50
13	0.35				
		6.10	7.25	8.25	1.15
67	2.25				
		8.35	8.35	9.35	0.00
37	0.95				
		9.30	9.35	10.35	0.05
57	1.75				
		11.05	11.05	12.05	0.00
08	0.20				
		11.25	12.05	13.05	0.80
36	0.90				
		12.15	13.05	14.05	0.90
13	0.35				
		12.50	14.05	15.05	1.55
01	0.05				
		12.55	15.05	16.05	2.50
45	1.25				
		13.80	16.05	17.05	2.25
12	0.30				
		14.10	17.05	18.05	2.95
70	2.45				
		16.55	18.05	19.05	1.50
98	8.00				
		24.55	24.55	25.55	0.00
12	0.30				
		24.85	25.55	26.55	0.70
11	0.25				
		25.10	26.55	27.55	1.45
82	3.50				
		28.60	28.60	29.60	0.00
55	1.65				
		30.25	30.25	31.25	0.00
76	2.85				
		33.10	33.10	34.10	0.00

Random Number	Time Between Arrivals	Time of Arrival	Serviced From	Until	Waiting Time in Queue
29	0.75				
		33.85	34.10	35.10	0.25
91	4.75				
		38.60	38.60	39.60	0.00
41	1.12				
		39.72	39.72	40.72	0.00
51	1.50				
		41.22	41.22	42.22	0.00
26	0.67				
		41.89	42.22	43.22	0.33
73	2.65				
		44.54	44.54	45.54	0.00
87	4.00				
		48.54	48.54	49.54	0.00
11	0.25				
		48.79	49.54	50.54	0.75
25	0.65				
		49.44	50.54	51.54	1.10
29	0.75				
		50.19	51.54	52.54	1.35
28	0.70				
		50.89	52.54	53.54	1.65
19	0.50				
		51.39	53.54	54.54	2.15
57	1.75				
		53.14	54.54	55.54	1.40
25	0.65				
		53.79	55.54	56.54	1.75
01	0.05				
		53.84	56.54	57.54	2.70
46	1.30				
		55.14	57.54	58.54	2.40
37	0.95				
		56.09	58.54	59.54	2.45
34	0.87				
		56.96	59.54	60.54	2.58
89	4.62				
		61.58	61.58	62.58	0.00
82	3.50				
		65.08	65.08	66.08	0.00
86	3.90				
		68.98	68.98	69.98	0.00
60	1.87				
		70.85	70.85	71.85	0.00

Random Number	Time Between Arrivals	Time of Arrival	Serviced From	Until	Waiting Time in Queue
72	2.62				
		73.47	73.47	74.47	0.00
00	0.00				
		73.47	74.47	75.47	1.00
66	2.20				
		75.67	75.67	76.67	0.00
15	0.37				
		76.04	76.67	77.67	0.63
71	2.50				
		78.54	78.54	79.54	0.00
34	0.87				
		79.41	79.54	80.54	0.13
66	2.20				
		81.61	81.61	82.61	0.00
37	0.95				
		82.56	82.61	83.61	0.05
33	0.85				
		83.41	83.61	84.61	0.20
22	0.58				
		83.99	84.61	85.61	0.62
95	6.00				
		89.99	89.99	90.99	0.00
08	0.20				
		90.19	90.99	91.99	0.80
56	1.70				
		91.89	91.99	92.99	0.10
49	1.40				
		93.29	93.29	94.29	0.00
95	6.00				
		99.29	99.29	0.29*	0.00
43	1.15				
		0.44	0.44	1.44	0.00
81	3.30				
		3.74	3.74	4.74	0.00
18	0.45				
		4.19	4.74	5.74	0.55
78	3.10				
		7.29	7.29	8.29	0.00
81	3.30				
		10.59	10.59	11.59	0.00
46	1.30				
		11.89	11.89	12.89	0.00
54	1.60				
		13.49	13.49	14.49	0.00

*To simplify the notation, we return to 0.00 after time 100.00.

Random Number	Time Between Arrivals	Time of Arrival	Serviced From	Serviced Until	Waiting Time in Queue
16	0.38				
		13.87	14.49	15.49	0.62
49	1.40				
		15.27	15.49	16.49	0.22
29	0.75				
		16.02	16.49	17.49	0.47
12	0.30				
		16.32	17.49	18.49	1.17
01	0.05				
		16.37	18.49	19.49	2.12
78	3.10				
		19.47	19.49	20.49	0.02
94	5.70				
		25.17	25.17	26.17	0.00
25	0.65				
		25.82	26.17	27.17	0.35
88	4.20				
		30.02	30.02	31.02	0.00
33	0.85				
		30.87	31.02	32.02	0.15
09	0.22				
		31.09	32.02	33.02	0.93
29	0.75				
		31.84	33.02	34.02	1.18
62	1.95				
		33.79	34.02	35.02	0.23
74	2.70				
		36.49	36.49	37.49	0.00
69	2.40				
		38.89	38.89	39.89	0.00
79	3.15				
		42.04	42.04	43.04	0.00
69	2.40				
		44.44	44.44	45.44	0.00
08	0.20				
		44.64	45.44	46.44	0.80
88	4.15				
		48.79	48.79	49.79	0.00
12	0.30				
		49.09	49.79	50.79	0.70
75	2.75				
		51.84	51.84	52.84	0.00
69	2.40				
		54.24	54.24	55.24	0.00

| Random Number | Time Between Arrivals | Time of Arrival | Serviced | | Waiting Time in Queue |
			From	Until	
36	0.90				
		55.14	55.24	56.24	0.10
52	1.55				
		56.69	56.69	57.69	0.00
25	0.65				
		57.34	57.69	58.69	0.35
61	1.90				
		59.24	59.24	60.24	0.00
94	5.70				
		64.94	64.94	65.94	0.00
58	1.80				
		66.74	66.74	67.74	0.00
54	1.60				
		68.34	68.34	69.34	0.00
29	0.75				
		69.09	69.34	70.34	0.25
28	0.70				
		69.79	70.34	71.34	0.55
75	2.75				
		72.54	72.54	73.54	0.00
86	3.95				
		74.49	76.49	77.49	0.00
98	8.00				
		84.49	84.49	85.49	0.00
90	4.70				
		89.19	89.19	90.19	0.00

of random numbers. To this end we draw two-digit random numbers, treating them as probabilities by placing a decimal point to the left of the first digit. Then we mark these probabilities on the vertical scale of Figure 16.5 and read off the times between arrivals by first going *horizontally* to the curve we fitted to the cumulative interarrival distribution and then *down vertically* to the "time between arrivals" scale. Thus, if we draw the random number 82, we mark 0.82 on the vertical scale and obtain a time between arrivals of 3.50 minutes as indicated in Figure 16.5.

In the table on pages 456–460, it is assumed that we start out with an arrival at time 0.00. The first column contains two-digit random numbers obtained from a table of random digits, the second column contains the corresponding times between arrivals obtained with the use of Figure 16.5 as has just been indicated, and the figures in the remaining columns are self-explanatory. Thus, the item with which we begin our simulated experiment is serviced from time 0.00 to time 1.00 and it did not have to wait; the second item arrives at time 0.75, it is serviced from time 1.00 to 2.00, and it had to wait 0.25 minutes from the time of its arrival until it began to get serviced. Note that time is measured in minutes in this table, beginning with the arrival of the first item, and that it is given to two decimals.

Not counting the first item which (by assumption) was served right away, we find that the average time spent waiting in the queue was $\frac{51.70}{100} = 0.517$ minutes, and this provides us with valuable information about the performance of the wash table, about the congestion of arrivals, and so on. We cannot check this value against the theory of the preceding section, since the service time was fixed and not exponentially distributed. However, appropriate theory shows that the *exact* value for the average waiting time in the queue is 0.50, and this shows that the approximation is very good even though the number of "trials" was fairly small. As in other sampling experiments, the precision of Monte Carlo results will, of course, increase with the number of trials.

So far we have used the sampling experiment only to estimate the average time spent waiting in the queue, but we might observe that there is additional information available on other aspects of the service system. Some of these are discussed in Exercises 2 and 3 below, and we might merely mention that the maximum time spent waiting was just under 3 minutes, the maximum time spent in the system was just under 4 minutes, and about 48 per cent of the "customers" did not have to wait for service.

EXERCISES

1. The following is an interesting and simple Monte Carlo technique for estimating π: With reference to Figure 16.6, suppose that we randomly select points inside the indicated unit square and that we note in each case whether or not the point falls inside the circle whose center is at the origin and whose radius is 1. The probability that such a point will fall inside the circle is the ratio of the area of the sector (quarter) of the circle to the area of the square, namely, $\pi/4$ divided by 1, or simply $\pi/4$. Thus, the proportion of points falling inside the circle provides an estimate of $\pi/4$ and hence four times this proportion is an estimate of π. The points are selected as follows: We take four consecutive random digits, say, 8016, and we look upon these digits as representing the point (0.80, 0.16), namely, the point whose x-coordinate is 0.80 and whose y-coordinate is 0.16. To check whether the point falls in or on the circle we have only to see whether $x^2 + y^2 \leq 1$; for the point (0.80, 0.16) we get $0.80^2 + 0.16^2 = 0.6656$, and it

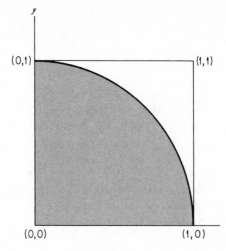

FIGURE 16.6

follows that the point lies in the circle. Similarly, the four random digits 8174 are interpreted as the point (0.81, 0.74) which lies outside the circle, since $0.81^2 + 0.74^2 = 1.2037$. Use this method to estimate π on the basis of 100 points selected in this fashion.

2. Considering the first 20 minutes of the Monte Carlo experiment of this section, it can be verified that there were no "customers" in the system for a total of 4.00 minutes, there was exactly one "customer" in the system for a total of 7.50 minutes, there were exactly two for a total of 4.30 minutes, three for a total of 2.50 minutes, and four for a total of 1.70 minutes. Use these figures to estimate the average number of "customers" in the system at any given time.

3. Using the figures of Exercise 2, show that for the given period the average length of the waiting line was 0.72 customers.

4. Suppose that we have an inventory item which costs $2 and sells for $4. Unsold items have no value and are destroyed at 0 cost at the end of the day. The probabilities that 0, 1, 2, 3, 4, 5, or 6 or more items will be demanded on any one day are 0.05, 0.15, 0.30, 0.25, 0.15, 0.05, and 0.05, respectively.

   (a) Devise a Monte Carlo sampling scheme for simulating the daily demand for this item in a 200-day period. (*Hint:* Proceed as we did in Technical Note 5 on page 232.)

   (b) Conduct the sampling experiment and construct a frequency table showing the number of days on which the demand was for 0, 1, 2, . . . , 6 or more items.

   (c) Referring to the sampling experiment of part (b) determine the respective profits over the 200-day period if the number of items stocked each day was 0, 1, 2, 3, 4, 5, or 6. Judging by this experiment, which stock level seems to maximize the expected profit?

5. As a prestige item, a pastry shop has decided to produce for general sale a very elegant kind of cake formerly available only by special order. The plan is to put the cake on sale each Saturday morning and destroy all unsold cakes after closing since they will not hold over the weekend. Due to the availability of ingredients, the number of cakes the shop can bake varies somewhat, and the probabilities that it will have 2, 3, 4, 5, or 6 cakes available on any given Saturday are 0.05, 0.30, 0.40, 0.15, and 0.10, respectively. Analysis of past orders leads the shop to believe that the weekly demand for the cakes can be approximated by a Poisson distribution with a mean of 4, so that the probabilities of a demand for 0, 1, 2, . . . , or 9 cakes are, respectively, 0.02, 0.07, 0.15, 0.20, 0.20, 0.16, 0.10, 0.06, 0.03, and 0.01.

   (a) Use the method described in Technical Note 5 on page 232 to simulate the production and sales of the cakes on 200 Saturdays.

   (b) If it costs 50 cents to produce one of these cakes, it sells for $1.75, and there is a 10-cent loss in good will for each customer turned away after the cakes are all sold, use the results of part (a) to estimate the pastry shop's average weekly profit on this item.

6. Suppose that in the quality control example of Exercise 25 on page 289 it is decided to use the medians of random samples of size 5 instead of their means.

   (a) Use Table I to draw a graph of the cumulative normal distribution having $\mu = 1.50$ and $\sigma = 0.02$ (similar to that of Figure 16.5).

   (b) Use the graph obtained in part (a) and the method discussed on page 454 to simulate taking 100 random samples of size 5 from the given normal population.

   (c) Calculate the medians of the 100 samples obtained in (b) and group them into a suitable frequency table. Then calculate the percentiles $P_{2.5}$ and $P_{97.5}$ (see page 54), which will serve as approximate 0.95 *control limits*. Using these limits, would the process be judged out of control if the median of a random sample of size 5 equaled 1.78?

A WORD OF CAUTION

The importance of creative management in the effective utilization of operations research cannot be overemphasized. Aside from such problems as organizing for operations research, managers must ask the right questions and suggest the proper criteria for evaluating results. Excessive concern with operating efficiency and the failure to think broadly and to plan wisely can be extremely damaging. Operating efficiency is always important, but unless it follows sound planning it is essentially useless: Nothing can be gained by producing most efficiently the wrong product for the wrong market.

The goal of operations research is not to make decisions itself, but it is definitely intended to serve as an aid in decision making. Indeed, unless an operations research study influences a decision, brings about some change, or leads to action of some sort, it is not fulfilling its primary function. In this sense, it is the function of operations research to tell an executive some of the consequences of a change in advertising or discount policy, or of building a new plant in a new area, but *not* what he must necessarily do about these matters.

BIBLIOGRAPHY

A. STATISTICS FOR THE LAYMAN

Bross, I. D. J., *Design for Decision*. New York: Free Press, 1965.

Huff, D., *How to Lie with Statistics*. New York: W. W. Norton, 1954.

Huff, D., and Geis, I., *How to Take a Chance*. New York: W. W. Norton, 1959.

Levinson, H. C., *Chance, Luck, and Statistics*. New York: Dover Publications, Inc., 1963.

Reichmann, W. J., *Use and Abuse of Statistics*. New York: Oxford University Press, 1962.

Sielaff, T. J. (ed.), *Statistics in Action*. San Jose, Calif.: Lansford Press, 1963.

Wallis, W. A., and Roberts, H. V., *The Nature of Statistics*. New York: Free Press, 1965.

Weaver, W., *Lady Luck—The Theory of Probability*. New York: Free Press, 1965.

Williams, J. D., *The Compleat Strategyst, revised*. New York: McGraw-Hill, 1965.

B. SOME GENERAL TEXTS ON BUSINESS STATISTICS

Bryant, E. C., *Statistical Analysis, 2nd ed.* New York: McGraw-Hill, 1966.

Croxton, F. E., and Cowden, D. J., *Practical Business Statistics, 3rd ed.* Englewood Cliffs, N.J.: Prentice-Hall, Inc., 1960.

Enrick, N. L., *Cases in Management Statistics.* New York: Holt, Rinehart, and Winston, 1962.

Hadley, G., *Introduction to Business Statistics.* San Francisco: Holden-Day, 1968.

L'Esperance, W. L., *Modern Statistics for Business and Economics.* New York: Macmillan, 1971.

Lewis, E. E., *Methods of Statistical Analysis in Economics and Business.* Boston: Houghton Mifflin, 1963.

Neter, J., and Wasserman, W., *Fundamental Statistics for Business and Economics, 3rd ed.* Boston: Allyn and Bacon, Inc., 1966.

Perles, B. M., and Sullivan, C. M., *Freund and Williams' Modern Business Statistics, rev. ed.* Englewood Cliffs, N.J.: Prentice-Hall, Inc., 1969.

Richmond, S. B., *Statistical Analysis, 2nd ed.* New York: Ronald Press, 1964.

Schlaifer, R., *Introduction to Statistics for Business Decisions.* New York: McGraw-Hill, 1961.

Spurr, W. A., and Bonini, C. P., *Statistical Analysis for Business Decisions.* Homewood, Ill.: R. D. Irwin, Inc., 1967.

Stockton, J. R., *Introduction to Business and Economic Statistics, 3rd ed.* Cincinnati: South-Western Publishing Co., 1966.

C. SOME BOOKS ON THE THEORY OF PROBABILITY AND STATISTICS

(*Many of these books require considerably more mathematical preparation than this text.*)

Brunk, H. D., *An Introduction to Mathematical Statistics, 2nd ed.* Boston: Ginn and Co., 1965.

Freund, J. E., *Mathematical Statistics, 2nd ed.* Englewood Cliffs, N.J.: Prentice-Hall, Inc., 1971.

Goldberg, S., *Probability—An Introduction.* Englewood Cliffs, N.J.: Prentice-Hall, Inc., 1960.

Hodges, J. L., and Lehmann, E. L., *Elements of Finite Probability*. San Francisco: Holden-Day, Inc., 1965.

Hoel, P., *Introduction to Mathematical Statistics, 3rd ed.* New York: John Wiley & Sons, Inc., 1962.

Keeping, E. S., *Introduction to Statistical Inference*. Princeton, N.J.: D. Van Nostrand Co., 1962.

Lindgreen, B. W., and McElrath, G. W., *Introduction to Probability and Statistics, 3rd ed.* New York: Macmillan, 1969.

Mosteller, F., Rourke, R. E. K., and Thomas, G. B., *Probability with Statistical Applications*. Reading, Mass.: Addison-Wesley, 1961.

D. SOME BOOKS DEALING WITH SPECIAL PROBLEMS OF STATISTICS AND SPECIAL APPLICATIONS

Cowden, D. J., *Statistical Methods in Quality Control*. Englewood Cliffs, N.J.: Prentice-Hall, Inc., 1957.

Cochran, W. G., *Sampling Techniques, 2nd ed.* New York: John Wiley & Sons, Inc. 1963.

Deming, W. E., *Sample Design in Business Research*. New York: John Wiley & Sons, Inc., 1960.

Duncan, A. J., *Quality Control and Industrial Statistics, 3rd ed.* Homewood, Ill.: R. D. Irwin, Inc., 1965.

Dresher, M., *Games of Strategy: Theory and Applications*. Englewood Cliffs, N.J.: Prentice-Hall, Inc., 1961.

Ferber, R., and Verdoorn, P. J., *Research Methods in Economics and Business*. New York: Macmillan, 1962.

Grant, E. L., *Statistical Quality Control, 3rd ed.* New York: McGraw-Hill, 1964.

Hicks, C. R., *Fundamental Concepts in the Design of Experiments*. New York: Holt, Rinehart, and Winston, 1964.

Li, C. C., *Introduction to Experimental Statistics*. New York: McGraw-Hill, 1964.

Mudgett, B. D., *Index Numbers*. New York: John Wiley & Sons, Inc., 1951.

Myers, J. H., *Statistical Presentation*. Littlefield, Adams and Co., 1956.

Noether, G. E., *Elements of Nonparametric Statistics*. New York: John Wiley & Sons, Inc., 1967.

Siegel, S., *Nonparametric Statistics for the Behavioral Sciences*. New York: McGraw-Hill, 1956.

Stuart, A., *Basic Ideas of Scientific Sampling*. New York: Hafner, 1962.

E. SOME BOOKS ON OPERATIONS RESEARCH

Ackoff, R. L., and Rivett, P., *A Manager's Guide to Operations Research*. New York: John Wiley & Sons, Inc., 1963.

Bennion, E. G., *Elementary Mathematics of Linear Programming and Game Theory*. East Lansing, Mich.: Michigan State University, 1960.

Buchan, J. F., and Koenigsberg, E., *Scientific Inventory Management*. Englewood Cliffs, N.J.: Prentice-Hall, Inc., 1962.

Charnes, A., and Cooper, W. W., *Management Models and Industrial Applications of Linear Programming*, Vols. I and II. New York: John Wiley & Sons, Inc., 1961.

Horowitz, I., *An Introduction to Quantitative Business Analysis*. New York: McGraw-Hill, 1965.

Luce, R. D., and Raiffa, H., *Games and Decisions*. New York: John Wiley & Sons, Inc., 1958.

Miller, D. W., and Starr, M. K., *Executive Decisions and Operations Research, 2nd ed.* Englewood Cliffs, N.J.: Prentice-Hall, Inc., 1969.

Shuchman, A., *Scientific Decision Making in Business: Readings in Operations Research for Nonmathematicians*. New York: Holt, Rinehart, and Winston, 1963.

Teichroew, D., *An Introduction to Management Science*. New York: John Wiley & Sons, Inc., 1964.

Wagner, H. M., *Principles of Operations Research*. Englewood Cliffs, N.J.: Prentice-Hall, Inc., 1970.

F. SOME STATISTICAL TABLES

Fisher, R. A., and Yates, F., *Statistical Tables for Biological, Agricultural and Medical Research, 6th ed.* Edinburgh: Oliver & Boyd, 1963.

National Bureau of Standards, *Tables of the Binomial Distribution*. Washington, D.C.: U.S. Government Printing Office, 1950.

Owen, D. B., *Handbook of Statistical Tables*. Reading, Mass.: Addison-Wesley, 1962.

Pearson, E. S., and Hartley, H. O., *Biometrika Tables for Statisticians, 3rd ed.* Cambridge: Cambridge University Press, 1966.

RAND Corporation, *A Million Random Digits with 100,000 Normal Deviates*. New York: Free Press, 1955.

Romig, H. G., *50–100 Binomial Tables*. New York: John Wiley & Sons, Inc., 1953.

G. SOME SOURCES OF STATISTICAL DATA

(*The following books list such various sources of statistical data as periodicals, year-books and annuals, books and pamphlets, directories and special reports.*)

Coman, E. T., Jr., *Sources of Business Information, 2nd ed.* Berkeley, Calif.: University of California Press, 1964.

Hauser, P. M., and Leonard, W. R., *Government Statistics for Business Use, 2nd ed.* New York: John Wiley & Sons, Inc., 1956.

Wasserman, P., *et al.*, eds., *Statistical Sources, 3rd ed.* Detroit: Gale Research Company, 1965.

(*The following are some annual publications devoted largely to presenting statistical data.*)

Agricultural Statistics. United States Department of Agriculture.

Commodity Yearbook. New York: Commodity Research Bureau, Inc.

Demographic Yearbook. Statistical Office of the United Nations.

Industrial Marketing (Annual Market Data and Directory Number). Chicago: Advertising Publications, Inc.

Statistical Abstract of the United States. United States Bureau of the Census.

Statistical Yearbook. Statistical Office of the United Nations.

Survey of Buying Power. New York: Sales Management.

The Economic Almanac. New York: National Industrial Conference Board.

The Market Guide. New York: Editor and Publishing Co., Inc.

The World Almanac. New York: Newspaper Enterprise Association, Inc.

STATISTICAL TABLES

TABLE I NORMAL CURVE AREAS

z	.00	.01	.02	.03	.04	.05	.06	.07	.08	.09
0.0	.0000	.0040	.0080	.0120	.0160	.0199	.0239	.0279	.0319	.0359
0.1	.0398	.0438	.0478	.0517	.0557	.0596	.0636	.0675	.0714	.0753
0.2	.0793	.0832	.0871	.0910	.0948	.0987	.1026	.1064	.1103	.1141
0.3	.1179	.1217	.1255	.1293	.1331	.1368	.1406	.1443	.1480	.1517
0.4	.1554	.1591	.1628	.1664	.1700	.1736	.1772	.1808	.1844	.1879
0.5	.1915	.1950	.1985	.2019	.2054	.2088	.2123	.2157	.2190	.2224
0.6	.2257	.2291	.2324	.2357	.2389	.2422	.2454	.2486	.2517	.2549
0.7	.2580	.2611	.2642	.2673	.2704	.2734	.2764	.2794	.2823	.2852
0.8	.2881	.2910	.2939	.2967	.2995	.3023	.3051	.3078	.3106	.3133
0.9	.3159	.3186	.3212	.3238	.3264	.3289	.3315	.3340	.3365	.3389
1.0	.3413	.3438	.3461	.3485	.3508	.3531	.3554	.3577	.3599	.3621
1.1	.3643	.3665	.3686	.3708	.3729	.3749	.3770	.3790	.3810	.3830
1.2	.3849	.3869	.3888	.3907	.3925	.3944	.3962	.3980	.3997	.4015
1.3	.4032	.4049	.4066	.4082	.4099	.4115	.4131	.4147	.4162	.4177
1.4	.4192	.4207	.4222	.4236	.4251	.4265	.4279	.4292	.4306	.4319
1.5	.4332	.4345	.4357	.4370	.4382	.4394	.4406	.4418	.4429	.4441
1.6	.4452	.4463	.4474	.4484	.4495	.4505	.4515	.4525	.4535	.4545
1.7	.4554	.4564	.4573	.4582	.4591	.4599	.4608	.4616	.4625	.4633
1.8	.4641	.4649	.4656	.4664	.4671	.4678	.4686	.4693	.4699	.4706
1.9	.4713	.4719	.4726	.4732	.4738	.4744	.4750	.4756	.4761	.4767
2.0	.4772	.4778	.4783	.4788	.4793	.4798	.4803	.4808	.4812	.4817
2.1	.4821	.4826	.4830	.4834	.4838	.4842	.4846	.4850	.4854	.4857
2.2	.4861	.4864	.4868	.4871	.4875	.4878	.4881	.4884	.4887	.4890
2.3	.4893	.4896	.4898	.4901	.4904	.4906	.4909	.4911	.4913	.4916
2.4	.4918	.4920	.4922	.4925	.4927	.4929	.4931	.4932	.4934	.4936
2.5	.4938	.4940	.4941	.4943	.4945	.4946	.4948	.4949	.4951	.4952
2.6	.4953	.4955	.4956	.4957	.4959	.4960	.4961	.4962	.4963	.4964
2.7	.4965	.4966	.4967	.4968	.4969	.4970	.4971	.4972	.4973	.4974
2.8	.4974	.4975	.4976	.4977	.4977	.4978	.4979	.4979	.4980	.4981
2.9	.4981	.4982	.4982	.4983	.4984	.4984	.4985	.4985	.4986	.4986
3.0	.4987	.4987	.4987	.4988	.4988	.4989	.4989	.4989	.4990	.4990

probability

TABLE II VALUES OF t^*

d.f.	$t_{.100}$	$t_{.050}$	$t_{.025}$	$t_{.010}$	$t_{.005}$	d.f.
1	3.078	6.314	12.706	31.821	63.657	1
2	1.886	2.920	4.303	6.965	9.925	2
3	1.638	2.353	3.182	4.541	5.841	3
4	1.533	2.132	2.776	3.747	4.604	4
5	1.476	2.015	2.571	3.365	4.032	5
6	1.440	1.943	2.447	3.143	3.707	6
7	1.415	1.895	2.365	2.998	3.499	7
8	1.397	1.860	2.306	2.896	3.355	8
9	1.383	1.833	2.262	2.821	3.250	9
10	1.372	1.812	2.228	2.764	3.169	10
11	1.363	1.796	2.201	2.718	3.106	11
12	1.356	1.782	2.179	2.681	3.055	12
13	1.350	1.771	2.160	2.650	3.012	13
14	1.345	1.761	2.145	2.624	2.977	14
15	1.341	1.753	2.131	2.602	2.947	15
16	1.337	1.746	2.120	2.583	2.921	16
17	1.333	1.740	2.110	2.567	2.898	17
18	1.330	1.734	2.101	2.552	2.878	18
19	1.328	1.729	2.093	2.539	2.861	19
20	1.325	1.725	2.086	2.528	2.845	20
21	1.323	1.721	2.080	2.518	2.831	21
22	1.321	1.717	2.074	2.508	2.819	22
23	1.319	1.714	2.069	2.500	2.807	23
24	1.318	1.711	2.064	2.492	2.797	24
25	1.316	1.708	2.060	2.485	2.787	25
26	1.315	1.706	2.056	2.479	2.779	26
27	1.314	1.703	2.052	2.473	2.771	27
28	1.313	1.701	2.048	2.467	2.763	28
29	1.311	1.699	2.045	2.462	2.756	29
inf.	1.282	1.645	1.960	2.326	2.576	inf.

* This table is abridged from Table IV of R. A. Fisher, *Statistical Methods for Research Workers*, published by Oliver and Boyd, Ltd., Edinburgh, by permission of the author and publishers.

TABLE III VALUES OF χ^{2*}

$d.f.$	$\chi^2_{.05}$	$\chi^2_{.025}$	$\chi^2_{.01}$	$\chi^2_{.005}$	$d.f.$
1	3.841	5.024	6.635	7.879	1
2	5.991	7.378	9.210	10.597	2
3	7.815	9.348	11.345	12.838	3
4	9.488	11.143	13.277	14.860	4
5	11.070	12.832	15.086	16.750	5
6	12.592	14.449	16.812	18.548	6
7	14.067	16.013	18.475	20.278	7
8	15.507	17.535	20.090	21.955	8
9	16.919	19.023	21.666	23.589	9
10	18.307	20.483	23.209	25.188	10
11	19.675	21.920	24.725	26.757	11
12	21.026	23.337	26.217	28.300	12
13	22.362	24.736	27.688	29.819	13
14	23.685	26.119	29.141	31.319	14
15	24.996	27.488	30.578	32.801	15
16	26.296	28.845	32.000	34.267	16
17	27.587	30.191	33.409	35.718	17
18	28.869	31.526	34.805	37.156	18
19	30.144	32.852	36.191	38.582	19
20	31.410	34.170	37.566	39.997	20
21	32.671	35.479	38.932	41.401	21
22	33.924	36.781	40.289	42.796	22
23	35.172	38.076	41.638	44.181	23
24	36.415	39.364	42.980	45.558	24
25	37.652	40.646	44.314	46.928	25
26	38.885	41.923	45.642	48.290	26
27	40.113	43.194	46.963	49.645	27
28	41.337	44.461	48.278	50.993	28
29	42.557	45.722	49.588	52.336	29
30	43.773	46.979	50.892	53.672	30

* This table is abridged from Table III of R. A. Fisher, *Statistical Methods for Research Workers*, published by Oliver and Boyd, Ltd., Edinburgh, by permission of the author and publishers.

Table IV(a) VALUES OF $F_{.05}$*

Degrees of freedom for numerator

d.f. denom.	1	2	3	4	5	6	7	8	9	10	12	15	20	24	30	40	60	120	∞
1	161	200	216	225	230	234	237	239	241	242	244	246	248	249	250	251	252	253	254
2	18.5	19.0	19.2	19.2	19.3	19.3	19.4	19.4	19.4	19.4	19.4	19.4	19.4	19.5	19.5	19.5	19.5	19.5	19.5
3	10.1	9.55	9.28	9.12	9.01	8.94	8.89	8.85	8.81	8.79	8.74	8.70	8.66	8.64	8.62	8.59	8.57	8.55	8.53
4	7.71	6.94	6.59	6.39	6.26	6.16	6.09	6.04	6.00	5.96	5.91	5.86	5.80	5.77	5.75	5.72	5.69	5.66	5.63
5	6.61	5.79	5.41	5.19	5.05	4.95	4.88	4.82	4.77	4.74	4.68	4.62	4.56	4.53	4.50	4.46	4.43	4.40	4.37
6	5.99	5.14	4.76	4.53	4.39	4.28	4.21	4.15	4.10	4.06	4.00	3.94	3.87	3.84	3.81	3.77	3.74	3.70	3.67
7	5.59	4.74	4.35	4.12	3.97	3.87	3.79	3.73	3.68	3.64	3.57	3.51	3.44	3.41	3.38	3.34	3.30	3.27	3.23
8	5.32	4.46	4.07	3.84	3.69	3.58	3.50	3.44	3.39	3.35	3.28	3.22	3.15	3.12	3.08	3.04	3.01	2.97	2.93
9	5.12	4.26	3.86	3.63	3.48	3.37	3.29	3.23	3.18	3.14	3.07	3.01	2.94	2.90	2.86	2.83	2.79	2.75	2.71
10	4.96	4.10	3.71	3.48	3.33	3.22	3.14	3.07	3.02	2.98	2.91	2.85	2.77	2.74	2.70	2.66	2.62	2.58	2.54
11	4.84	3.98	3.59	3.36	3.20	3.09	3.01	2.95	2.90	2.85	2.79	2.72	2.65	2.61	2.57	2.53	2.49	2.45	2.40
12	4.75	3.89	3.49	3.26	3.11	3.00	2.91	2.85	2.80	2.75	2.69	2.62	2.54	2.51	2.47	2.43	2.38	2.34	2.30
13	4.67	3.81	3.41	3.18	3.03	2.92	2.83	2.77	2.71	2.67	2.60	2.53	2.46	2.42	2.38	2.34	2.30	2.25	2.21
14	4.60	3.74	3.34	3.11	2.96	2.85	2.76	2.70	2.65	2.60	2.53	2.46	2.39	2.35	2.31	2.27	2.22	2.18	2.13
15	4.54	3.68	3.29	3.06	2.90	2.79	2.71	2.64	2.59	2.54	2.48	2.40	2.33	2.29	2.25	2.20	2.16	2.11	2.07
16	4.49	3.63	3.24	3.01	2.85	2.74	2.66	2.59	2.54	2.49	2.42	2.35	2.28	2.24	2.19	2.15	2.11	2.06	2.01
17	4.45	3.59	3.20	2.96	2.81	2.70	2.61	2.55	2.49	2.45	2.38	2.31	2.23	2.19	2.15	2.10	2.06	2.01	1.96
18	4.41	3.55	3.16	2.93	2.77	2.66	2.58	2.51	2.46	2.41	2.34	2.27	2.19	2.15	2.11	2.06	2.02	1.97	1.92
19	4.38	3.52	3.13	2.90	2.74	2.63	2.54	2.48	2.42	2.38	2.31	2.23	2.16	2.11	2.07	2.03	1.98	1.93	1.88
20	4.35	3.49	3.10	2.87	2.71	2.60	2.51	2.45	2.39	2.35	2.28	2.20	2.12	2.08	2.04	1.99	1.95	1.90	1.84
21	4.32	3.47	3.07	2.84	2.68	2.57	2.49	2.42	2.37	2.32	2.25	2.18	2.10	2.05	2.01	1.96	1.92	1.87	1.81
22	4.30	3.44	3.05	2.82	2.66	2.55	2.46	2.40	2.34	2.30	2.23	2.15	2.07	2.03	1.98	1.94	1.89	1.84	1.78
23	4.28	3.42	3.03	2.80	2.64	2.53	2.44	2.37	2.32	2.27	2.20	2.13	2.05	2.01	1.96	1.91	1.86	1.81	1.76
24	4.26	3.40	3.01	2.78	2.62	2.51	2.42	2.36	2.30	2.25	2.18	2.11	2.03	1.98	1.94	1.89	1.84	1.79	1.73
25	4.24	3.39	2.99	2.76	2.60	2.49	2.40	2.34	2.28	2.24	2.16	2.09	2.01	1.96	1.92	1.87	1.82	1.77	1.71
30	4.17	3.32	2.92	2.69	2.53	2.42	2.33	2.27	2.21	2.16	2.09	2.01	1.93	1.89	1.84	1.79	1.74	1.68	1.62
40	4.08	3.23	2.84	2.61	2.45	2.34	2.25	2.18	2.12	2.08	2.00	1.92	1.84	1.79	1.74	1.69	1.64	1.58	1.51
60	4.00	3.15	2.76	2.53	2.37	2.25	2.17	2.10	2.04	1.99	1.92	1.84	1.75	1.70	1.65	1.59	1.53	1.47	1.39
120	3.92	3.07	2.68	2.45	2.29	2.18	2.09	2.02	1.96	1.91	1.83	1.75	1.66	1.61	1.55	1.50	1.43	1.35	1.25
∞	3.84	3.00	2.60	2.37	2.21	2.10	2.01	1.94	1.88	1.83	1.75	1.67	1.57	1.52	1.46	1.39	1.32	1.22	1.00

Degrees of freedom for denominator

* This table is reproduced from M. Merrington and C. M. Thompson, "Tables of percentage points of the inverted beta (F) distribution," *Biometrika*, vol. 33 (1943), by permission of the *Biometrika* trustees.

TABLE IV(b)   VALUES OF $F_{.01}$*

Degrees of freedom for numerator

Den.	1	2	3	4	5	6	7	8	9	10	12	15	20	24	30	40	60	120	∞
1	4,052	5,000	5,403	5,625	5,764	5,859	5,928	5,982	6,023	6,056	6,106	6,157	6,209	6,235	6,261	6,287	6,313	6,339	6,366
2	98.5	99.0	99.2	99.2	99.3	99.3	99.4	99.4	99.4	99.4	99.4	99.4	99.4	99.5	99.5	99.5	99.5	99.5	99.5
3	34.1	30.8	29.5	28.7	28.2	27.9	27.7	27.5	27.3	27.2	27.1	26.9	26.7	26.6	26.5	26.4	26.3	26.2	26.1
4	21.2	18.0	16.7	16.0	15.5	15.2	15.0	14.8	14.7	14.5	14.4	14.2	14.0	13.9	13.8	13.7	13.7	13.6	13.5
5	16.3	13.3	12.1	11.4	11.0	10.7	10.5	10.3	10.2	10.1	9.89	9.72	9.55	9.47	9.38	9.29	9.20	9.11	9.02
6	13.7	10.9	9.78	9.15	8.75	8.47	8.26	8.10	7.98	7.87	7.72	7.56	7.40	7.31	7.23	7.14	7.06	6.97	6.88
7	12.2	9.55	8.45	7.85	7.46	7.19	6.99	6.84	6.72	6.62	6.47	6.31	6.16	6.07	5.99	5.91	5.82	5.74	5.65
8	11.3	8.65	7.59	7.01	6.63	6.37	6.18	6.03	5.91	5.81	5.67	5.52	5.36	5.28	5.20	5.12	5.03	4.95	4.86
9	10.6	8.02	6.99	6.42	6.06	5.80	5.61	5.47	5.35	5.26	5.11	4.96	4.81	4.73	4.65	4.57	4.48	4.40	4.31
10	10.0	7.56	6.55	5.99	5.64	5.39	5.20	5.06	4.94	4.85	4.71	4.56	4.41	4.33	4.25	4.17	4.08	4.00	3.91
11	9.65	7.21	6.22	5.67	5.32	5.07	4.89	4.74	4.63	4.54	4.40	4.25	4.10	4.02	3.94	3.86	3.78	3.69	3.60
12	9.33	6.93	5.95	5.41	5.06	4.82	4.64	4.50	4.39	4.30	4.16	4.01	3.86	3.78	3.70	3.62	3.54	3.45	3.36
13	9.07	6.70	5.74	5.21	4.86	4.62	4.44	4.30	4.19	4.10	3.96	3.82	3.66	3.59	3.51	3.43	3.34	3.25	3.17
14	8.86	6.51	5.56	5.04	4.70	4.46	4.28	4.14	4.03	3.94	3.80	3.66	3.51	3.43	3.35	3.27	3.18	3.09	3.00
15	8.68	6.36	5.42	4.89	4.56	4.32	4.14	4.00	3.89	3.80	3.67	3.52	3.37	3.29	3.21	3.13	3.05	2.96	2.87
16	8.53	6.23	5.29	4.77	4.44	4.20	4.03	3.89	3.78	3.69	3.55	3.41	3.26	3.18	3.10	3.02	2.93	2.84	2.75
17	8.40	6.11	5.19	4.67	4.34	4.10	3.93	3.79	3.68	3.59	3.46	3.31	3.16	3.08	3.00	2.92	2.83	2.75	2.65
18	8.29	6.01	5.09	4.58	4.25	4.01	3.84	3.71	3.60	3.51	3.37	3.23	3.08	3.00	2.92	2.84	2.75	2.66	2.57
19	8.19	5.93	5.01	4.50	4.17	3.94	3.77	3.63	3.52	3.43	3.30	3.15	3.00	2.92	2.84	2.76	2.67	2.58	2.49
20	8.10	5.85	4.94	4.43	4.10	3.87	3.70	3.56	3.46	3.37	3.23	3.09	2.94	2.86	2.78	2.69	2.61	2.52	2.42
21	8.02	5.78	4.87	4.37	4.04	3.81	3.64	3.51	3.40	3.31	3.17	3.03	2.88	2.80	2.72	2.64	2.55	2.46	2.36
22	7.95	5.72	4.82	4.31	3.99	3.76	3.59	3.45	3.35	3.26	3.12	2.98	2.83	2.75	2.67	2.58	2.50	2.40	2.31
23	7.88	5.66	4.76	4.26	3.94	3.71	3.54	3.41	3.30	3.21	3.07	2.93	2.78	2.70	2.62	2.54	2.45	2.35	2.26
24	7.82	5.61	4.72	4.22	3.90	3.67	3.50	3.36	3.26	3.17	3.03	2.89	2.74	2.66	2.58	2.49	2.40	2.31	2.21
25	7.77	5.57	4.68	4.18	3.86	3.63	3.46	3.32	3.22	3.13	2.99	2.85	2.70	2.62	2.53	2.45	2.36	2.27	2.17
30	7.56	5.39	4.51	4.02	3.70	3.47	3.30	3.17	3.07	2.98	2.84	2.70	2.55	2.47	2.39	2.30	2.21	2.11	2.01
40	7.31	5.18	4.31	3.83	3.51	3.29	3.12	2.99	2.89	2.80	2.66	2.52	2.37	2.29	2.20	2.11	2.02	1.92	1.80
60	7.08	4.98	4.13	3.65	3.34	3.12	2.95	2.82	2.72	2.63	2.50	2.35	2.20	2.12	2.03	1.94	1.84	1.73	1.60
120	6.85	4.79	3.95	3.48	3.17	2.96	2.79	2.66	2.56	2.47	2.34	2.19	2.03	1.95	1.86	1.76	1.66	1.53	1.38
∞	6.63	4.61	3.78	3.32	3.02	2.80	2.64	2.51	2.41	2.32	2.18	2.04	1.88	1.79	1.70	1.59	1.47	1.32	1.00

Degrees of freedom for denominator

* This table is reproduced from M. Merrington and C. M. Thompson, "Tables of percentage points of the inverted beta (F) distribution," *Biometrika*, vol. 33 (1943), by permission of the *Biometrika* trustees.

TABLE V(a)　0.95 CONFIDENCE INTERVALS FOR PROPORTIONS*

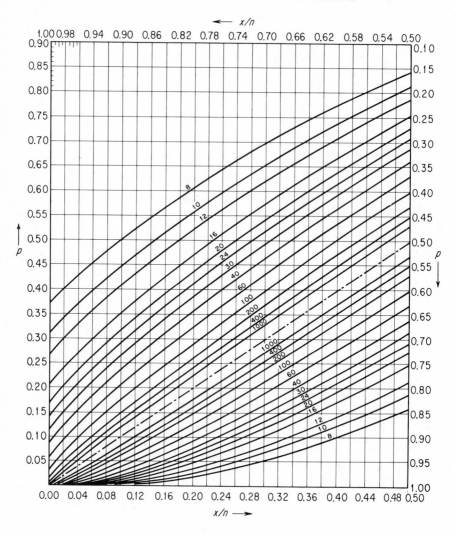

TABLE V(b) 0.99 CONFIDENCE INTERVALS FOR PROPORTIONS*

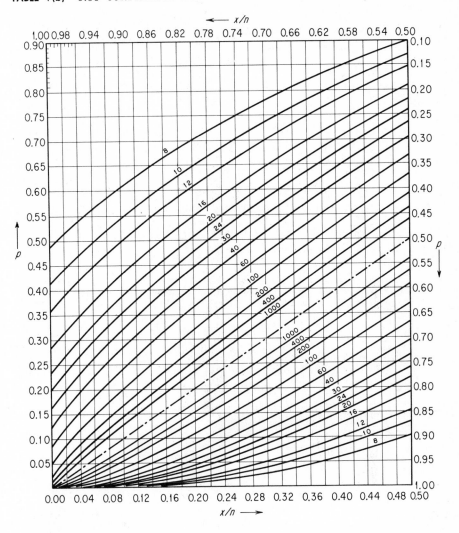

TABLE VI CRITICAL VALUES OF r*

n	$r_{.025}$	$r_{.010}$	$r_{.005}$	n	$r_{.025}$	$r_{.010}$	$r_{.005}$
3	0.997			18	0.468	0.543	0.590
4	0.950	0.990	0.999	19	0.456	0.529	0.575
5	0.878	0.934	0.959	20	0.444	0.516	0.561
6	0.811	0.882	0.917	21	0.433	0.503	0.549
7	0.754	0.833	0.875	22	0.423	0.492	0.537
8	0.707	0.789	0.834	27	0.381	0.445	0.487
9	0.666	0.750	0.798	32	0.349	0.409	0.449
10	0.632	0.715	0.765	37	0.325	0.381	0.418
11	0.602	0.685	0.735	42	0.304	0.358	0.393
12	0.576	0.658	0.708	47	0.288	0.338	0.372
13	0.553	0.634	0.684	52	0.273	0.322	0.354
14	0.532	0.612	0.661	62	0.250	0.295	0.325
15	0.514	0.592	0.641	72	0.232	0.274	0.302
16	0.497	0.574	0.623	82	0.217	0.256	0.283
17	0.482	0.558	0.606	92	0.205	0.242	0.267

* This table is abridged from Table VI of R. A. Fisher and F. Yates, *Statistical Tables for Biological, Agricultural, and Medical Research*, published by Oliver and Boyd, Ltd., Edinburgh, by permission of the author and publishers.

TABLE VII CRITICAL VALUES OF u (RUNS)*

Number of a's = Number of b's	$u_{.05}$	$u_{.01}$
5	3	2
6	3	2
7	4	3
8	5	4
9	6	4
10	6	5
11	7	6
12	8	6
13	9	7
14	10	8
15	11	9
16	11	10
17	12	10
18	13	11
19	14	12
20	15	13
25	19	17
30	24	21
35	28	25
40	33	30
45	37	34
50	42	38
60	51	47
80	70	65
100	88	84

* This table is based on F. S. Swed and C. Eisenhart, "Tables for testing randomness of grouping in a sequence of alternatives," *Annals of Mathematical Statistics*, vol. 14, p. 66.

TABLE VIII RANDOM NUMBERS*

6063	2353	8531	8892	4109	5782	2283	1385	0699	5927
6305	1326	4551	2815	8937	2908	0698	5509	4303	9911
0143	0187	8127	2026	8313	8341	2479	4722	6602	2236
1031	0754	7989	4948	1804	3025	0997	9562	3674	7876
2022	3227	2147	5613	2857	8859	4941	7274	9412	0620
9149	0806	9751	8870	9677	9676	1854	8094	7658	7012
5863	0513	1402	3866	8696	9142	6063	2252	7818	2477
8724	0806	9644	8284	7010	0868	9076	4915	5751	9214
6783	4207	2958	5295	3175	3396	8117	5918	1037	4319
0862	1620	4690	0036	9654	4078	1918	8721	8454	7671
9394	2466	6427	5395	9393	0520	7074	0634	5578	4023
3220	3058	7787	7706	4094	5603	3303	8300	6185	8705
1491	3503	0584	7221	6176	0116	0309	1975	0910	3535
4368	5705	8579	5790	7244	6547	8495	7973	1805	7251
2325	4026	2919	8327	0267	2616	6572	8620	8245	6257
0591	1775	5134	8709	7373	3332	0507	5525	7640	2840
3471	1461	1149	6798	6070	9930	1862	3672	6718	3849
2600	9885	6219	3668	1005	5418	5832	0416	4220	4692
9572	7874	6034	4514	2628	1693	0628	2200	9006	3795
0822	2790	9386	5783	2689	2565	1565	0349	3410	5216
4329	3028	2549	2529	9434	3083	6800	8569	9290	8298
9289	5212	2355	9367	1297	1638	9282	3720	7178	2695
3932	9960	3399	1700	8253	1375	4594	6024	1223	5383
2282	0648	7561	7528	5870	7907	0713	8608	9682	8576
9933	3416	5957	2574	5553	5534	4707	3206	0963	2459
9015	6416	6603	2967	7591	5013	2878	8424	5452	4659
1539	0719	2637	9969	8450	4489	3528	3364	1459	9708
6849	5595	7969	2582	5627	1920	9772	8560	0892	6500
2523	7769	3536	9611	1079	1694	1254	4195	5799	5928
0701	7355	0587	8878	3446	1137	7690	0647	1407	6362
2163	8543	4594	6022	0496	8648	2999	1262	6702	0811
0327	5727	1070	5996	8660	9024	2135	9799	8414	9136
2169	3160	8707	6361	6339	4054	3251	7397	3480	5805
8393	8147	5360	4150	2990	3380	1789	7436	4781	0337
9726	9151	2064	0609	5878	9095	9737	2897	6510	8891
0515	2296	2636	9756	5313	7754	0916	6066	3905	1298
0649	8398	5614	0140	3155	2211	4988	3674	7663	0620
0026	9426	8005	8579	5774	7962	5092	5856	1626	0980
3422	0092	1626	1298	2475	1997	9796	7076	1541	1731
8191	1983	9164	1885	5468	8216	4327	8109	5880	9804
7408	0486	7654	4829	2711	6592	4785	5901	7147	9314
8261	9440	8118	6338	8157	9052	9093	8449	4066	4894
9274	8838	8342	3114	0455	6212	8862	6701	0099	0501
2699	0383	1400	3484	1492	4683	5369	3851	5870	0903
8740	0349	3502	3971	9960	6325	6727	4715	2945	9938
0247	2372	0424	0578	0036	1619	4479	7108	8520	1487
5136	9444	8343	1152	3615	1420	8923	7307	3978	5724
4844	8931	0964	2878	8212	9328	2656	1965	4805	0634
0205	8457	4333	2555	5353	9201	1606	2715	4014	1877
2517	5061	7642	3891	7713	7066	5435	1200	7455	5562

* From Donald B. Owen, *Handbook of Statistical Tables*. Reading, Mass.: Addison-Wesley, 1962

TABLE VIII RANDOM NUMBERS (*Continued*)

2271	2572	8665	3272	9033	8256	2822	3646	7599	0270
3025	0788	5311	7792	1837	4739	4552	3234	5572	9885
3382	6151	1011	3778	9951	7709	8060	2258	8536	2290
7870	5799	6032	9043	4526	8100	1957	9539	5370	0046
1697	0002	2340	6959	1915	1626	1297	1533	6572	3835
3395	3381	1862	3250	8614	5683	6757	5628	2551	6971
6081	6526	3028	2338	5702	8819	3679	4829	9909	4712
3470	9879	2935	1141	6398	6387	5634	9589	3212	7963
0432	8641	5020	6612	1038	1547	0948	4278	0020	6509
4995	5596	8286	8377	8567	8237	3520	8244	5694	3326
8246	6718	3851	5870	1216	2107	1387	1621	5509	5772
7825	8727	2849	3501	3551	1001	0123	7873	5926	6078
6258	2450	2962	1183	3666	4156	4454	8239	4551	2920
3235	5783	2701	2378	7460	3398	1223	4688	3674	7872
2525	9008	5997	0885	1053	2340	7066	5328	6412	5054
5852	9739	1457	8999	2789	9068	9829	1336	3148	7875
0440	3769	7864	4029	4494	9829	1339	4910	1303	9161
0820	4641	2375	2542	4093	5364	1145	2848	2792	0431
7114	2842	8554	6881	6377	9427	8216	1193	8042	8449
6558	9301	9096	0577	8520	5923	4717	0188	8545	8745
0345	9937	5569	0279	8951	6183	7787	7808	5149	2185
7430	2074	9427	8422	4082	5629	2971	9456	0649	7981
8030	7345	3389	4739	5911	1022	9189	2565	1982	8577
6272	6718	3849	4715	3156	2823	4174	8733	5600	7702
4894	9847	5611	4763	8755	3388	5114	3274	6681	3657
2676	5984	6806	2692	4012	0934	2436	0869	9557	2490
9305	2074	9378	7670	8284	7431	7361	2912	2251	7395
5138	2461	7213	1905	7775	9881	8782	6272	0632	4418
2452	4200	8674	9202	0812	3986	1143	7343	2264	9072
8882	3033	8746	7390	8609	1144	2531	6944	8869	1570
1087	9336	8020	9166	4472	8293	2904	7949	3165	7400
5666	2841	8134	9588	2915	4116	2802	6917	3993	8764
9790	2228	9702	1690	7170	7511	1937	0723	4505	7155
3250	8860	3294	2684	6572	3415	5750	8726	2647	6596
5450	3922	0950	0890	6434	2306	2781	1066	3681	2404
5765	0765	7311	5270	5910	7009	0240	7435	4568	6484
8408	1939	0599	5347	2160	7376	4696	6969	0787	3838
8460	7658	6906	9177	1492	4680	3719	3456	8681	6736
4198	7244	3849	4819	1008	6781	3388	5253	7041	6712
9872	4441	6712	9614	2736	5533	9062	2534	0855	7946
6485	0487	0004	5563	1481	1546	8245	6116	6920	0990
2064	0512	9509	0341	8131	7778	8609	9417	1216	4189
9927	8987	5321	3125	9992	9449	5951	5872	2057	5731
4918	9690	6121	8770	6053	6931	7252	5409	1869	4229
8099	5821	3899	2685	6781	3178	0096	2986	8878	8991
1901	4974	1262	6810	4673	8772	6616	2632	7891	9970
8273	6675	4925	3924	2274	3860	1662	7480	8674	4503
2878	8213	3170	5126	0434	9481	7029	8688	4027	3340
6088	1182	3242	0835	1765	8819	3462	9820	5759	4189
5773	6600	5306	0354	8295	0148	6608	9064	3421	8570

* From Donald B. Owen, *Handbook of Statistical Tables*. Reading, Mass.: Addison-Wesley, 1962.

TABLE VIII RANDOM NUMBERS (*Continued*)

5500	2276	6307	2346	1285	7000	5306	0414	3383	2137
3251	8902	8843	2112	8567	8131	8116	5270	5994	7445
4675	1435	2192	0874	2897	0262	5092	5541	4014	2086
3543	6130	4247	4859	2660	7852	9096	0578	0097	4746
3521	8772	6612	0721	3899	2999	1263	7017	8057	4983
5573	9396	3464	1702	9204	3389	5678	2589	0288	4633
7478	7569	7551	3380	2152	5411	2647	7242	2800	6183
3339	2854	9691	9562	3252	9848	6030	8472	2266	1270
5505	8474	3167	8552	5409	1556	4247	4652	2953	5394
6381	2086	5457	7703	2758	2963	8167	6712	9820	5654
6975	5239	0762	5846	2431	0543	4956	8787	9651	2605
7185	4019	7332	2820	4853	8636	9505	6575	0365	6648
4510	1658	5615	2194	1901	4975	1895	4383	0415	3771
7752	0105	4769	2994	7445	0781	4960	4253	9451	6518
4834	4043	6591	3646	8918	4603	1970	9145	7615	3905
8866	6036	9755	4508	9061	2080	3406	9856	1298	6281
6622	4612	2030	7299	8414	8822	5176	9443	6054	6462
9094	8973	3335	2183	5192	1630	0959	8143	9182	8012
5618	6445	2983	0375	2540	2735	4901	5515	4787	7058
2705	2693	1944	8074	2015	3261	5529	7193	5401	9531
1797	4334	3293	2632	3770	1675	9363	7795	3331	8995
9448	5174	5869	0448	8613	4400	6938	5161	8691	2838
3461	1304	9682	8577	4449	1896	8328	1698	7138	1141
7092	5007	5596	8522	2580	4495	4728	8948	4434	2438
5533	4294	0939	4050	1225	6414	5895	0148	7053	5935
7852	8988	5951	4919	7404	2426	4450	2358	3082	4561
8313	8456	9892	0981	6736	8021	6226	5573	1664	9489
1158	2241	9661	7588	2669	5480	9160	4267	1690	7278
9338	7226	0025	8844	8181	5565	2418	9394	0837	3106
7711	1336	3251	8902	8425	5766	3262	5848	3545	7073
2656	1863	3884	6516	6922	1808	1896	8853	0964	3089
7980	9370	2850	3818	7281	8352	9637	0618	2430	6525
1409	7865	5908	4296	1888	2792	4014	1667	1295	0814
7657	6630	5000	1493	5459	5869	0315	8134	9587	2184
2863	5450	1329	8787	8795	4604	2615	0075	1433	7707
3988	2042	2906	8995	0818	9288	1650	0803	8319	2533
4551	2815	8941	4893	8612	4844	0042	3890	7069	8512
5772	4732	2829	3931	9540	6256	5420	2179	9448	5489
9150	1435	3817	8975	4276	9569	0175	6663	0045	5549
5764	7914	8280	1337	3779	8197	9105	5985	1054	2866
5895	0044	5021	3846	7599	0398	5212	9509	0134	4656
6857	1174	8085	6503	5355	3027	1708	3626	7059	0167
2538	2669	3746	3270	1214	9983	8434	1344	1160	3292
9983	1387	1410	8891	2523	8705	9190	2986	7654	5142
5061	9529	2922	2199	8310	6954	8090	5371	0672	6281
9999	4226	2815	8817	5606	5190	0495	7867	9968	5951
9078	5936	2393	7875	6871	3163	9203	2863	5693	9973
4823	2291	8925	6306	1717	0320	2549	3107	5488	0303
1232	1384	5698	9313	3501	3238	7227	0220	6118	7655
7694	6484	0279	8528	7214	1750	0577	8418	0698	5403

* From Donald B. Owen, *Handbook of Statistical Tables*. Reading, Mass.: Addison-Wesley, 1962.

TABLE VIII RANDOM NUMBERS (*Continued*)

0366	6390	2107	3875	4488	2911	1727	8108	3484	6370
3686	8812	8754	2758	3079	2994	3642	1580	1475	0366
4195	4602	1481	7324	8570	6913	6228	1934	6165	0554
8180	5460	0134	4469	8619	7723	8084	3293	1895	4886
1498	7883	5280	0692	7202	1273	3334	1554	3303	8569
9428	8633	9606	7679	4182	4035	6849	5593	6712	9822
9630	5879	9342	9618	8513	4399	9734	7744	4600	0224
1086	8918	7713	5909	2620	6612	0616	1298	2476	2386
2478	3551	1247	8004	0301	6672	6176	0682	2493	6381
2808	1133	5853	8737	9804	2404	7400	5904	8803	0377
8934	2047	4963	4531	6391	9064	3526	2482	9328	5556
1156	1191	1182	3032	8640	4681	3932	6975	4926	4870
5677	7494	0987	8870	4837	5267	4119	4163	1953	3553
3719	3586	5775	7309	5111	0919	7721	7032	1164	2105
6556	8472	1848	1056	3670	7509	0854	7210	9336	8127
1246	3476	4027	3654	2444	9040	5331	2363	4738	9822
6591	3387	4109	7956	5837	6914	6435	2624	8610	4005
8197	9026	4868	6372	2695	7143	2783	1925	3383	9060
5035	4569	7158	8531	8891	0975	6329	1329	8746	0989
1563	9650	2139	7696	7511	1725	7292	0664	8440	8593
6034	4512	1505	3857	0290	3270	8389	9612	1892	8707
2435	0238	6478	5727	0862	1621	5228	5038	2000	0433
9418	4486	5992	7172	8353	6516	6605	6387	8126	1603
3116	1295	0563	6475	4382	9902	6621	9209	8060	1787
5426	5517	5603	3722	4965	5892	8135	5214	9877	6429
2494	6696	5881	1198	2055	4624	4592	4788	7477	7149
1362	2650	8867	6503	5250	7622	5989	5909	2623	7875
5622	8415	9553	7882	1402	4723	7101	1917	8305	0440
6687	5386	9837	9111	8123	3859	1134	6321	4756	1325
0045	5546	2340	7068	6692	3802	8740	0563	8253	1589
3441	4562	1126	6427	7674	6564	1996	9167	4995	6200
9354	3914	6037	7309	5111	3080	3616	2152	2426	4450
8655	6422	1264	7859	3622	8979	7253	4257	5523	4808
0143	0292	0220	2205	4773	4964	5055	5460	0240	7505
5860	4714	6437	3670	5881	1131	7609	9690	3736	7266
8400	6939	5684	7116	3472	4006	1069	5272	5209	8271
3262	4214	5901	1064	7064	4286	1038	1178	3658	4628
2220	1426	2920	8956	8142	4642	3008	9816	5548	7753
9734	7954	9700	1489	3213	8400	7043	7552	4019	9938
3178	1061	8942	8397	4898	3793	6603	2864	6014	5225
4189	6015	5328	8242	0427	1270	1992	4789	8075	7632
4774	5282	1202	5496	8949	8940	9032	6872	3581	6631
9541	6606	6881	4916	5257	3207	9530	4546	9880	0479
4560	8877	8779	1690	6959	1916	2049	7214	0761	5111
2719	2098	7631	2574	5660	8600	2922	1570	6442	8082
7081	8366	4236	6582	9193	4328	8842	1588	1391	7714
2300	5410	2186	6846	4440	6180	6021	5258	3080	3723
4090	3091	2193	1295	0563	6579	6249	9151	1959	8949
2656	1861	2833	0067	2726	3697	5862	6058	8434	1240
9465	8924	6068	1461	0656	2718	1468	5401	9638	0931

* From Donald B. Owen, *Handbook of Statistical Tables*. Reading, Mass.: Addison-Wesley, 1962.

TABLE IX VALUES OF e^{-x}

x	e^{-x}	x	e^{-x}	x	e^{-x}	x	e^{-x}
0.0	1.000	2.5	0.082	5.0	0.0067	7.5	0.00055
0.1	0.905	2.6	0.074	5.1	0.0061	7.6	0.00050
0.2	0.819	2.7	0.067	5.2	0.0055	7.7	0.00045
0.3	0.741	2.8	0.061	5.3	0.0050	7.8	0.00041
0.4	0.670	2.9	0.055	5.4	0.0045	7.9	0.00037
0.5	0.607	3.0	0.050	5.5	0.0041	8.0	0.00034
0.6	0.549	3.1	0.045	5.6	0.0037	8.1	0.00030
0.7	0.497	3.2	0.041	5.7	0.0033	8.2	0.00028
0.8	0.449	3.3	0.037	5.8	0.0030	8.3	0.00025
0.9	0.407	3.4	0.033	5.9	0.0027	8.4	0.00023
1.0	0.368	3.5	0.030	6.0	0.0025	8.5	0.00020
1.1	0.333	3.6	0.027	6.1	0.0022	8.6	0.00018
1.2	0.301	3.7	0.025	6.2	0.0020	8.7	0.00017
1.3	0.273	3.8	0.022	6.3	0.0018	8.8	0.00015
1.4	0.247	3.9	0.020	6.4	0.0017	8.9	0.00014
1.5	0.223	4.0	0.018	6.5	0.0015	9.0	0.00012
1.6	0.202	4.1	0.017	6.6	0.0014	9.1	0.00011
1.7	0.183	4.2	0.015	6.7	0.0012	9.2	0.00010
1.8	0.165	4.3	0.014	6.8	0.0011	9.3	0.00009
1.9	0.150	4.4	0.012	6.9	0.0010	9.4	0.00008
2.0	0.135	4.5	0.011	7.0	0.0009	9.5	0.00008
2.1	0.122	4.6	0.010	7.1	0.0008	9.6	0.00007
2.2	0.111	4.7	0.009	7.2	0.0007	9.7	0.00006
2.3	0.100	4.8	0.008	7.3	0.0007	9.8	0.00006
2.4	0.091	4.9	0.007	7.4	0.0006	9.9	0.00005

TABLE X BINOMIAL COEFFICIENTS

n	$\binom{n}{0}$	$\binom{n}{1}$	$\binom{n}{2}$	$\binom{n}{3}$	$\binom{n}{4}$	$\binom{n}{5}$	$\binom{n}{6}$	$\binom{n}{7}$	$\binom{n}{8}$	$\binom{n}{9}$	$\binom{n}{10}$
0	1										
1	1	1									
2	1	2	1								
3	1	3	3	1							
4	1	4	6	4	1						
5	1	5	10	10	5	1					
6	1	6	15	20	15	6	1				
7	1	7	21	35	35	21	7	1			
8	1	8	28	56	70	56	28	8	1		
9	1	9	36	84	126	126	84	36	9	1	
10	1	10	45	120	210	252	210	120	45	10	1
11	1	11	55	165	330	462	462	330	165	55	11
12	1	12	66	220	495	792	924	792	495	220	66
13	1	13	78	286	715	1287	1716	1716	1287	715	286
14	1	14	91	364	1001	2002	3003	3432	3003	2002	1001
15	1	15	105	455	1365	3003	5005	6435	6435	5005	3003
16	1	16	120	560	1820	4368	8008	11440	12870	11440	8008
17	1	17	136	680	2380	6188	12376	19448	24310	24310	19448
18	1	18	153	816	3060	8568	18564	31824	43758	48620	43758
19	1	19	171	969	3876	11628	27132	50388	75582	92378	92378
20	1	20	190	1140	4845	15504	38760	77520	125970	167960	184756

If necessary, use the identity $\binom{n}{k} = \binom{n}{n-k}$.

$$\frac{n(n-1) \cdot (n-f+1)}{f!}$$

indep trials

TABLE XI LOGARITHMS

N	0	1	2	3	4	5	6	7	8	9
10	0000	0043	0086	0128	0170	0212	0253	0294	0334	0374
11	0414	0453	0492	0531	0569	0607	0645	0682	0719	0755
12	0792	0828	0864	0899	0934	0969	1004	1038	1072	1106
13	1139	1173	1206	1239	1271	1303	1335	1367	1399	1430
14	1461	1492	1523	1553	1584	1614	1644	1673	1703	1732
15	1761	1790	1818	1847	1875	1903	1931	1959	1987	2014
16	2041	2068	2095	2122	2148	2175	2201	2227	2253	2279
17	2304	2330	2355	2380	2405	2430	2455	2480	2504	2529
18	2553	2577	2601	2625	2648	2672	2695	2718	2742	2765
19	2788	2810	2833	2856	2878	2900	2923	2945	2967	2989
20	3010	3032	3054	3075	3096	3118	3139	3160	3181	3201
21	3222	3243	3263	3284	3304	3324	3345	3365	3385	3404
22	3424	3444	3464	3483	3502	3522	3541	3560	3579	3598
23	3617	3636	3655	3674	3692	3711	3729	3747	3766	3784
24	3802	3820	3838	3856	3874	3892	3909	3927	3945	3962
25	3979	3997	4014	4031	4048	4065	4082	4099	4116	4133
26	4150	4166	4183	4200	4216	4232	4249	4265	4281	4298
27	4314	4330	4346	4362	4378	4393	4409	4425	4440	4456
28	4472	4487	4502	4518	4533	4548	4564	4579	4594	4609
29	4624	4639	4654	4669	4683	4698	4713	4728	4742	4757
30	4771	4786	4800	4814	4829	4843	4857	4871	4886	4900
31	4914	4928	4942	4955	4969	4983	4997	5011	5024	5038
32	5051	5065	5079	5092	5105	5119	5132	5145	5159	5172
33	5185	5198	5211	5224	5237	5250	5263	5276	5289	5302
34	5315	5328	5340	5353	5366	5378	5391	5403	5416	5428
35	5441	5453	5465	5478	5490	5502	5514	5527	5539	5551
36	5563	5575	5587	5599	5611	5623	5635	5647	5658	5670
37	5682	5694	5705	5717	5729	5740	5752	5763	5775	5786
38	5798	5809	5821	5832	5843	5855	5866	5877	5888	5899
39	5911	5922	5933	5944	5955	5966	5977	5988	5999	6010
40	6021	6031	6042	6053	6064	6075	6085	6096	6107	6117
41	6128	6138	6149	6160	6170	6180	6191	6201	6212	6222
42	6232	6243	6253	6263	6274	6284	6294	6304	6314	6325
43	6335	6345	6355	6365	6375	6385	6395	6405	6415	6425
44	6435	6444	6454	6464	6474	6484	6493	6503	6513	6522
45	6532	6542	6551	6561	6571	6580	6590	6599	6609	6618
46	6628	6637	6646	6656	6665	6675	6684	6693	6702	6712
47	6721	6730	6739	6749	6758	6767	6776	6785	6794	6803
48	6812	6821	6830	6839	6848	6857	6866	6875	6884	6893
49	6902	6911	6920	6928	6937	6946	6955	6964	6972	6981
50	6990	6998	7007	7016	7024	7033	7042	7050	7059	7067
51	7076	7084	7093	7101	7110	7118	7126	7135	7143	7152
52	7160	7168	7177	7185	7193	7202	7210	7218	7226	7235
53	7243	7251	7259	7267	7275	7284	7292	7300	7308	7316
54	7324	7332	7340	7348	7356	7364	7372	7380	7388	7396

TABLE XI   LOGARITHMS (*Continued*)

N	0	1	2	3	4	5	6	7	8	9
55	7404	7412	7419	7427	7435	7443	7451	7459	7466	7474
56	7482	7490	7497	7505	7513	7520	7528	7536	7543	7551
57	7559	7566	7574	7582	7589	7597	7604	7612	7619	7627
58	7634	7642	7649	7657	7664	7672	7679	7686	7694	7701
59	7709	7716	7723	7731	7738	7745	7752	7760	7767	7774
60	7782	7789	7796	7803	7810	7818	7825	7832	7839	7846
61	7853	7860	7868	7875	7882	7889	7896	7903	7910	7917
62	7924	7931	7938	7945	7952	7959	7966	7973	7980	7987
63	7993	8000	8007	8014	8021	8028	8035	8041	8048	8055
64	8062	8069	8075	8082	8089	8096	8102	8109	8116	8122
65	8129	8136	8142	8149	8156	8162	8169	8176	8182	8189
66	8195	8202	8209	8215	8222	8228	8235	8241	8248	8254
67	8261	8267	8274	8280	8287	8293	8299	8306	8312	8319
68	8325	8331	8338	8344	8351	8357	8363	8370	8376	8382
69	8388	8395	8401	8407	8414	8420	8426	8432	8439	8445
70	8451	8457	8463	8470	8476	8482	8488	8494	8500	8506
71	8513	8519	8525	8531	8537	8543	8549	8555	8561	8567
72	8573	8579	8585	8591	8597	8603	8609	8615	8621	8627
73	8633	8639	8645	8651	8657	8663	8669	8675	8681	8686
74	8692	8698	8704	8710	8716	8722	8727	8733	8739	8745
75	8751	8756	8762	8768	8774	8779	8785	8791	8797	8802
76	8808	8814	8820	8825	8831	8837	8842	8848	8854	8859
77	8865	8871	8876	8882	8887	8893	8899	8904	8910	8915
78	8921	8927	8932	8938	8943	8949	8954	8960	8965	8971
79	8976	8982	8987	8993	8998	9004	9009	9015	9020	9025
80	9031	9036	9042	9047	9053	9058	9063	9069	9074	9079
81	9085	9090	9096	9101	9106	9112	9117	9122	9128	9133
82	9138	9143	9149	9154	9159	9165	9170	9175	9180	9186
83	9191	9196	9201	9206	9212	9217	9222	9227	9232	9238
84	9243	9248	9253	9258	9263	9269	9274	9279	9284	9289
85	9294	9299	9304	9309	9315	9320	9325	9330	9335	9340
86	9345	9350	9355	9360	9365	9370	9375	9380	9385	9390
87	9395	9400	9405	9410	9415	9420	9425	9430	9435	9440
88	9445	9450	9455	9460	9465	9469	9474	9479	9484	9489
89	9494	9499	9504	9509	9513	9518	9523	9528	9533	9538
90	9542	9547	9552	9557	9562	9566	9571	9576	9581	9586
91	9590	9595	9600	9605	9609	9614	9619	9624	9628	9633
92	9638	9643	9647	9652	9657	9661	9666	9671	9675	9680
93	9685	9689	9694	9699	9703	9708	9713	9717	9722	9727
94	9731	9736	9741	9745	9750	9754	9759	9763	9768	9773
95	9777	9782	9786	9791	9795	9800	9805	9809	9814	9818
96	9823	9827	9832	9836	9841	9845	9850	9854	9859	9863
97	9868	9872	9877	9881	9886	9890	9894	9899	9903	9908
98	9912	9917	9921	9926	9930	9934	9939	9943	9948	9952
99	9956	9961	9965	9969	9974	9978	9983	9987	9991	9996

TABLE XII SQUARES AND SQUARE ROOTS*

n	n^2	$\sqrt{n}$	$\sqrt{10n}$	n	n^2	$\sqrt{n}$	$\sqrt{10n}$
1.00	1.0000	1.000000	3.162278	1.50	2.2500	1.224745	3.872983
1.01	1.0201	1.004988	3.178050	1.51	2.2801	1.228821	3.885872
1.02	1.0404	1.009950	3.193744	1.52	2.3104	1.232883	3.898718
1.03	1.0609	1.014889	3.209361	1.53	2.3409	1.236932	3.911521
1.04	1.0816	1.019804	3.224903	1.54	2.3716	1.240967	3.924283
1.05	1.1025	1.024695	3.240370	1.55	2.4025	1.244990	3.937004
1.06	1.1236	1.029563	3.255764	1.56	2.4336	1.249000	3.949684
1.07	1.1449	1.034408	3.271085	1.57	2.4649	1.252996	3.962323
1.08	1.1664	1.039230	3.286335	1.58	2.4964	1.256981	3.974921
1.09	1.1881	1.044031	3.301515	1.59	2.5281	1.260952	3.987480
1.10	1.2100	1.048809	3.316625	1.60	2.5600	1.264911	4.000000
1.11	1.2321	1.053565	3.331666	1.61	2.5921	1.268858	4.012481
1.12	1.2544	1.058301	3.346640	1.62	2.6244	1.272792	4.024922
1.13	1.2769	1.063015	3.361547	1.63	2.6569	1.276715	4.037326
1.14	1.2996	1.067708	3.376389	1.64	2.6896	1.280625	4.049691
1.15	1.3225	1.072381	3.391165	1.65	2.7225	1.284523	4.062019
1.16	1.3456	1.077033	3.405877	1.66	2.7556	1.288410	4.074310
1.17	1.3689	1.081665	3.420526	1.67	2.7889	1.292285	4.086563
1.18	1.3924	1.086278	3.435113	1.68	2.8224	1.296148	4.098780
1.19	1.4161	1.090871	3.449638	1.69	2.8561	1.300000	4.110961
1.20	1.4400	1.095445	3.464102	1.70	2.8900	1.303840	4.123106
1.21	1.4641	1.100000	3.478505	1.71	2.9241	1.307670	4.135215
1.22	1.4884	1.104536	3.492850	1.72	2.9584	1.311488	4.147288
1.23	1.5129	1.109054	3.507136	1.73	2.9929	1.315295	4.159327
1.24	1.5376	1.113553	3.521363	1.74	3.0276	1.319091	4.171331
1.25	1.5625	1.118034	3.535534	1.75	3.0625	1.322876	4.183300
1.26	1.5876	1.122497	3.549648	1.76	3.0976	1.326650	4.195235
1.27	1.6129	1.126943	3.563706	1.77	3.1329	1.330413	4.207137
1.28	1.6384	1.131371	3.577709	1.78	3.1684	1.334166	4.219005
1.29	1.6641	1.135782	3.591657	1.79	3.2041	1.337909	4.230839
1.30	1.6900	1.140175	3.605551	1.80	3.2400	1.341641	4.242641
1.31	1.7161	1.144552	3.619392	1.81	3.2761	1.345362	4.254409
1.32	1.7424	1.148913	3.633180	1.82	3.3124	1.349074	4.266146
1.33	1.7689	1.153256	3.646917	1.83	3.3489	1.352775	4.277850
1.34	1.7956	1.157584	3.660601	1.84	3.3856	1.356466	4.289522
1.35	1.8255	1.161895	3.674235	1.85	3.4225	1.360147	4.301163
1.36	1.8496	1.166190	3.687818	1.86	3.4596	1.363818	4.312772
1.37	1.8769	1.170470	3.701351	1.87	3.4969	1.367479	4.324350
1.38	1.9044	1.174734	3.714835	1.88	3.5344	1.371131	4.335897
1.39	1.9321	1.178983	3.728270	1.89	3.5721	1.374773	4.347413
1.40	1.9600	1.183216	3.741657	1.90	3.6100	1.378405	4.358899
1.41	1.9881	1.187434	3.754997	1.91	3.6481	1.382027	4.370355
1.42	2.0164	1.191638	3.768289	1.92	3.6864	1.385641	4.381780
1.43	2.0449	1.195826	3.781534	1.93	3.7249	1.389244	4.393177
1.44	2.0736	1.200000	3.794733	1.94	3.7636	1.392839	4.404543
1.45	2.1025	1.204159	3.807887	1.95	3.8025	1.396424	4.415880
1.46	2.1316	1.208305	3.820995	1.96	3.8416	1.400000	4.427189
1.47	2.1609	1.212436	3.834058	1.97	3.8809	1.403567	4.438468
1.48	2.1904	1.216553	3.847077	1.98	3.9204	1.407125	4.449719
1.49	2.2201	1.220656	3.860052	1.99	3.9601	1.410674	4.460942

TABLE XII   SQUARES AND SQUARE ROOTS (*Continued*)

n	n^2	$\sqrt{n}$	$\sqrt{10n}$	n	n^2	$\sqrt{n}$	$\sqrt{10n}$
2.00	4.0000	1.414214	4.472136	2.50	6.2500	1.581139	5.000000
2.01	4.0401	1.417745	4.483302	2.51	6.3001	1.584298	5.009990
2.02	4.0804	1.421267	4.494441	2.52	6.3504	1.587451	5.019960
2.03	4.1209	1.424781	4.505552	2.53	6.4009	1.590597	5.029911
2.04	4.1616	1.428286	4.516636	2.54	6.4516	1.593738	5.039841
2.05	4.2025	1.431782	4.527693	2.55	6.5025	1.596872	5.049752
2.06	4.2436	1.435270	4.538722	2.56	6.5536	1.600000	5.059644
2.07	4.2849	1.438749	4.549725	2.57	6.6049	1.603122	5.069517
2.08	4.3264	1.442221	4.560702	2.58	6.6564	1.606238	5.079370
2.09	4.3681	1.445683	4.571652	2.59	6.7081	1.609348	5.089204
2.10	4.4100	1.449138	4.582576	2.60	6.7600	1.612452	5.099020
2.11	4.4521	1.452584	4.593474	2.61	6.8121	1.615549	5.108816
2.12	4.4944	1.456022	4.604346	2.62	6.8644	1.618641	5.118594
2.13	4.5369	1.459452	4.615192	2.63	6.9169	1.621727	5.128353
2.14	4.5796	1.462874	4.626013	2.64	6.9696	1.624808	5.138093
2.15	4.6225	1.466288	4.636809	2.65	7.0225	1.627882	5.147815
2.16	4.6656	1.469694	4.647580	2.66	7.0756	1.630951	5.157519
2.17	4.7089	1.473092	4.658326	2.67	7.1289	1.634013	5.167204
2.18	4.7524	1.476482	4.669047	2.68	7.1824	1.637071	5.176872
2.19	4.7961	1.479865	4.679744	2.69	7.2361	1.640122	5.186521
2.20	4.8400	1.483240	4.690416	2.70	7.2900	1.643168	5.196152
2.21	4.8841	1.486607	4.701064	2.71	7.3441	1.646208	5.205766
2.22	4.9284	1.489966	4.711688	2.72	7.3984	1.649242	5.215362
2.23	4.9729	1.493318	4.722288	2.73	7.4529	1.652271	5.224940
2.24	5.0176	1.496663	4.732864	2.74	7.5076	1.655295	5.234501
2.25	5.0625	1.500000	4.743416	2.75	7.5625	1.658312	5.244044
2.26	5.1076	1.503330	4.753946	2.76	7.6176	1.661325	5.253570
2.27	5.1529	1.506652	4.764452	2.77	7.6729	1.664332	5.263079
2.28	5.1984	1.509967	4.774935	2.78	7.7284	1.667333	5.272571
2.29	5.2441	1.513275	4.785394	2.79	7.7841	1.670329	5.282045
2.30	5.2900	1.516575	4.795832	2.80	7.8400	1.673320	5.291503
2.31	5.3361	1.519868	4.806246	2.81	7.8961	1.676305	5.300943
2.32	5.3824	1.523155	4.816638	2.82	7.9524	1.679286	5.310367
2.33	5.4289	1.526434	4.827007	2.83	8.0089	1.682260	5.319774
2.34	5.4756	1.529706	4.837355	2.84	8.0656	1.685230	5.329165
2.35	5.5225	1.532971	4.847680	2.85	8.1225	1.688194	5.338539
2.36	5.5696	1.536229	4.857983	2.86	8.1796	1.691153	5.347897
2.37	5.6169	1.539480	4.868265	2.87	8.2369	1.694107	5.357238
2.38	5.6644	1.542725	4.878524	2.88	8.2944	1.697056	5.366563
2.39	5.7121	1.545962	4.888763	2.89	8.3521	1.700000	5.375872
2.40	5.7600	1.549193	4.898979	2.90	8.4100	1.702939	5.385165
2.41	5.8081	1.552417	4.909175	2.91	8.4681	1.705872	5.394442
2.42	5.8564	1.555635	4.919350	2.92	8.5264	1.708801	5.403702
2.43	5.9049	1.558846	4.929503	2.93	8.5849	1.711724	5.412947
2.44	5.9536	1.562050	4.939636	2.94	8.6436	1.714643	5.422177
2.45	6.0025	1.565248	4.949747	2.95	8.7025	1.717556	5.431390
2.46	6.0516	1.568439	4.959839	2.96	8.7616	1.720465	5.440588
2.47	6.1009	1.571623	4.969909	2.97	8.8209	1.723369	5.449771
2.48	6.1504	1.574802	4.979960	2.98	8.8804	1.726268	5.458938
2.49	6.2001	1.577973	4.989990	2.99	8.9401	1.729162	5.468089

TABLE XII SQUARES AND SQUARE ROOTS (*Continued*)

n	n^2	$\sqrt{n}$	$\sqrt{10n}$	n	n^2	$\sqrt{n}$	$\sqrt{10n}$
3.00	9.0000	1.732051	5.477226	3.50	12.2500	1.870829	5.916080
3.01	9.0601	1.734935	5.486347	3.51	12.3201	1.873499	5.924525
3.02	9.1204	1.737815	5.495453	3.52	12.3904	1.876166	5.932959
3.03	9.1809	1.740690	5.504544	3.53	12.4609	1.878829	5.941380
3.04	9.2416	1.743560	5.513620	3.54	12.5316	1.881489	5.949790
3.05	9.3025	1.746425	5.522681	3.55	12.6025	1.884144	5.958188
3.06	9.3636	1.749286	5.531727	3.56	12.6736	1.886796	5.966574
3.07	9.4249	1.752142	5.540758	3.57	12.7449	1.889444	5.974948
3.08	9.4864	1.754993	5.549775	3.58	12.8164	1.892089	5.983310
3.09	9.5481	1.757840	5.558777	3.59	12.8881	1.894730	5.991661
3.10	9.6100	1.760682	5.567764	3.60	12.9600	1.897367	6.000000
3.11	9.6721	1.763519	5.576737	3.61	13.0321	1.900000	6.008328
3.12	9.7344	1.766352	5.585696	3.62	13.1044	1.902630	6.016644
3.13	9.7969	1.769181	5.594640	3.63	13.1769	1.905256	6.024948
3.14	9.8596	1.772005	5.603570	3.64	13.2496	1.907878	6.033241
3.15	9.9225	1.774824	5.612486	3.65	13.3225	1.910497	6.041523
3.16	9.9856	1.777639	5.621388	3.66	13.3956	1.913113	6.049793
3.17	10.0489	1.780449	5.630275	3.67	13.4689	1.915724	6.058052
3.18	10.1124	1.783255	5.639149	3.68	13.5424	1.918333	6.066300
3.19	10.1761	1.786057	5.648008	3.69	13.6161	1.920937	6.074537
3.20	10.2400	1.788854	5.656854	3.70	13.6900	1.923538	6.082763
3.21	10.3041	1.791647	5.665686	3.71	13.7641	1.926136	6.090977
3.22	10.3684	1.794436	5.674504	3.72	13.8384	1.928730	6.099180
3.23	10.4329	1.797220	5.683309	3.73	13.9129	1.931321	6.107373
3.24	10.4976	1.800000	5.692100	3.74	13.9876	1.933908	6.115554
3.25	10.5625	1.802776	5.700877	3.75	14.0625	1.936492	6.123724
3.26	10.6276	1.805547	5.709641	3.76	14.1376	1.939072	6.131884
3.27	10.6929	1.808314	5.718391	3.77	14.2129	1.941649	6.140033
3.28	10.7584	1.811077	5.727128	3.78	14.2884	1.944222	6.148170
3.29	10.8241	1.813836	5.735852	3.79	14.3641	1.946792	6.156298
3.30	10.8900	1.816590	5.744563	3.80	14.4400	1.949359	6.164414
3.31	10.9561	1.819341	5.753260	3.81	14.5161	1.951922	6.172520
3.32	11.0224	1.822087	5.761944	3.82	14.5924	1.954483	6.180615
3.33	11.0889	1.824829	5.770615	3.83	14.6689	1.957039	6.188699
3.34	11.1556	1.827567	5.779273	3.84	14.7456	1.959592	6.196773
3.35	11.2225	1.830301	5.787918	3.85	14.8225	1.962142	6.204837
3.36	11.2896	1.833030	5.796551	3.86	14.8996	1.964688	6.212890
3.37	11.3569	1.835756	5.805170	3.87	14.9769	1.967232	6.220932
3.38	11.4244	1.838478	5.813777	3.88	15.0544	1.969772	6.228965
3.39	11.4921	1.841195	5.822371	3.89	15.1321	1.972308	6.236986
3.40	11.5600	1.843909	5.830952	3.90	15.2100	1.974842	6.244998
3.41	11.6281	1.846619	5.839521	3.91	15.2881	1.977372	6.252999
3.42	11.6964	1.849324	5.848077	3.92	15.3664	1.979899	6.260990
3.43	11.7649	1.852026	5.856620	3.93	15.4449	1.982423	6.268971
3.44	11.8336	1.854724	5.865151	3.94	15.5236	1.984943	6.276942
3.45	11.9025	1.857418	5.873670	3.95	15.6025	1.987461	6.284903
3.46	11.9716	1.860108	5.882176	3.96	15.6816	1.989975	6.292853
3.47	12.0409	1.862794	5.890671	3.97	15.7609	1.992486	6.300794
3.48	12.1104	1.865476	5.899152	3.98	15.8404	1.994994	6.308724
3.49	12.1801	1.868154	5.907622	3.99	15.9201	1.997498	6.316645

TABLE XII SQUARES AND SQUARE ROOTS (*Continued*)

n	n^2	$\sqrt{n}$	$\sqrt{10n}$	n	n^2	$\sqrt{n}$	$\sqrt{10n}$
4.00	16.0000	2.000000	6.324555	4.50	20.2500	2.121320	6.708204
4.01	16.0801	2.002498	6.332456	4.51	20.3401	2.123676	6.715653
4.02	16.1604	2.004994	6.340347	4.52	20.4304	2.126029	6.723095
4.03	16.2409	2.007486	6.348228	4.53	20.5209	2.128380	6.730527
4.04	16.3216	2.009975	6.356099	4.54	20.6116	2.130728	6.737952
4.05	16.4025	2.012461	6.363961	4.55	20.7025	2.133073	6.745369
4.06	16.4836	2.014944	6.371813	4.56	20.7936	2.135416	6.752777
4.07	16.5649	2.017424	6.379655	4.57	20.8849	2.137756	6.760178
4.08	16.6464	2.019901	6.387488	4.58	20.9764	2.140093	6.767570
4.09	16.7281	2.022375	6.395311	4.59	21.0681	2.142429	6.774954
4.10	16.8100	2.024846	6.403124	4.60	21.1600	2.144761	6.782330
4.11	16.8921	20.27313	6.410928	4.61	21.2521	2.147091	6.789698
4.12	16.9744	2.029778	6.418723	4.62	21.3444	2.149419	6.797058
4.13	17.0569	2.032240	6.426508	4.63	21.4369	2.151743	6.804410
4.14	17.1396	2.034699	6.434283	4.64	21.5296	2.154066	6.811755
4.15	17.2225	2.037155	6.442049	4.65	21.6225	2.156386	6.819091
4.16	17.3056	2.039608	6.449806	4.66	21.7156	2.158703	6.826419
4.17	17.3889	2.042058	6.457554	4.67	21.8089	2.161018	6.833740
4.18	17.4724	2.044505	6.465292	4.68	21.9024	2.163331	6.841053
4.19	17.5561	2.046949	6.473021	4.69	21.9961	2.165641	6.848357
4.20	17.6400	2.049390	6.480741	4.70	22.0900	2.167948	6.855655
4.21	17.7241	2.051828	6.488451	4.71	22.1841	2.170253	6.862944
4.22	17.8084	2.054264	6.496153	4.72	22.2784	2.172556	6.870226
4.23	17.8929	2.056696	6.503845	4.73	22.3729	2.174856	6.877500
4.24	17.9776	2.059126	6.511528	4.74	22.4676	2.177154	6.884766
4.25	18.0625	2.061553	6.519202	4.75	22.5625	2.179449	6.892024
4.26	18.1476	2.063977	6.526868	4.76	22.6576	2.181742	6.899275
4.27	18.2329	2.066398	6.534524	4.77	22.7529	2.184033	6.906519
4.28	18.3184	2.068816	6.542171	4.78	22.8484	2.186321	6.913754
4.29	18.4041	2.071232	6.549809	4.79	22.9441	2.188607	6.920983
4.30	18.4900	2.073644	6.557439	4.80	23.0400	2.190890	6.928203
4.31	18.5761	2.076054	6.565059	4.81	23.1361	2.193171	6.935416
4.32	18.6624	2.078461	6.572671	4.82	23.2324	2.195450	6.942622
4.33	18.7489	2.080865	6.580274	4.83	23.3289	2.197726	6.949820
4.34	18.8356	2.083267	6.587868	4.84	23.4256	2.200000	6.957011
4.35	18.9225	2.085665	6.595453	4.85	23.5225	2.202272	6.964194
4.36	19.0096	2.088061	6.603030	4.86	23.6196	2.204541	6.971370
4.37	19.0969	2.090454	6.610598	4.87	23.7169	2.206808	6.978539
4.38	19.1844	2.092845	6.618157	4.88	23.8144	2.209072	6.985700
4.39	19.2721	2.095233	6.625708	4.89	23.9121	2.211334	6.992853
4.40	19.3600	2.097618	6.633250	4.90	24.0100	2.213594	7.000000
4.41	19.4481	2.100000	6.640783	4.91	24.1081	2.215852	7.007139
4.42	19.5364	2.102380	6.648308	4.92	24.2064	2.218107	7.014271
4.43	19.6249	2.104757	6.655825	4.93	24.3049	2.220360	7.021396
4.44	19.7136	2.107131	6.663332	4.94	24.4036	2.222611	7.028513
4.45	19.8025	2.109502	6.670832	4.95	24.5025	2.224860	7.035624
4.46	19.8916	2.111871	6.678323	4.96	24.6016	2.227106	7.042727
4.47	19.9809	2.114237	6.685806	4.97	24.7009	2.229350	7.049823
4.48	20.0704	2.116601	6.693280	4.98	24.8004	2.231591	7.056912
4.49	20.1601	2.118962	6.700746	4.99	24.9001	2.233831	7.063993

TABLE XII SQUARES AND SQUARE ROOTS (*Continued*)

n	n^2	$\sqrt{n}$	$\sqrt{10n}$	n	n^2	$\sqrt{n}$	$\sqrt{10n}$
5.00	25.0000	2.236068	7.071068	5.50	30.2500	2.345208	7.416198
5.01	25.1001	2.238303	7.078135	5.51	30.3601	2.347339	7.422937
5.02	25.2004	2.240536	7.085196	5.52	30.4704	2.349468	7.429670
5.03	25.3009	2.242766	7.092249	5.53	30.5809	2.351595	7.436397
5.04	25.4016	2.244994	7.099296	5.54	30.6916	2.353720	7.443118
5.05	25.5025	2.247221	7.106335	5.55	30.8025	2.355844	7.449832
5 06	25.6036	2.249444	7.113368	5.56	30.9136	2.357965	7.456541
5.07	25.7049	2.251666	7.120393	5.57	31.0249	2.360085	7.463243
5.08	25.8064	2.253886	7.127412	5.58	31.1364	2.362202	7.469940
5.09	25.9081	2.256103	7.134424	5.59	31.2481	2.364318	7.476630
5.10	26.0100	2.258318	7.141428	5.60	31.3600	2.366432	7.483315
5.11	26.1121	2.260531	7.148426	5.61	31.4721	2.368544	7.489993
5.12	26.2144	2.262742	7.155418	5.62	31.5844	2.370654	7.496666
5.13	26.3169	2.264950	7.162402	5.63	31.6969	2.372762	7.503333
5.14	26.4196	2.267157	7.169379	5.64	31.8096	2.374868	7.509993
5.15	26.5225	2.269361	7.176350	5.65	31.9225	2.376973	7.516648
5.16	26.6256	2.271563	7.183314	5.66	32.0356	2.379075	7.523297
5.17	26.7289	2.273763	7.190271	5.67	32.1489	2.381176	7.529940
5.18	26.8324	2.275961	7.197222	5.68	32.2624	2.383275	7.536577
5.19	26.9361	2.278157	7.204165	5.69	32.3761	2.385372	7.543209
5.20	27.0400	2.280351	7.211103	5.70	32.4900	2.387467	7.549834
5.21	27.1441	2.282542	7.218033	5.71	32.6041	2.389561	7.556454
5.22	27.2484	2.284732	7.224957	5.72	32.7184	2.391652	7.563068
5.23	27.3529	2.286919	7.231874	5.73	32.8329	2.393742	7.569676
5.24	27.4576	2.289105	7.238784	5.74	32.9476	2.395830	7.576279
5.25	27.5625	2.291288	7.245688	5.75	33.0625	2.397916	7.582875
5.26	27.6676	2.293469	7.252586	5.76	33.1776	2.400000	7.589466
5.27	27.7729	2.295648	7.259477	5.77	33.2929	2.402082	7.596052
5.28	27.8784	2.297825	7.266361	5.78	33.4084	2.404163	7.602631
5.29	27.9841	2.300000	7.273239	5.79	33.5241	2.406242	7.609205
5.30	28.0900	2.302173	7.280110	5.80	33.6400	2.408319	7.615773
5.31	28.1961	2.304344	7.286975	5.81	33.7561	2.410394	7.622336
5.32	28.3024	2.306513	7.293833	5.82	33.8724	2.412468	7.628892
5.33	28.4089	2.308679	7.300685	5.83	33.9889	2.414539	7.635444
5.34	28.5156	2.310844	7.307530	5.84	34.1056	2.416609	7.641989
5.35	28.6225	2.313007	7.314369	5.85	34.2225	2.418677	7.648529
5.36	28.7296	2.315167	7.321202	5.86	34.3396	2.420744	7.655064
5.37	28.8369	2.317326	7.328028	5.87	34.4569	2.422808	7.661593
5.38	28.9444	2.319483	7.334848	5.88	34.5744	2.424871	7.668116
5.39	29.0521	2.321637	7.341662	5.89	34.6921	2.426932	7.674634
5.40	29.1600	2.323790	7.348469	5.90	34.8100	2.428992	7.681146
5.41	29.2681	2.325941	7.355270	5.91	34.9281	2.431049	7.687652
5.42	29.3764	2.328089	7.362065	5.92	35.0464	2.433105	7.694154
5.43	29.4849	2.330236	7.368853	5.93	35.1649	2.435159	7.700649
5.44	29.5936	2.332381	7.357636	5.94	35.2836	2.437212	7.707140
5.45	29.7025	2.334524	7.382412	5.95	35.4025	2.439262	7.713624
5.46	29.8116	2.336664	7.389181	5.96	35.5216	2.441311	7.720104
5.47	29.9209	2.338803	7.395945	5.97	35.6409	2.443358	7.726578
5.48	30.0304	2.340940	7.402702	5.98	35.7604	2.445404	7.733046
5.49	30.1401	2.343075	7.409453	5.99	35.8801	2.447448	7.739509

TABLE XII SQUARES AND SQUARE ROOTS (*Continued*)

n	n^2	$\sqrt{n}$	$\sqrt{10n}$	n	n^2	$\sqrt{n}$	$\sqrt{10n}$
6.00	36.0000	2.449490	7.745967	6.50	42.2500	2.549510	8.062258
6.01	36.1201	2.451530	7.752419	6.51	42.3801	2.551470	8.068457
6.02	36.2404	2.453569	7.758866	6.52	42.5104	2.553429	8.074652
6.03	36.3609	2.455606	7.765307	6.53	42.6409	2.555386	8.080842
6.04	36.4816	2.457641	7.771744	6.54	42.7716	2.557342	8.087027
6.05	36.6025	2.459675	7.778175	6.55	42.9025	2.559297	8.093207
6.06	36.7236	2.461707	7.784600	6.56	43.0336	2.561250	8.099383
6.07	36.8449	2.463737	7.791020	6.57	43.1649	2.563201	8.105554
6.08	36.9664	2.465766	7.797435	6.58	43.2964	2.565151	8.111720
6.09	37.0881	2.467793	7.803845	6.59	43.4281	2.567100	8.117881
6.10	37.2100	2.469818	7.810250	6.60	43.5600	2.569047	8.124038
6.11	37.3321	2.471841	7.816649	6.61	43.6921	2.570992	8.130191
6.12	37.4544	2.473863	7.823043	6.62	43.8244	2.572936	8.136338
6.13	37.5769	2.475884	7.829432	6.63	43.9569	2.574879	8.142481
6.14	37.6996	2.477902	7.835815	6.64	44.0896	2.576820	8.148620
6.15	37.8225	2.479919	7.842194	6.65	44.2225	2.578759	8.154753
6.16	37.9456	2.481935	7.848567	6.66	44.3556	2.580698	8.160882
6.17	38.0689	2.483948	7.854935	6.67	44.4889	2.582634	8.167007
6.18	38.1924	2.485961	7.861298	6.68	44.6224	2.584570	8.173127
6.19	38.3161	2.487971	7.867655	6.69	44.7561	2.586503	8.179242
6.20	38.4400	2.489980	7.874008	6.70	44.8900	2.588436	8.185353
6.21	38.5641	2.491987	7.880355	6.71	45.0241	2.590367	8.191459
6.22	38.6884	2.493993	7.886698	6.72	45.1584	2.592296	8.197561
6.23	38.8129	2.495997	7.893035	6.73	45.2929	2.594224	8.203658
6.24	38.9376	2.497999	7.899367	6.74	45.4276	2.596151	8.209750
6.25	39.0625	2.500000	7.905694	6.75	45.5625	2.598076	8.215838
6.26	39.1876	2.501999	7.912016	6.76	45.6976	2.600000	8.221922
6.27	39.3129	2.503997	7.918333	6.77	45.8329	2.601922	8.228001
6.28	39.4384	2.505993	7.924645	6.78	45.9684	2.603843	8.234076
6.29	39.5641	2.507987	7.930952	6.79	46.1041	2.605763	8.240146
6.30	39.6900	2.509980	7.937254	6.80	46.2400	2.607681	8.246211
6.31	39.8161	2.511971	7.943551	6.81	46.3761	2.609598	8.242272
6.32	39.9424	2.513961	7.949843	6.82	46.5124	2.611513	8.258329
6.33	40.0689	2.515949	7.956130	6.83	46.6489	2.613427	8.264381
6.34	40.1956	2.517936	7.962412	6.84	46.7856	2.615339	8.270429
6.35	40.3225	2.519921	7.968689	6.85	46.9225	2.617250	8.276473
6.36	40.4496	2.521904	7.974961	6.86	47.0596	2 619160	8.282512
6.37	40.5769	2.523886	7.981228	6.87	47.1969	2.621068	8.288546
6.38	40.7044	2.525866	7.987490	6.88	47.3344	2.622975	8.294577
6.39	40.8321	2.527845	7.993748	6.89	47.4721	2.624881	8.300602
6.40	40.9600	2.529822	8.000000	6.90	47.6100	2.626785	8.306624
6.41	41.0881	2.531798	8.006248	6.91	47.7481	2.628688	8.312641
6.42	41.2164	2.533772	8.012490	6.92	47.8864	2.630589	8.318654
6.43	41.3449	2.535744	8.018728	6.93	48.0249	2.632489	8.324662
6.44	41.4736	2.537716	8.024961	6.94	48.1636	2.634388	8.330666
6.45	41.6025	2.539685	8.031189	6.95	48.3025	2.636285	8.336666
6.46	41.7316	2.541653	8.037413	6.96	48.4416	2.638181	8.342661
6.47	41.8609	2.543619	8.043631	6.97	48.5809	2.640076	8.348653
6.48	41.9904	2.545584	8.049845	6.98	48.7204	2.641969	8.354639
6.49	42.1201	2.547548	8.056054	6.99	48.8601	2.643861	8.360622

TABLE XII   SQUARES AND SQUARE ROOTS (*Continued*)

n	n^2	$\sqrt{n}$	$\sqrt{10n}$	n	n^2	$\sqrt{n}$	$\sqrt{10n}$
7.00	49.0000	2.645751	8.366600	7.50	56.2500	2.738613	8.660254
7.01	49.1401	2.647640	8.372574	7.51	56.4001	2.740438	8.660026
7.02	49.2804	2.649528	8.378544	7.52	56.5504	2.742262	8.671793
7.03	49.4209	2.651415	8.384510	7.53	56.7009	2.744085	8.677557
7.04	49.5616	2.653300	8.390471	7.54	56.8516	2.745906	8.683317
7.05	49.7025	2.655184	8.396428	7.55	57.0025	2.747726	8.689074
7.06	49.8436	2.657066	8.402381	7.56	57.1536	2.749545	8.694826
7.07	49.9849	2.658947	8.408329	7.57	57.3049	2.751363	8.700575
7.08	50.1264	2.660827	8.414274	7.58	57.4564	2.753180	8.706320
7.09	50.2681	2.662705	8.420214	7.59	57.6081	2.754995	8.712061
7.10	50.4100	2.664583	8.426150	7.60	57.7600	2.756810	8.717798
7.11	50.5521	2.666458	8.432082	7.61	57.9121	2.758623	8.723531
7.12	50.6944	2.668333	8.438009	7.62	58.0644	2.760435	8.729261
7.13	50.8369	2.670206	8.443933	7.63	58.2169	2.762245	8.734987
7.14	50.9796	2.672078	8.449852	7.64	58.3696	2.764055	8.740709
7.15	51.1225	2.673948	8.455767	7.65	58.5225	2.765863	8.746428
7.16	51.2656	2.675818	8.461678	7.66	58.6756	2.767671	8.752143
7.17	51.4089	2.677686	8.467585	7.67	58.8289	2.769476	8.757854
7.18	51.5524	2.679552	8.473488	7.68	58.9824	2.771281	8.763561
7.19	51.6961	2.681418	8.479387	7.69	59.1361	2.773085	8.769265
7.20	51.8400	2.683282	8.485281	7.70	59.2900	2.774887	8.774964
7.21	51.9841	2.685144	8.491172	7.71	59.4441	2.776689	8.780661
7.22	52.1284	2.687006	8.497058	7.72	59.5984	2.778489	8.786353
7.23	52.2729	2.688866	8.502941	7.73	59.7529	2.780288	8.792042
7.24	52.4176	2.690725	8.508819	7.74	59.9076	2.782086	8.797727
7.25	52.5625	2.692582	8.514693	7.75	60.0625	2.783882	8.803408
7.26	52.7076	2.694439	8.520563	7.76	60.2176	2.785678	8.809086
7.27	52.8529	2.696294	8.526429	7.77	60.3729	2.787472	8.814760
7.28	52.9984	2.698148	8.532292	7.78	60.5284	2.789265	8.820431
7.29	53.1441	2.700000	8.538150	7.79	60.6841	2.791057	8.826098
7.30	53.2900	2.701851	8.544004	7.80	60.8400	2.792848	8.831761
7.31	53.4361	2.703701	8.549854	7.81	60.9961	2.794638	8.837420
7.32	53.5824	2.705550	8.555700	7.82	61.1524	2.796426	8.843076
7.33	53.7289	2.707397	8.561542	7.83	61.3089	2.798214	8.848729
7.34	53.8756	2.709243	8.567380	7.84	61.4656	2.800000	8.854377
7.35	54.0225	2.711088	8.573214	7.85	61.6225	2.801785	8.860023
7.36	54.1696	2.712932	8.579044	7.86	61.7796	2.803569	8.865664
7.37	54.3169	2.714774	8.584870	7.87	61.9369	2.805352	8.871302
7.38	54.4644	2.716616	8.590693	7.88	62.0944	2.807134	8.876936
7.39	54.6121	2.718455	8.596511	7.89	62.2521	2.808914	8.882567
7.40	54.7600	2.720294	8.602325	7.90	62.4100	2.810694	8.888194
7.41	54.9081	2.722132	8.608136	7.91	62.5681	2.812472	8.893818
7.42	55.0564	2.723968	8.613942	7.92	62.7264	2.814249	8.899438
7.43	55.2049	2.725803	8.619745	7.93	62.8849	2.816026	8.905055
7.44	55.3536	2.727636	8.625543	7.94	63.0436	2.817801	8.910668
7.45	55.5025	2.729469	8.631338	7.95	63.2025	2.819574	8.916277
7.46	55.6516	2.731300	8.637129	7.96	63.3616	2.821347	8.921883
7.47	55.8009	2.733130	8.642916	7.97	63.5209	2.823119	8.927486
7.48	55.9504	2.734959	8.648699	7.98	63.6804	2.824889	8.933085
7.49	56.1001	2.736786	8.654479	7.99	63.8401	2.826659	8.938680

TABLE XII SQUARES AND SQUARE ROOTS (*Continued*)

n	n^2	$\sqrt{n}$	$\sqrt{10n}$	n	n^2	$\sqrt{n}$	$\sqrt{10n}$
8.00	64.0000	2.828427	8.944272	8.50	72.2500	2.915476	9.219544
8.01	64.1601	2.830194	8.949860	8.51	72.4201	2.917190	9.224966
8.02	64.3204	2.831960	8.955445	8.52	72.5904	2.918904	9.230385
8.03	64.4809	2.833725	8.961027	8.53	72.7609	2.920616	9.235800
8.04	64.6416	2.835489	8.966605	8.54	72.9316	2.922328	9.241212
8.05	64.8025	2.837252	8.972179	8.55	73.1025	2.924038	9.246621
8.06	64.9636	2.839014	8.977750	8.56	73.2736	2.925748	9.252027
8.07	65.1249	2.840775	8.983318	8.57	73.4449	2.927456	9.257429
8.08	65.2864	2.842534	8.988882	8.58	73.6164	2.929164	9.262829
8.09	65.4481	2.844293	8.994443	8.59	73.7881	2.930870	9.268225
8.10	65.6100	2.846050	9.000000	8.60	73.9600	2.932576	9.273618
8.11	65.7721	2.847806	9.005554	8.61	74.1321	2.934280	9.279009
8.12	65.9344	2.849561	9.011104	8.62	74.3044	2.935984	9.284396
8.13	66.0969	2.851315	9.016651	8.63	74.4769	2.937686	9.289779
8.14	66.2596	2.853069	9.022195	8.64	74.6496	2.939388	9.295160
8.15	66.4225	2.854820	9.027735	8.65	74.8225	2.941088	9.300538
8.16	66.5856	2.856571	9.033272	8.66	74.9956	2.942788	9.305912
8.17	66.7489	2.858321	9.038805	8.67	75.1689	2.944486	9.311283
8.18	66.9124	2.860070	9.044335	8.68	75.3424	2.946184	9.316652
8.19	67.0761	2.861818	9.049862	8.69	75.5161	2.947881	9.322017
8.20	67.2400	2.863564	9.055385	8.70	75.6900	2.949576	9.327379
8.21	67.4041	2.865310	9.060905	8.71	75.8641	2.951271	9.332738
8.22	67.5684	2.867054	9.066422	8.72	76.0384	2.952965	9.338094
8.23	67.7329	2.868798	9.071935	8.73	76.2129	2.954657	9.343447
8.24	67.8976	2.870540	9.077445	8.74	76.3876	2.956349	9.348797
8.25	68.0625	2.872281	9.082951	8.75	76.5625	2.958040	9.354143
8.26	68.2276	2.874022	9.088454	8.76	76.7376	2.959730	9.359487
8.27	68.3929	2.875761	9.093954	8.77	76.9129	2.961419	9.364828
8.28	68.5584	2.877499	9.099451	8.78	77.0884	2.963106	9.370165
8.29	68.7241	2.879236	9.104944	8.79	77.2641	2.964793	9.375500
8.30	68.8900	2.880972	9.110434	8.80	77.4400	2.966479	9.380832
8.31	69.0561	2.882707	9.115920	8.81	77.6161	2.968164	9.386160
8.32	69.2224	2.884441	9.121403	8.82	77.7924	2.969848	9.391486
8.33	69.3889	2.886174	9.126883	8.83	77.9689	2.971532	9.396808
8.34	69.5556	2.887906	9.132360	8.84	78.1456	2.973214	9.402127
8.35	69.7225	2.889637	9.137833	8.85	78.3225	2.974895	9.407444
8.36	69.8896	2.891366	9.143304	8.86	78.4996	2.976575	9.412757
8.37	7.00569	2.893095	9.148770	8.87	78.6769	2.978255	9.418068
8.38	70.2244	2.894823	9.154234	8.88	78.8544	2.979933	9.423375
8.39	70.3921	2.896550	9.159694	8.89	79.0321	2.981610	9.428680
8.40	70.5600	2.898275	9.165151	8.90	79.2100	2.983287	9.433981
8.41	70.7281	2.900000	9.170605	8.91	79.3881	2.984962	9.439280
8.42	70.8964	2.901724	9.176056	8.92	79.5664	2.986637	9.444575
8.43	71.0649	2.903446	9.181503	8.93	79.7449	2.988311	9.449868
8.44	71.2336	2.905168	9.186947	8.94	79.9236	2.989983	9.455157
8.45	71.4025	2.906888	9.192388	8.95	80.1025	2.991655	9.460444
8.46	71.5716	2.908608	9.197826	8.96	80.2816	2.993326	9.465728
8.47	71.7409	2.910326	9.203260	8.97	80.4609	2.994996	9.471008
8.48	71.9104	2.912044	9.208692	8.98	80.6404	2.996665	9.476286
8.49	72.0801	2.913760	9.214120	8.99	80.8201	2.998333	9.481561

TABLE XII   SQUARES AND SQUARE ROOTS (*Continued*)

n	n^2	$\sqrt{n}$	$\sqrt{10n}$	n	n^2	$\sqrt{n}$	$\sqrt{10n}$
9.00	81.0000	3.000000	9.486833	9.50	90.2500	3.082207	9.746794
9.01	81.1801	3.001666	9.492102	9.51	90.4401	3.083829	9.751923
9.02	81.3604	3.003331	9.497368	9.52	90.6304	3.085450	9.757049
9.03	81.5409	3.004996	9.502631	9.53	90.8209	3.087070	9.762172
9.04	81.7216	3.006659	9.507891	9.54	91.0116	3.088689	9.767292
9.05	81.9025	3.008322	9.513149	9.55	91.2025	3.090307	9.772410
9.06	82.0836	3.009983	9.518403	9.56	91.3936	3.091925	9.777525
9.07	82.2649	3.011644	9.523655	9.57	91.5849	3.093542	9.782638
9.08	82.4464	3.013304	9.528903	9.58	91.7764	3.095158	9.787747
9.09	82.6281	3.014963	9.534149	9.59	91.9681	3.096773	9.792855
9.10	82.8100	3.016621	9.539392	9.60	92.1600	3.098387	9.797959
9.11	82.9921	3.018278	9.544632	9.61	92.3521	3.100000	9.803061
9.12	83.1744	3.019934	9.549869	9.62	92.5444	3.101612	9.808160
9.13	83.3569	3.021589	9.555103	9.63	92.7369	3.103224	9.813256
9.14	83.5396	3.023243	9.560335	9.64	92.9296	3.104835	9.818350
9.15	83.7225	3.024897	9.565563	9.65	93.1225	3.106445	9.823441
9.16	83.9056	3.026549	9.570789	9.66	93.3156	3.108054	9.828530
9.17	84.0889	3.028201	9.576012	9.67	93.5089	3.109662	9.833616
9.18	84.2724	3.029851	9.581232	9.68	93.7024	3.111270	9.838699
9.19	84.4561	3.031501	9.586449	9.69	93.8961	3.112876	9.843780
9.20	84.6400	3.033150	9.591663	9.70	94.0900	3.114482	9.848858
9.21	84.8241	3.034798	9.596874	9.71	94.2841	3.116087	9.853933
9.22	85.0084	3.036445	9.602083	9.72	94.4784	3.117691	9.859006
9.23	85.1929	3.038092	9.607289	9.73	94.6729	3.119295	9.864076
9.24	85.3776	3.039737	9.612492	9.74	94.8676	3.120897	9.869144
9.25	85.5625	3.041381	9.617692	9.75	95.0625	3.122499	9.874209
9.26	85.7476	3.043025	9.622889	9.76	95.2576	3.124100	9.879271
9.27	85.9329	3.044667	9.628084	9.77	95.4529	3.125700	9.884331
9.28	86.1184	3.046309	9.633276	9.78	95.6484	3.127299	9.889388
9.29	86.3041	3.047950	9.638465	9.79	95.8441	3.128898	9.894443
9.30	86.4900	3.049590	9.643651	9.80	96.0400	3.130495	9.899495
9.31	86.6761	3.051229	9.648834	9.81	96.2361	3.132092	9.904544
9.32	86.8624	3.052868	9.654015	9.82	96.4324	3.133688	9.909591
9.33	87.0489	3.054505	9.659193	9.83	96.6289	3.135283	9.914636
9.34	87.2356	3.056141	9.664368	9.84	96.8256	3.136877	9.919677
9.35	87.4225	3.057777	9.669540	9.85	97.0225	3.138471	9.924717
9.36	87.6096	3.059412	9.674709	9.86	97.2196	3.140064	9.929753
9.37	87.7969	3.061046	9.679876	9.87	97.4169	3.141656	9.934787
9.38	87.9844	3.062679	9.685040	9.88	97.6144	3.143247	9.939819
9.39	88.1721	3.064311	9.690201	9.89	97.8121	3.144837	9.944848
9.40	88.3600	3.065942	9.695360	9.90	98.0100	3.146427	9.949874
9.41	88.5481	3.067572	9.700515	9.91	98.2081	3.148015	9.954898
9.42	88.7364	3.069202	9.705668	9.92	98.4064	3.149603	9.959920
9.43	88.9249	3.070831	9.710819	9.93	98.6049	3.151190	9.964939
9.44	89.1136	3.072458	9.715966	9.94	98.8036	3.152777	9.969955
9.45	89.3025	3.074085	9.721111	9.95	99.0025	3.154362	9.974969
9.46	89.4916	3.075711	9.726253	9.96	99.2016	3.155947	9.979980
9.47	89.6809	3.077337	9.731393	9.97	99.4009	3.157531	9.984989
9.48	89.8704	3.078961	9.736529	9.98	99.6004	3.159114	9.989995
9.49	90.0601	3.080584	9.741663	9.99	99.8001	3.160696	9.994999

ANSWERS
TO ODD-NUMBERED
EXERCISES

Page 14

1. (a) 246; (b) no; (c) no; (d) 317; (e) no; (f) 1,058.

3. (a) Yes; (b) no; (c) no; (d) yes.

5. The class boundaries are 9.5, 14.5, 19.5, 24.5, and 29.5; the class marks are 12.0, 17.0, 22.0, 27.0, and 32.0.

7. There are various possibilities. One is 150.0–159.0, 160.0–169.0, . . . , and 220.0–229.0.

9. (a) The class frequencies are 5, 8, 9, 30, 21, 10, 8, 7, and 2. The cumulative frequencies for "less than 3.0," "less than 4.0," and so on, are 5, 13, 22, 52, 73, 83, 91, 98, and 100.

11. The class frequencies are 2, 6, 11, 16, 9, 5, and 1. The corresponding percentages are 4, 12, 22, 32, 18, 10, and 2. The cumulative percentages for the classes "2 or more," "5 or more," and so on, are 100, 96, 84, 62, 30, 12, and 2.

13. The class frequencies are 8, 16, 12, 4, 4, 3, 2, and 1.

17. The class frequencies are 31, 25, 17, 9, 5, and 3.

Page 34

3. 10.4.

5. Sour cream: mean 59.7, median 59, mode 59; cream cheese: mean 29.4, median 29, mode 29; fruit pies: mean 60.7, median 61, mode 62.

7. The mean is 1.15, the median is 0.98, and the mode is not unique (three values, 0.80, 0.98, and 1.00 occur twice). The mode is not suitable, and the mean is unduly influenced by the extreme value 4.20 (the mean is higher than all the other values than this one); the median seems to be the best measure.

9. (a) Yes; (b) no; (c) yes.

13. $21.19.

15. (a) 16; (b) 6; (c) 4; (d) 131.2.

17. (a) $\sum_{i=1}^{30} y_i^2$; (b) $\sum_{i=3}^{9} x_i y_i$; (c) $\sum_{i=1}^{n} x_i f_i$; (d) $\sum_{i=5}^{10} A_i$;

(e) $2 \cdot \sum_{i=1}^{20} x_i$; (f) $\sum_{i=1}^{n} z_i - \sum_{i=1}^{n} y_i$.

19. (a) 9; (b) 19; (c) -1; (d) 3; (e) 17; (f) 289.

21. (a) 3; (b) 41; (c) 218.

23. No.

Page 47

1. 1.66.

3. For the prices of sour cream, cream cheese, and fruit pies, respectively, the standard deviations are 1.3, 1.2, and 1.5 cents.

5. 2.3.

7. For the syrup densities of the fancy, choice, and standard grades, respectively, the ranges are 5, 6, and 6 per cent.

9. (a) The mean, range, and standard deviation are 51, 7, and 2.94 minutes, respectively; (b) the mean, range, and standard deviation are 1, 7, and 2.94 minutes, respectively; (c) the mean, range, and standard deviation are 61, 7, and 2.94 minutes, respectively; (d) the mean, range, and standard deviation are 102, 14, and 5.89 minutes, respectively; (e) adding a constant to each sample value increases (if the constant is positive) or decreases (if the constant is negative) the mean by the amount of the constant, but it has no effect on the range or standard deviation; multiplying each sample value by a positive constant has the effect of multiplying the mean, range, and standard deviation by the constant.

11. 560.

13. A is selling at 1.68 standard deviations above average, and B is selling at 3.00 standard deviations above average. Leaving all other considerations aside, the investor might well sell B.

15. A is relatively most expensive since his, B's, and C's charges are, respectively, 1.52, 1.38, and 0.74 standard deviations above average.

Page 55

1. (a) 16.36 and 0.34; (b) 16.36 and 0.34.

3. For Exercise 9 the mean and standard deviation are 6.08 and 1.86; for Exercise 11 they are 11.6 and 4.1; for Exercise 12 they are 66.3 and 15.2.

5. The values for the data of Exercise 15 depend on the way in which the data are grouped. The mean and standard deviation of the wages of Exercise 16 are $43.11 and $13.89.

7. $109.55.

9. (a) 44.4; (b) 28.6 and 58.6; (c) 17.1 and 74.3; (d) 35.3 and 52.9.

11. (a) Midquartile 43.4, interquartile range 22.2, semi-interquartile range 11.1, and coefficient of quartile variation 25.6; (b) midquartile 16.34, interquartile range 0.44; (c) midquartile 43.6, semi-interquartile range 15.0, and coefficient of quartile variation 34.4.

Page 59

1. 0.26.

Page 76

1. (a) With 1966 = 100 the values for 1967, 1968, and 1969 are 101.7, 103.0, and 104.0; (c) 105.1.

3. (a) With 1960 = 100 the values for 1965, 1966, 1967, and 1968 are 139.7, 138.2, 140.4, and 144.1; (b) 104.3; (c) 143.3 for 1967, and 151.0 for 1968; (d) 116.6; (e) 131.2.

5. (a) The index values are 74.7, 83.4, 156.0, 51.9, and 85.6 for Ohio, Indiana, Illinois, Michigan, and Wisconsin, respectively; (b) 59.9.

11. 129.4.

13. (a) 98.8 for 1967, and 100.6 for 1968; (b) 98.9 for 1967, and 100.6 for 1968;

(c) 100.7; (d) 100.6; (e) the index number is 100.6, the same as the one calculated for 1968 in part (a); (f) 100.7.

Page 86

5. No.

7. In constant 1957–59 dollars the earnings are 86.86, 85.80, 86.22, 85.35, and 85.71.

11. With 1966 = 100 the index values are 90, 91, 93, 96, 100, 104, 105, and 108.

13. The simple aggregative index, the geometric mean of price relatives, and the Ideal Index satisfy the time reversal test, but the weighted aggregative index and the arithmetic mean of price relatives do not.

Page 97

1. (b) E is the event that the house chosen has a combined total of 4 bedrooms and baths; F is the event that the house chosen does not have the same number of bedrooms as baths; G is the event that the house chosen has at most 2 bedrooms; (c) (i) $F' = \{(1, 1), (2, 2), (3, 3)\}$ is the event that the house chosen has the same number of bedrooms as baths, (ii) $G' = \{(3, 1), (3, 2), (3, 3)\}$ is the event that the house chosen has 3 bedrooms, (iii) $E \cap F = \{(3, 1)\}$ is the event that the house chosen has 3 bedrooms and 1 bath, (iv) $F \cup G = \{(1, 1), (2, 1), (2, 2), (3, 1), \text{and } (3, 2)\}$ is the event that the house chosen does not have 3 bedrooms and 3 baths, (v) $E \cap G = \{(2, 2)\}$ is the event that the house chosen has 2 bedrooms and 2 baths, (vi) $E \cup G' = \{(2, 2), (3, 1), (3, 2), (3, 3)\}$ is the event that the house chosen has either 3 bedrooms or has 2 bedrooms and 2 baths; (d) only the pair E and $F \cap G$.

3. (a) 2; (b) 7; (c) 4 and 1; (d) 6 and 3; (e) 8.

5. HHH, HHT, HTH, HTT, THH, THT, TTH, and TTT.

9. (1, 1) corresponds to (2, 0, 0), (2, 2) corresponds to (0, 2, 0), (3, 3) corresponds to (0, 0, 2), (1, 2) and (2, 1) correspond to (1, 1, 0), (2, 3) and (3, 2) correspond to (0, 1, 1), and (1, 3) and (3, 1) correspond to (1, 0, 1).

11. The pairs of events given in parts (c), (d), (g), and (h) are mutually exclusive.

Page 107

5. 12.

7. (a) 20; (b) 5; (c) 25.

9. 1,048,576.

11. (a) When we arrange n things in a circle we arbitrarily fix the position of one

thing, then calculate the number of ways in which the rest may be arranged in a straight line. Thus, the number of circular permutations of n objects is $(n - 1)!$; (b) 120; (c) 6.

13. (a) Each time we take r objects from n objects we leave $n - r$ objects, so there is the same number of ways of selecting n objects to be taken as there is of selecting $n - r$ objects to be left.

15. (a) 28; (b) 6; (c) 16; (d) 6.

17. (a) 84; (b) 70.

19. 25,025.

21. (a) 1,260; (b) 140; (c) 280; (d) 15,120.

Page 113

1. 703/925.

3. 2,014/5,300.

5. (a) 125/375; (b) 2 to 1; (c) Mr. G.

7. 5/8.

9. 13/20.

11. His personal probability is at least 9/10.

Page 123

1. The three events are mutually exclusive and one of them must occur; therefore, the sum of the measures assigned to the events must equal 1 (by Postulate 2). We cannot call A's and C's assignments probabilities at all, but B's assignment satisfies our probability model.

3. According to the student, the probability of an A is 1/11, the probability of a B is 1/9, and the probability of either an A or a B is 1/7; the probabilities are inconsistent since the sum of the first two does not equal the third.

5. According to the businessman, the probability that business conditions will improve is 2/3, the probability that they will remain the same is 1/4, and the probability that they will either improve or remain the same is 5/6; the probabilities are inconsistent since the sum of the first two does not equal the third.

7. (a) 0.76; (b) 0.48; (c) 0.64; (d) 0.40; (e) 0.12; (f) 0.36.

9. (a) 3/9; (b) 6/9; (c) 4/9; (d) 3/9; (e) 5/9; (f) 2/9; (g) 8/9; (h) 1/9; (i) 6/9.

11. 19/80.

13. (a) 0.04; (b) 0.27; (c) 0.25; (d) 0.73.

15. (a) 0.50; (b) 0.50; (c) 0.95; (d) 0.90.

Page 130

1. $P(E/T') = 32/160$ is the probability of selecting someone with experience given that he has had no training. $P(T/E') = 24/152$ is the probability of selecting someone with training given that he has had no experience.
3. (a) 0.35; (b) 0.21; (c) 0.74; (d) 0.26.
7. (a) 20/40; (b) 20/40; (c) 24/40; (d) 9/40; (e) 4/13; (f) 11/27; (g) 4/13; (h) 9/20; (i) 16/20.
9. (a) 480/1,000; (b) 520/1,000; (c) 80/1,000; (d) 80/200.
11. We must assume that the events that the first partner will live another 20 years and that the second partner will live another 20 years are independent. It is impossible to say whether or not this is reasonable, but if it is the probability that both will be alive 20 years hence is $(0.66)(0.47) = 0.31$.
13. (a) 2/3; (b) 80/81.

Page 136

1. 10/13.
3. 288/360.
5. 3/11.
7. 9/43.

Page 146

1. (a) 4¢; (b) not on the basis of expectation alone.
3. $18,800.
5. The expectation is $3.50, so it is not rational on the basis of expectation alone.
7. $-2.8¢$.
9. (a) $p < 78/117$; (b) $p = 64/400$; (c) $p > 4/19$.
11. (a) $-$56,000 if the program is continued, and $0 if it is terminated; (b) $-$13,333 if the program is continued, and $-$13,333 if it is terminated.
13. (b) The expected profits are $80,000 from the old stadium and $70,000 from a new stadium, so the regents should not authorize construction of a new stadium; (c) the expected profits are $104,000 from a new stadium and $92,000

from the old stadium, so the regents should authorize construction of a new stadium; (d) to use the old stadium because he would prefer a profit of $20,000 to a loss of $100,000; (e) to authorize construction of a new stadium because he would prefer a profit of $410,000 to a profit of $200,000; (f) to minimize the maximum opportunity loss they should authorize construction of a new stadium; to minimize the expected opportunity loss they should not authorize construction of a new stadium; (g) the expected value of perfect information is $150,000 and it would be worthwhile to spend the $10,000.

15. (a) Continue drilling; (b) stop drilling; (c) continue drilling; (d) continue drilling; (e) 700 units.

17. (a) He should choose job 1 since the expectation is $55,000 as against $45,000 for job 2; (b) he might choose job 2 because it offers a prospect of making $120,000.

19. (a) None; (b) the expected profits from stocking 0, 1, 2, 3, or 4 items are, respectively, $0, $1.60, $2.00, $0.80, and $-$0.80, so the retailer should stock 2 items.

Page 159

1. (a) Strategy 1 dominates strategy 2; (b) strategy 1 dominates strategy 2; (c) there is no dominance.

3. (a) A should choose strategies I, II, III, and IV with probabilities 0, 2/5, 3/5, and 0, respectively; B should choose strategies 1 and 2 with probabilities 4/5 and 1/5, respectively; the value of the game is $-2/5$; (b) A should choose strategies I, II, and III with probabilities 0, 13/18, and 5/18, respectively; B should choose strategies 1, 2, and 3 with probabilities 0, 2/3, and 1/3, respectively; the value of the game is 10/3.

9. A should choose strategies I, II, and III with probabilities 0, 3/4, and 1/4, respectively; B should choose strategies 1, 2, and 3 with probabilities 0, 7/12, and 5/12, respectively; the value of the game is 3/4.

11. Both should introduce a new model.

13. (a) The first player should write 1 and 4 with probabilities 3/4 and 1/4; the second player should write 0 and 3 with probabilities 3/4 and 1/4; the value of the game is 25¢; (c) the best strategies are the same as those given in part (a), but the value of the game is $1.25.

15. He should choose to stop drilling with probability 13/19 and to continue drilling with probability 6/19.

17. He should choose between methods 1, 2, and 3 with probabilities 5/22, 0, and 17/22, respectively.

19. (a) The owner of the first station should lower his prices; (b) the two owners should take turns lowering their prices.

Page 173

1. (a) 9,375/46,656; (b) 43,750/46,656; (c) 84/512; (d) 130/512.
3. (a) 27,436/160,000; (b) 157,757/160,000.
5. (a) 3,093/3,125; (b) 1,080/3,125; (c) 2,133/3,125.
7. The respective probabilities of winning $1, $2, $3, and losing $1 are 75/216, 15/216, 1/216, and 125/216; the player's expectation is −7.9 cents.
9. 0.765.
11. (a) 126/495; (b) 108/495; (c) 117/495.
13. 0.301; the binomial probability is 0.296.
15. 13,101/18,564.
17. 0.174.
19. 0.175.
21. (a) 0.018; (b) 0.144; (c) 0.154.
23. (a) 0.144; (b) 0.234.

Page 182

3. (a) 0.50, 0.45, and 0.05, respectively, to 20, 10, and 50 per cent.
5. 5/70, 4/70, 3/70, 8/70, and 50/70.
7. The posterior probabilities would be 0.76, 0.17, and 0.07; the expected cost would be minimized by not checking the batch.
9. (a) No, his expected profit is −$4; (b) no, his expected profit is −$3.08; (c) yes, his expected profit is $7.62.
11. (a) $1; (b) yes, his expected profit is $41 if he buys the policy and $30 if he does not; (c) no, his expected profit is $60 if he does not buy the policy and $44 if he does.

Page 190

1. 0.88 and 0.85.
3. (a) 5/2 and $\sqrt{5}/2$; (b) same as (a).
5. (a) 338 and 13; (b) 101 and 9.2; (c) 24 and 4.8; (d) 90 and 9; (e) 120 and 10.1.
7. (a) 9/22 and 1/22; (b) 1/2.
11. At least 0.84.

Page 203

1. (a) 0.0392; (b) 0.8531; (c) 0.5517; (d) 0.1271; (e) 0.0367; (f) 0.2971; (g) 0.8241; (h) 0.1008.
3. (a) 0.6826; (b) 0.9544; (c) 0.9974; (d) 0.9500; (e) 0.9802; (f) 0.9902.
5. (a) 1.28; (b) 1.64; (c) 1.96; (d) 2.05; (e) 2.33; (f) 2.58.
7. (a) 0.9332; (b) 0.2266; (c) 0.3891; (d) 0.9104.
9. 94.7.

Page 208

1. (a) 0.0294; (b) 0.4483; (c) $211.49.
3. 32.05.
7. 1,456.
9. (a) 0.0611; (b) 0.0594.
11. 0.9732.
13. 0.0154.

Page 216

1. (a) 10; (b) 66; (c) 300.

Page 227

1. (b) The samples are 6,7; 6,8; 6,9; 6,10; 6,11; 7,8; 7,9; 7,10; 7,11; 8,9; 8,10; 8,11; 9,10; 9,11; and 10,11; their means are 6.5, 7.0, 7.5, 8.0, 8.5, 7.5, 8.0, 8.5, 9.0, 8.5, 9.0, 9.5, 9.5, 10.0, and 10.5, respectively; (c) the probabilities of means of 6.5, 7.0, 7.5, 8.0, 8.5, 9.0, 9.5, 10.0, and 10.5 are, respectively, 1/15, 1/15, 2/15, 2/15, 3/15, 2/15, 2/15, 1/15, and 1/15; (d) the mean of the distribution is 8.5 and the variance is 7/6.
3. (a) The samples are 6,6; 6,7; 6,8; 6,9; 6,10; 6,11; 7,6; 7,7; 7,8; 7,9; 7,10; 7,11; 8,6; 8,7; 8,8; 8,9; 8,10; 8,11; 9,6; 9,7; 9,8; 9,9; 9,10; 9,11; 10,6; 10,7; 10,8; 10,9; 10,10; 10,11; 11,6; 11,7; 11,8; 11,9; 11,10; and 11,11; (b) the probabilities of means of 6.0, 6.5, 7.0, 7.5, 8.0, 8.5, 9.0, 9.5, 10.0, 10.5, and 11.0 are, respectively, 1/36, 2/36, 3/36, 4/36, 5/36, 6/36, 5/36, 4/36, 3/36, 2/36, and 1/36; (c) the mean is 8.5 and the standard deviation is $\sqrt{35/24}$ which are identical with the values expected according to the theorem.

5. The mean is 15.68; the standard deviation is 1.80 which is slightly larger than the standard deviation of the sampling distribution of the mean.

7. 100.

11. (a) The probability is 0.8889; (b) the probability is 0.9974.

13. (a) 0.0668; (b) 0.0062; (c) 0.6687; (d) 0.0026.

Page 244

1. The error is less than $5.25.

3. The error is less than $13.43.

5. (a) The error is less than $1.44; (b) $25.76–$29.54.

7. 0.8230.

9. (a) The error is less than $4.93; (b) with the correction the error is less than 21.70, and without the correction it is less than 25.03; (c) 234.58–245.42.

11. 107.

13. The error is less than $15.37.

15. 63.2–65.8.

17. (a) The error is less than 0.47 pounds; (b) 24.72–26.28.

19. The error is less than 0.82 per cent.

21. 70.7.

Page 253

1. 0.62–0.71.

3. 0.48–0.80.

5. (a) 0.61–0.71; (b) 0.22–0.46; (c) 0.29–0.43.

7. (a) 0.83–0.87; (b) 0.828–0.872; (c) the error is less than 0.029.

9. The error is less than 0.052.

11. 666.

13. (a) 1,849; (b) 1,184.

Page 257

1. $15.62–$23.44.

3. $6.44–$8.52.

Page 265

1. We would commit a Type I error if passenger cars are in fact driven on the average 12,000 miles a year and we concluded that the average mileage is not 12,000; we would commit a Type II error if passenger cars are in fact not driven on the average 12,000 miles a year and we concluded that the average mileage is 12,000.

5. (a) The probabilities of accepting 300 as the true value when the true value is in fact 294, 295, . . . , 306 are, respectively, 0.0000, 0.0062, 0.0668, 0.3085, 0.6915, 0.9332, 0.9876, 0.9332, 0.6915, 0.3085, 0.0668, 0.0062, and 0.0000. (Here the value 0.9876 is the probability of avoiding a Type I error, and each other value is a probability of committing a Type II error.); (c) the values of the power function (the probabilities of rejecting 300 as the true value when the true value is in fact 294, 295, . . . , 306) are 1 minus the corresponding probabilities of the OC-curve given in part (a). (Here the value $1 - 0.9876 = 0.0124$ is the probability of committing a Type I error, and each other value is a probability of avoiding a Type II error.)

7. (a) The probabilities of accepting 30.0 as the true value when the true value is in fact 27.5, 28.0, . . . , 32.5 are, respectively, 0.0013, 0.0228, 0.1587, 0.5000, 0.8400, 0.9544, 0.8400, 0.5000, 0.1587; 0.0228, and 0.0013. (Here the value 0.9544 is the probability of avoiding a Type I error, and each other value is a probability of committing a Type II error.)

Page 271

1. (a) The alternative is "less than 65 per cent of our policy holders will be interested in the new policy." The consequences of a Type I error are those that result from proceeding in the belief that less than 65 per cent will be interested when in fact 65 per cent or more will be interested. The consequences of a Type II error are those that result from proceeding in the belief that 65 per cent or more will be interested when in fact less than 65 per cent will be interested. (b) The alternative is "at least 65 per cent of our policy holders will be interested in the new policy." The consequences of a Type I error are those that result from proceeding in the belief that 65 per cent or more will be interested when in fact less than 65 per cent will be interested. The consequences of a Type II error are those that result from proceeding in the belief that less than 65 per cent will be interested when in fact 65 per cent or more will be interested.

3. (a) The alternative is that the belts are not effective. The consequences of a Type I error are those which result from deciding that the belts are not effective when in fact they are effective. The consequences of a Type II error are

those which result from deciding that the belts are effective when in fact they are not. (b) The alternative is that the belts are effective. The consequences of a Type I error are those which result from deciding that the belts are effective when in fact they are not effective. The consequences of a Type II error are those which result from concluding that the belts are not effective when in fact they are effective.

5. (a) Alternative: $r < 0.04$; (b) Alternative: $r > 0.04$.

7. (a) The hypothesis should be that the clerk makes the average number of mistakes, and the alternative that the clerk makes more than the average number of mistakes; (b) the hypothesis should be that the clerk makes the average number of mistakes, and the alternative that the clerk makes less than the average number of mistakes.

Page 283

1. $z = -4.67$; yes.

3. $t = 2.44$; yes.

5. (a) $z = -3.50$, which is less than the critical value $z = -2.33$.

7. $z = 8.00$; yes.

9. (a) For possible mean weights of 246, 247, . . . , 254 pounds the probabilities are 0.000, 0.000, 0.000, 0.004, 0.050, 0.261, 0.641, 0.913, and 0.991 using the right-hand one-tail test; 0.991, 0.913, 0.641, 0.261, 0.050, 0.004, 0.000, 0.000, and 0.000 using the left-hand one-tail test; and 0.979, 0.851, 0.516, 0.170, 0.050, 0.170, 0.516, 0.851, and 0.979 using the two-tail test; (c) the left-hand one-tail test is the best test to use against a specific alternative of 247 pounds because it is the "most powerful" of the three tests against this alternative (that is, it gives the greatest protection against committing a Type II error); the right-hand one-tail test is the best test to use against a specific alternative of 252 pounds because it is the "most powerful" of the three tests against this alternative; if the mean weight is 250 pounds (if the hypothesis is true) there is no difference between the three tests because the "power of the test" (the probability of rejecting the hypothesis) is the same for all three tests; (d) clearly, the right-hand one-tail test is the best test to use against an alternative that specifies the true value is greater than the hypothetical value, the left-hand one-tail test is best when the alternative specifies that the true value is less than the hypothetical value, and the two-tail test is best when the alternative specifies that the true value is not equal to the hypothetical value.

11. Reject the hypothesis for all values of $\bar{x} < 14.31$; the probability of failing to detect a lot whose mean is 14.3 is 0.464.

13. $t = -2.45$; yes.

15. (a) $z = -1.37$; no real difference; (b) 0.58–0.87.

17. $t = -3.57$; yes.

19. In the second example of the exercise show first that the hypothesis will be rejected if $\bar{x} < 496.5$ or if $\bar{x} > 503.5$; then show that the probability of rejecting the hypothesis that the mean is 500 when it is true is 0.05, and that the probability of rejecting the hypothesis that the mean is 500 when the mean is in fact 495 (or 505) is 0.80.

21. The buyer should reject the hypothesis for values of $\bar{x} < 3.975$.

23. $t = -2.66$; the diet is not effective.

25. The lower control limit is 1.473, and the upper control limit is 1.527; the process is out of control with samples 8, 9, 10, 11, and 14.

27. 1,020/4,060, or 0.251.

Page 297

1. $z = 1.25$, so the proportion receiving beneficial results falls just short of significance at the 0.10 level. The results are certainly "encouraging," if not significant, and the doctor may well conclude that there is a need for further study of the new relaxant's effect.

3. $z = 1.73$; the proportion is greater than 0.75.

5. $z = 5.86$; yes.

7. $\chi^2 = 0.77$; the null hypothesis cannot be rejected.

9. $\chi^2 = 8.14$ and $z = 2.86$; there is a real difference between the two proportions. (z^2 and χ^2 are equal except for a slight rounding error.)

11. (a) $z = -0.35$; yes; (b) 0.58–0.77; (c) $z = 3.33$; yes.

13. (a) $z = 2.18$; there is evidence to support the claim; (b) $z = 2.40$; the hypothesis is rejected.

15. $z = 0.58$; the hypothesis cannot be rejected.

17. The lower control limit is 0.08, and the upper control limit is 0.32; the process is out of control at the time of the last sample.

Page 306

1. $\chi^2 = 4.99$; no.

3. $\chi^2 = 43.06$; there is strong evidence of a real relationship.

5. $C = 0.30$.

7. $\chi^2 = 27.36$; evidently the dice are not properly balanced or the tossing is not random.

9. $\chi^2 = 4.59$; there is no evidence that the dice are not balanced.

11. $\chi^2 = 5.82$; yes, it is reasonable.

Page 320

1. $F = 15.8$; the differences are significant.
5. $F = 2.35$; the differences are not significant.
7. (a) $F = 1.05$; we cannot reject the null hypothesis; (b) $F = 9.55$; reject the null hypothesis.

Page 325

1. The differences among detergents are significant ($F = 9.85$), but the differences among water temperatures are not significant ($F = 2.38$).
3. The differences among salesmen are not significant ($F = 0.17$), but the differences among weeks are significant ($F = 4.63$).

Page 330

1. The estimates of the mean and standard deviation are 38.45 and 8.59, respectively.
3. 0.29.
5. 20.95.
9. The median is 14.05 and $\bar{x}$ is 14.04; the estimate of s is 0.105 and s is 0.106.

Page 339

1. $z = 2.02$; the amount of waste is significantly less.
3. $z = 2.94$; the gain is significant.
5. (a) $z = 0.67$; we cannot reject the hypothesis that the population mean time is 32 days even at a 0.20 level of significance; (b) $z = 1.57$; no.
7. $z = 0.46$; there is no evidence of a change.
9. $z = 1.33$; there is no significant difference.
11. (a) $z = 0.15$; there is no significant difference between the means; (b) $z = 0.62$; there is no significant difference between the dispersions.
15. $H = 4.91$; the null hypothesis cannot be rejected.

Page 354

1. (b) $y' = 2.19 + 1.16x$; (c) 8.0 months.
3. (b) $y' = -0.1689 + 0.2451x$; (c) 12.1 per cent.
5. (b) $y' = 0.96 + 1.2485x$; (c) 3.5 per cent.
7. (b) $y' = -10.61 + 3.7043x$; (c) 19.0.
9. (b) $y' = 24.17 + 5.9456x$; (c) 59.8.

Page 360

1. (a) 8.3–13.5; (b) 10.1–11.7.
3. (a) 0.04–7.66; (b) 2.56–5.14.

Page 367

1. (a) 0.90; there is a relationship; (b) 81 per cent.
3. (a) 0.91; the relationship is significant; (b) 83 per cent.
5. 0.96; there is a relationship.
7. (a) $r = 1$; (b) $r = -1$. (There are only two points in each case, and one can always draw a straight line through two distinct points.)
9. (a) Not significant; (b) not significant; (c) significant; (d) not significant.
13. (a) 0.17–0.37; (b) β lies on the interval from -1.84 to -1.24.

Page 371

1. $r' = -0.95$; yes, there is evidence.
3. (a) $r = 0.16$; there is no relationship; (b) $r' = -0.19$; there is no relationship.
5. $r' = 0.29$; no.
7. The difference is accounted for by the ties.
9. $r' = -0.93$; there is a significant relationship.

Page 389

1. (b) $y' = 4.83 + 0.5571x$ (origin, 1968–69; x units, 6 months; y, number of projects begun); (c) $y' = 2.04 + 0.5571x$.

3. (b) $y' = 4.02 + 0.1900x$ (origin, 1967; x units, 1 year; y, net sales in hundreds of millions of dollars); (c) $y' = 0.296 + 0.0013x$.

5. (b) $y' = 2.86 + 0.4050x$ (origin, 1966–67; x units, 1 year; y, net investments in billions of dollars); (c) $y' = 3.87 + 0.4050x$; (d) $y' = 2.05 + 0.4050x$.

7. (b) $y' = 1,257.20 + 71.1515x$ (origin, 1964–65; x units, 6 months; y, total annual sales in millions of dollars); (c) $y' = 75.37 + 0.4941x$.

9. (a) The three-year moving average for the years 1951 through 1968 is: 4.53, 5.07, 4.17, 4.63, 4.83, 4.43, 4.13, 4.23, 4.43, 4.67, 4.97, 5.20, 4.67, 4.00, 3.40, 3.13, 3.50, and 3.73; (b) the five-year moving average for the years 1952 through 1967 is: 4.36, 5.10, 4.48, 4.30, 4.54, 4.52, 4.18, 4.44, 4.84, 4.82, 4.70, 4.58, 4.06, 3.60, 3.56, and 3.50.

11. (a) The three-year moving average for the years 1940 through 1967 is: 15.9, 15.2, 15.5, 16.4, 18.9, 20.3, 20.0, 20.1, 20.6, 20.5, 20.3, 19.7, 20.5, 19.5, 19.2, 19.3, 20.6, 20.8, 21.4, 21.8, 22.5, 23.0, 23.5, 23.7, 22.6, 22.1, 22.5, and 22.9; (b) the five-year moving average for the years 1941 through 1966 is: 15.5, 16.4, 17.4, 18.1, 19.3, 20.1, 20.6, 20.2, 20.1, 20.4, 20.4, 19.4, 19.7, 19.9, 20.0, 19.8, 21.0, 21.5, 21.7, 22.5, 23.2, 23.1, 22.9, 22.9, 22.3, and 22.6.

13. $u = 6$; there is no significant trend.

15. $u = 2$; there is a real trend.

17. (a) $u = 11$; there is no significant trend; (b) $u = 8$; there is a significant trend; (c) $u = 2$; there is a significant trend.

Page 404

1. The seasonal index is 79.6, 76.6, 94.9, 97.5, 105.5, 97.9, 85.1, 89.2, 102.1, 121.9, 113.4, and 136.3.

3. The deseasonalized data are as follows. *1964*: 8.6, 8.0, 7.3, 8.8, 6.3, 5.7, 7.8, 7.4, 7.8, 8.6, 6.2, and 8.7; *1965*: 7.6, 8.3, 9.7, 9.4, 6.2, 8.2, 8.3, 8.3, 7.3, 8.3, 7.7, and 8.3; *1966*: 8.3, 8.4, 7.6, 9.1, 8.5, 8.8, 7.8, 8.3, 9.2, 8.6, 8.8, and 8.5; *1967*: 8.0, 8.6, 8.4, 8.5, 8.9, 7.7, 9.3, 8.6, 8.7, 8.3, 8.6, and 7.4; *1968*: 9.5, 8.5, 8.5, 8.4, 8.5, 8.6, 7.3, 8.8, 8.1, 8.5, 8.6, and 8.8; *1969*: 9.1, 8.3, 9.0, 8.7, 7.9, 8.1, 9.4, 7.4, 8.9, 7.7, 7.7, and 8.2.

5. The seasonal index is 66.6, 67.3, 100.9, 125.3, 125.0, 119.2, 111.4, 107.5, 100.4, 111.2, 92.9, and 72.3.

7. The projected monthly housing starts (in thousands) for 1974 are 77, 78, 117, 145, 144, 138, 129, 124, 116, 128, 107, and 83.

9. (a) The seasonal index is 102.5, 80.1, 82.6, 77.0, 81.3, 86.8, 93.8, 101.8, 111.1, 139.4, 132.5, and 110.9; (c) the deseasonalized data (in billions of dollars) for 1969 are 3.7, 3.9, 4.0, 4.2, 4.1, 3.9, 4.5, 5.1, 5.2, 4.2, 3.8, and 4.1; (d) 48 billion dollars; (e) the projected monthly cash receipts (in billions of dollars) for 1975 are 5.3, 4.2, 4.3, 4.0, 4.3, 4.6, 5.0, 5.4, 5.9, 7.4, 7.1, and 5.9.

11. (b) The monthly sales forecast (in millions of dollars) for 1972 is 1.50, 1.45, 1.95, 2.48, 2.73, 2.41, 2.03, 2.10, 2.02, 2.11, 1.65, and 1.56.
13. (a) 250; (b) 1974; (c) 750 machines.

Page 413

1. (a) Primary; (b) secondary; (c) secondary; (d) primary; (e) primary; (f) primary; (g) secondary.

Page 419

1. (a) 51, 38, 67, and 61; 20, 68, 55, and 90; 43, 27, 76, and 52; (b) the means are 54.25, 58.25, 49.50, and 54.00; the mean of the sampling distribution is 54.0 which is equal to the mean of the 12 amounts.
5. (a) 3,719,250; (b) 7,714,000.
7. (a) With proportional allocation $n_1 = 50$ and $n_2 = 150$; (b) with optimum allocation $n_1 = 40$ and $n_2 = 160$.

Page 425

1. AIS, AIF, AIIS, AIIF, BIS, BIF, BIIS, BIIF, CIS, CIF, CIIS and CIIF.
7. Personnel: Flynn; Purchasing: Gordon; Insurance: Duncan or Evans; Research: Evans or Duncan.

Page 438

3. (a) The processor should use 12 ounces of Compound A and 4 ounces of Compound B; the minimum cost is 24¢; (b) the processor can buy at a cost of 18¢ *either* 12 ounces of Compound A and 4 ounces of Compound B *or* 36 ounces of Compound A and no Compound B; there is, in fact, a line segment solution and any point (x, y) on the line segment joining $(12, 4)$ and $(36, 0)$ leads to a minimum cost of 18¢.
5. The company should schedule 20 dozen of Toy A and 10 dozen of Toy B; the maximum profit is $490.
7. There are many possible procurement policies which lead to a minimum cost of $31,900. If Supplier 1 ships no cans to Warehouse A it can then ship any number of cans between 1,000 and 3,000 (both included) to Warehouse B; all such schedules lead to a cost of $31,900.

Page 445

1. The optimum order quantity is 200 boxes, and lots of this size should be bought five times during the year. Total inventory cost for the year is $100 of which $50 is the carrying cost and $50 is the order cost.

5. (a) The optimum order quantity is 240 parts and the minimum cost is $180; (b) if the company buys 200 parts (two standard lots) nine times, the cost will be $183 of which $75 is the carrying cost and $108 is the order cost.

7. The expected profits from stock levels of 0, 1, 2, 3, 4, 5, and 6 corsages are $0, $2.80, $5.00, $6.00, $6.00, $5.40, and $4.60; the vendor should stock 4 corsages.

9. The optimum stock level is 1 booster.

Page 452

1. (a) 5/3; (b) 40/3; (c) 25/24; (d) 25/3.

5. Yes, the total cost would be reduced by $145.

7. The respective probabilities are 0.375, 0.234, 0.146, 0.092, 0.057, 0.036, 0.022, and 0.014. The exact value of the mean is 1.67 and the approximation is 1.44.

INDEX